Molecular Pathogenesis of Colorectal Cancer

Kevin M. Haigis
Editor

Molecular Pathogenesis of Colorectal Cancer

Springer

Editor
Kevin M. Haigis, Ph.D.
Molecular Pathology Unit
Harvard Medical School
Charlestown, MA, USA

ISBN 978-1-4614-8411-0 ISBN 978-1-4614-8412-7 (eBook)
DOI 10.1007/978-1-4614-8412-7
Springer New York Heidelberg Dordrecht London

Library of Congress Control Number: 2013949281

Printed on acid-free paper

Springer is part of Springer Science+Business Media (www.springer.com)

Preface

Colorectal cancer (CRC) represents a major healthcare burden worldwide for two simple reasons: it is common and it is deadly. And while improvements in early detection have helped to reduce the incidence of CRC-related death over the past several decades, the overall frequency of the disease is likely to increase steadily due to its connection to western style diet, which is spreading across the globe, and to obesity and chronic inflammation (i.e., inflammatory bowel disease), which are also rapidly increasing in incidence. As a result, a new generation of effective CRC therapies is greatly needed. The search for new therapies to treat CRC is intertwined with the identification of the molecular etiology of the disease.

Even prior to the advent of whole genome sequencing, many of the mutant genes that contribute to CRC (*APC*, *KRAS*, *TP53*) were known from targeted sequencing efforts. As a result, CRC has become the paradigm for multistage tumorigenesis, where the histologically defined transition states from normal tissue to malignancy can be associated with mutations in specific genes or pathways. As we enter the post-genomic era, it is possible that most, if not all, of the genes that contribute to CRC in a meaningful way have been identified. Now it is time to leverage the extensive mutational information to establish new therapeutic strategies. This will require a combination of functional genomics (i.e., genetics), medicinal chemistry, and preclinical and clinical efforts.

The goal of this book is to provide a broad overview of the state of understanding of the molecular pathogenesis of CRC. This book is organized as a timeline of the study of CRC. Chapters 1 and 2 provide a general and historical viewpoint of the role for genetic changes and genomic instability in CRC. Chapters 3–8 discuss the roles of specific pathways (RAS, PI3K, TGF-β) or environmental conditions (inflammation). And Chaps. 9–12 look toward the future, focusing on the potential for genome-wide analyses to find new genes/pathways that contribute to CRC. I hope that the reader will get an appreciation for the rich history of CRC research and a fresh perspective on the possibilities for emerging therapeutic options.

Charlestown, MA — Kevin M. Haigis, Ph.D.

Contents

1 **The Genetics of Colorectal Cancer** 1
William Hankey and Joanna Groden

2 **Molecular Mechanisms of Colorectal Carcinogenesis** 25
Jatin Roper and Kenneth E. Hung

3 **The Association Between Inflammation and Colorectal Cancer** 67
Maria José Oliveira and Sérgia Velho

4 **Importance of the Niche: Wnt Signaling and Stem Cell Plasticity in Intestinal Homeostasis and Disease** 107
Owen J. Sansom and Inke Näthke

5 **Mutational Activation of *KRAS* and *BRAF* in Colorectal Cancer** 121
Katherine H. Pedone, Jennifer L. Sells, and Channing J. Der

6 **The PI3K Pathway in Colorectal Cancers** 157
Jihye Yun, George Poulogiannis, Evan T. Brower, Samuel Klempner, and Lewis L. Cantley

7 **TGF-ß Signaling Pathway and Colorectal Cancer** 201
William M. Grady

8 **The Clinical Significance of Mutations in Colorectal Cancer** 231
Franklin W. Huang, Laura B. Kleiman, and Theodore S. Hong

9 **Colorectal Cancer Genome and Its Implications** 247
Nickolas Papadopoulos

10 **Copy-Number Alterations in the Colorectal Cancer Genome** 267
Jihun Kim and Adam J. Bass

11 Genome-Wide Association Studies in Colorectal Cancer 289
Ian Tomlinson

12 Future Prospects for Leveraging Molecular Information in the Fight Against Colorectal Cancer 303
Laura B. Kleiman and Kevin M. Haigis

Index 309

Chapter 1
The Genetics of Colorectal Cancer

William Hankey and Joanna Groden

Abstract Colorectal cancer (CRC) develops over a period of years through a defined progression from a single aberrant crypt to a benign adenoma and ultimately to an invasive malignancy. These phenotypic steps parallel a series of underlying changes at the DNA level. Many of the critical tumor suppressor loci have been identified through cytogenetic or genetic linkage studies of inherited disorders that predispose affected family members to the development of benign or malignant lesions in the colorectal epithelium. Inactivating mutations in the *APC* gene not only were first identified in the germline of individuals with familial adenomatous polyposis coli but also are present in most sporadic CRCs. Germline mutations in *MSH2*, *MLH1*, *MSH6*, or *PMS2* predispose individuals with Lynch syndrome to CRCs with DNA mismatch repair defects; these genes can be mutated or silenced in sporadic CRCs as well. Other inherited mutations are responsible for benign colorectal lesions that rarely progress to malignancy, including those found in the *SMAD4*, *BMPR1A*, and *PTEN* genes. Sporadic changes in these genes are found in malignant rather than premalignant lesions, suggesting that these mutations promote rather than initiate tumorigenesis. Genetic analysis of CRCs will permit stratification for improved prognosis and treatment.

1.1 Introduction

Colorectal carcinoma has been observed in humans as far back in history as the time of the ancient Egyptians (Zimmerman 2003) while the term carcinoma has been used to describe the broader classification of solid tumors since the time of

W. Hankey • J. Groden (✉)
Department of Molecular Immunology, Virology & Medical Genetics,
The Ohio State University College of Medicine, 460 West 12th Avenue,
986 Biomedical Research Tower, Columbus, OH, USA
e-mail: groden.2@osu.edu

K.M. Haigis (ed.), *Molecular Pathogenesis of Colorectal Cancer*,
DOI 10.1007/978-1-4614-8412-7_1,

Hippocrates (Tsuchiya and Fujisawa 1999). Current pharmacological interventions for colorectal carcinoma also draw from discoveries that date back long before the modern era. For example, the topoisomerase I inhibitor irinotecan, a common component of current therapeutic regimens, has its roots in traditional Chinese medicine (Wall and Wani 1995). More recent targeted therapies for colorectal carcinoma, particularly humanized monoclonal antibodies, have emerged in response to the discovery of oncogenes frequently mutated or aberrantly expressed in solid tumors (Soker et al. 1998; Zhang et al. 2007). The current molecular era of biomedical research and drug discovery is increasingly oriented toward the development of such targeted therapeutics and therefore increasingly informed by the study of tumor genetics.

Colorectal tumor development occurs in a stepwise fashion from lesions involving a single crypt, to a small benign tumor, to a malignant or invasive tumor described histopathologically as an adenocarcinoma. This progression is thought to take place over a period of years, such that 8 % of detected adenomatous polyps develop into adenocarcinomas within a 10-year period (Stryker et al. 1987). Genetic studies comparing normal epithelial tissue and sporadic tumors of various stages reveal a sequence of events occurring at the DNA level that underlie and parallel the clinical and histological stages of tumor development (Kinzler and Vogelstein 1996). Before these studies at the genomic scale became technically feasible, other types of experiments enabled critical discoveries about colorectal cancer (CRC) genetics. Cytogenetic studies revealed key chromosomal loci frequently lost or gained in CRCs of various stages (Dutrillaux 1988; Migliore et al. 2011). In addition, specific genes that play critical roles in CRC development were identified through genetic mapping of familial cancer syndromes that predispose patients to development of CRC. Based on these accumulated studies, this chapter reviews the genetics of CRC and addresses its implications for disease progression, treatment, and prognosis.

1.2 Genetic vs. Environmental Components of Colorectal Cancer Risk

The majority of CRC cases occur sporadically in individuals without a family history of the disease. A strong environmental component has been suggested by epidemiological findings of significant geographic and lifestyle contributions to CRC risk, with particularly influential contributions of age and diet (Haggar and Boushey 2009). Between one quarter and one-third of CRCs occur in patients with a family history of the disease. Having one first-degree relative diagnosed with CRC confers a 1.72-fold increase in age-adjusted risk. Having multiple first-degree relatives with CRC confers either a 2.75- or 5.37-fold increase in age-adjusted risk, depending on whether these relatives were above or below the age of 45 at the time of tumor onset (Fuchs et al. 1994). In many such cases, low-penetrance genetic susceptibility factors are thought to work in conjunction with environmental influences, while 5–6 % of all CRC cases are estimated to be the result of high-penetrance genetic mutations (Migliore et al. 2011).

1.3 Early History of Colorectal Cancer Genetics

The history of CRC genetics began with an 1882 report of "disseminated polyps of the rectum" observed in two siblings by W. Harrison Cripps, an English rectal specialist and abdominal surgeon (Cripps 1882). This observation was made during an era when the relationship between heredity and carcinoma was still a matter of debate. Cripps is credited with recognizing multiple intestinal polyposis as both heritable and potentially premalignant in nature (Gardner 1951). A subsequent study published in 1913 by the American pathologist Aldred Scott Warthin (Fig. 1.1) was the first to apply statistics to identify a hereditary component in carcinoma development. Warthin accumulated records on 1,600 cases of carcinoma treated at the University of Michigan hospitals over a 19-year period and assembled detailed family histories for approximately 500 patients (Warthin 1913). Warthin's idea was conceived during an 1895 conversation with his seamstress, Pauline Gross, when she predicted to him that she would die at an early age from cancer of the stomach, colon, or the female organs. Gross' prediction was based on an unusually potent family history, and she ultimately did succumb to metastatic endometrial carcinoma at a relatively young age. Warthin's publication included her genealogy under the name "Family G," mapping an overwhelming predisposition to gastrointestinal and endometrial carcinomas over three generations (Warthin 1913).

The efforts of later human geneticists subsequently led to the characterization of Warthin's Family G as a Lynch syndrome pedigree (Douglas et al. 2005) and to the identification of Cripps' sibling patients as suffering from familial adenomatous polyposis coli (FAP) (Gardner 1951). The next section discusses Lynch syndrome, FAP, and other familial cancer syndromes and addresses their implications for the development of sporadic colorectal carcinomas.

Fig. 1.1 Aldred Scott Warthin, M.D., Ph.D. Image courtesy of the A.S. Warthin Papers, Bentley Historical Library University of Michigan

1.4 *APC* and Familial Adenomatous Polyposis

Following the initial report by Cripps in (1882), several pedigrees were published in the early half of the twentieth century (reviewed in Dukes 1930) in which affected family members exhibited numerous polyps or polyposis of the colon and rectum, as well as predisposition to the development of CRC (Gardner 1951). Eldon J. Gardner (Fig. 1.2), a Professor and human geneticist at the University of Utah, subsequently conducted a larger study of patients exhibiting multiple polyps at the Salt Lake General Hospital and constructed large kindreds in order to link conclusively this "intestinal polyposis" syndrome with predisposition to carcinoma (Gardner and Stephens 1950). While all of Gardner's kindreds showed colorectal phenotypes, one kindred in particular exhibited additional enhanced risk for certain extracolonic phenotypes. These included osteomas (Gardner and Plenk 1952), soft tissue tumors, and dental abnormalities (Gardner and Richards 1953). These findings led to the characterization of Gardner's syndrome, in which affected family members develop hundreds to thousands of polyps in the colon and rectum over their lifetime that confer predisposition to the development of CRC and also exhibit an increased risk of developing osteomas of the jawbone and skull, cutaneous and subcutaneous cysts, desmoid tumors, congenital hypertrophy of the retinal pigment epithelium (CHRPE), as well as cancers at various other sites such as the thyroid, brain, liver, small intestine, and stomach (Groen et al. 2008).

More than 30 years after Gardner's initial findings, cytogenetics provided the first evidence linking polyposis and CRC predisposition to a specific region of the genome. In 1986, a male patient with mental retardation was diagnosed with intestinal polyposis, multiple CRCs, and soft-tissue neoplasms that included a desmoid tumor (Herrera et al. 1986). Although neither of his parents exhibited Gardner's

Fig. 1.2 Eldon J. Gardner M.D. Image courtesy of Special Collections, Merrill-Cazier Library, Utah State University

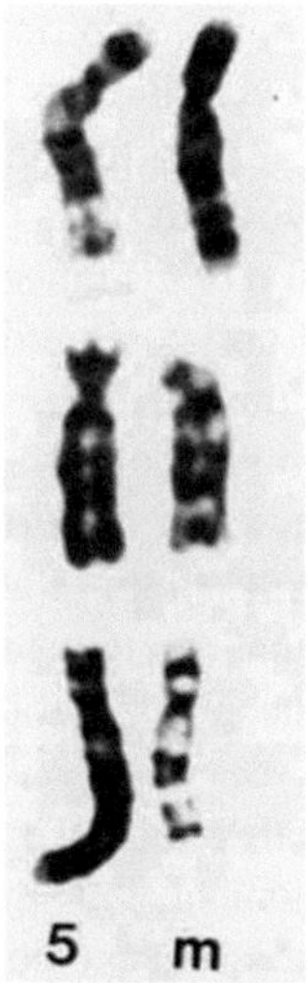

Fig. 1.3 Partial karyotype from peripheral blood cells of a patient with Gardner's syndrome and mental retardation. It shows an interstitial deletion within the long or q arm of chromosome 5. The *left side* of the figure displays the patient's normal chromosome 5. The *right side* of the figure displays the abnormal (or marker) chromosome in which the G-band between 5q13 and 5q31 is missing. This image was first published in Herrera et al. (1986). Gardner's syndrome in a man with an interstitial deletion of 5q. *American Journal of Medical Genetics* 25: 473–476. Image courtesy of John Wiley and Sons, Inc

syndrome phenotypes or mental retardation, the patient's cells contained an apparently germline, *de novo* interstitial deletion within the long (q) arm of chromosome 5 (Herrera et al. 1986). This cytogenetic abnormality (Fig. 1.3) appeared to be the cause of both Gardner syndrome and mental retardation in the patient and suggested the localization of a tumor suppressor gene on chromosome 5q. Following this discovery, large-scale cytogenetic studies of sporadic tumors detected biallelic loss of markers within this genomic region in a significant proportion of CRCs (Solomon et al. 1987). Subsequent genetic linkage studies of FAP kindreds mapped the causal locus to chromosome 5q21 (Leppert et al. 1987; Bodmer et al. 1987). Positional cloning identified the adenomatous polyposis coli or *APC* gene in this region of 5q21 (Joslyn et al. 1991; Kinzler et al. 1991) and the presence of nonsense mutations in its open reading frame in a significant number of individuals with FAP (Groden et al. 1991; Nishisho et al. 1991).

Since the cloning of the *APC* gene, Gardner's syndrome has been assimilated into the broader condition of FAP, in which mutations in the *APC* gene cause affected family members to develop hundreds to thousands of adenomatous polyps, resulting in onset of CRC typically by the age of 40 (Giardiello 1995). Affected individuals from most FAP families experience elevated risks of extraintestinal abnormalities as well (Galiatsatos and Foulkes 2006), although Gardner's syndrome is considered to be a severe variant in which the incidence of extraintestinal manifestations is elevated. Attenuated forms of FAP have also been identified, in which affected individuals exhibit fewer polyps (Spirio et al. 1993). Also included within FAP are some families previously classified as having Turcot's syndrome. In this variant, affected individuals present with tumors of the central nervous system in addition to multiple intestinal polyps and CRC (Hamilton et al. 1995). Among the multiple variants of FAP, the location of the inherited mutation within the *APC* gene strongly influences the severity of polyposis and age of CRC onset, as well as the site and frequency of extracolonic manifestations.

While individuals suffering from FAP inherit a mutation in a single allele of *APC*, the adenomas and adenocarcinomas that develop in these individuals also bear a mutation in the second allele of APC (Ichii et al. 1993; Levy et al. 1994; Luongo et al. 1994). The ability of *APC* mutations to precipitate the formation of precancerous lesions indicates that the gene plays a role in suppressing the initiation of colorectal tumors (Ichii et al. 1993; Levy et al. 1994; Luongo et al. 1994). How does *APC* perform this function in normal colorectal epithelium? Its protein product functions to negatively regulate the canonical Wnt signaling pathway by participating in a cytoplasmic complex that mediates the proteolytic degradation of the transcriptional coregulator β-catenin (Munemitsu et al. 1995; Behrens et al. 1998). The great majority of *APC* mutations produce a truncated APC protein and either abolish or impair its ability to interact with β-catenin (Miyoshi et al. 1992; Powell et al. 1992; Nagase and Nakamura 1993). The result is that APC-deficient cells accumulate β-catenin in the cytoplasm as well as in the nucleus (Korinek et al. 1997; Morin et al. 1997), where its interactions with the transcription factors Tcf4 and Lef1 modify the expression of numerous genes controlling growth, proliferation (Shtutman et al. 1999; Tetsu and McCormick 1999), differentiation (He et al. 1998), survival (Zhang et al. 2001), and other processes related to cancer development. Thus, constitutively active Wnt signaling leads to the dysregulation of critical genes whose expression is tightly regulated in normal colorectal epithelium. The *MYC* proto-oncogene is an important example of such a gene whose transcription is directly upregulated by β-catenin and activated Wnt signaling in APC-deficient CRC (He et al. 1998). Recent studies have demonstrated that the APC protein possesses other functions that protect against tumor development, in addition to its role in negatively regulating canonical Wnt signaling. The APC protein sensitizes cells to apoptosis (Steigerwald et al. 2005; Qian et al. 2007), regulates components of the cytoskeleton (Smith et al. 1994; Munemitsu et al. 1994; Hulsken et al. 1994), and shuttles in and out of the nucleus to regulate the nuclear pool of β-catenin (Henderson 2000; Neufeld et al. 2000; Rosin-Arbesfeld et al. 2000). Understanding of the relative importance of these diverse functions of APC in preventing tumor development is still evolving.

FAP occurs in the general population at a frequency of 1 in 7,000–22,000 people (Half et al. 2009), and therefore germline mutations in *APC* account for less than 1 % of CRC diagnoses in the Western world. However, somatic *APC* mutations occur in most sporadic CRCs (Miyoshi et al. 1992; Powell et al. 1992) and usually in the earliest detectable neoplasms known as dysplastic aberrant crypt foci, thought to be early intermediates in the development of adenomas (Jen et al. 1994). The prevalence of these somatic *APC* mutations suggests that *APC* inactivation is an early and rate-limiting step in the majority of CRCs. Sporadic tumors without *APC* mutations frequently carry an activating mutation in *CTNNB1*, the gene encoding β-catenin (Jen et al. 1994). Evidence that *CTNNB1* mutations occur more frequently in small adenomas than in more advanced adenomas and adenocarcinomas suggests that *APC* mutations may initiate CRC development more efficiently than *CTNNB1* mutations (Samowitz et al. 1999). Mutations in the *AXIN1* gene, encoding another critical component of the cytoplasmic complex that mediates the proteolytic degradation of β-catenin, have been proposed as another source of aberrant Wnt signaling in CRC, based on their ability to promote nuclear accumulation of β-catenin and tumorigenesis in hepatocellular carcinomas (Satoh et al. 2000). Overall, nuclear accumulation of β-catenin and constitutive activation of Wnt signaling can be characterized as the early events in the development of most CRCs. The prevalence of abnormal Wnt signaling speaks to the nature of CRC as a disease driven by aberrant gene expression, and in particular the activation of a stem cell-like transcriptional program of proliferation.

1.5 *MUTYH* and Familial Adenomatous Polyposis

An autosomal recessive variant of FAP occurs in the absence of germline *APC* mutations and with affected individuals exhibiting less severe polyposis, often with fewer than 100 polyps, and delayed or adult onset (Sieber et al. 2003; Jones et al. 2002). Consistent with the attenuated form of FAP, approximately 35–53 % of the affected individuals go on to develop CRC (Aretz et al. 2006). Germline mutations inherited by these individuals have been mapped to the *MUTYH* gene, the homolog of *E. coli mutY* that encodes a component of the base excision repair pathway (Al-Tassan et al. 2002). *MUTYH* encodes a DNA glycosylase that removes purine bases damaged by oxidation. In particular, one of its primary functions is to remove adenine bases misincorporated opposite oxidized guanine bases known as 8-oxoG (Slupska et al. 1999; Takao et al. 1999; Ohtsubo et al. 2000). In individuals who inherit biallelic *MUTYH* mutations, bases misincorporated during DNA replication become fixed, and G:C to A:T mutations accumulate throughout the genome (Al-Tassan et al. 2002). This variant of FAP is referred to as *MUTYH*-associated polyposis (MAP) (Sampson et al. 2003; Dolwani et al. 2003). As part of their mutator phenotype, CRCs from these patients exhibit G:C to A:T transversions within the *APC* gene in particular, often leading to truncation of the APC protein (Halford et al. 2003). Sporadic mutations in *MUTYH* do not appear to play a significant role in CRC development (Halford et al. 2003).

1.6 DNA Mismatch Repair Genes and Lynch Syndrome

More than 50 years after Aldred Scott Warthin's 1913 construction of family pedigrees, Henry T. Lynch and colleagues published two large Midwestern kindreds (Lynch et al. 1966) and updated the original "Family G" pedigree (Lynch and Krush 1971).

Their work defined a "cancer family syndrome" now known as Lynch syndrome, previously referred to as hereditary nonpolyposis CRC or hereditary nonpolyposis colon cancer (HNPCC). This autosomal dominant inherited condition confers a predisposition to CRC with an 80 % lifetime risk and an average age of 44–45 years old at the time of onset (Lynch and Lynch 2000). Lynch syndrome comprises roughly 3–5 % of all CRCs, making it the most commonly inherited form of CRC. Shared across all Lynch syndrome families is an enhanced risk of developing CRC in the absence of widespread premalignant polyposis. Some families also exhibit a predisposition for developing certain extracolonic tumors. This originally led to the classification of families into one of two variants: Lynch syndrome I families exhibiting site-specific manifestation in the colon and rectum and Lynch syndrome II families exhibiting additional predisposition to malignancies of the endometrium, stomach, ovaries, small intestine, hepatobiliary tract, urinary tract, and/or nervous system (Vasen et al. 1996). Approximately 60–85 % of CRCs in Lynch syndrome patients occur proximal to the splenic flexure, compared to 34 % for overall CRC (Lynch et al. 1993).

Henry T. Lynch coined the alternative designation of HNPCC to describe Lynch syndrome (Lynch et al. 1985), referring to the fact that those affected do not develop adenomatous polyps at an elevated rate compared to the general population (Lynch and Lynch 2000). This term has fallen out of favor with the discovery that the adenomatous polyps that develop in these patients are in fact the precancerous benign lesions that ultimately progress into CRC (Lanspa et al. 1990). Lynch syndrome individuals develop adenomatous polyps at a relatively early age; these lesions have a greater likelihood of progressing to malignancy relative to their counterparts in the general population (Lynch and Lynch 2000). This first suggested that Lynch syndrome may result from the inheritance of a genetic factor or factors that enable accelerated progression of adenomas to adenocarcinomas. What are these genetic factors, and what is the mechanism by which they accelerate carcinogenesis?

The molecular basis of Lynch syndrome further emerged with the critical observation of mutations of simple repetitive DNA sequences in primarily right-sided colon cancers (Ionov et al. 1993). Mapping of this genetic susceptibility identified a locus on chromosome 2p (Aaltonen et al. 1993; Peltomaki et al. 1993) in which causative mutations were found specifically in the *MSH2* gene, a homolog of the *E. coli mutS* gene, which encodes a critical mismatch repair protein (Fishel et al. 1993). In general, the DNA mismatch repair pathway protects the genome from single-base mispairings that can result from DNA damage (Duncan and Miller 1980), or errors in replication (Modrich 1991) or recombination (Holliday 1964). By impairing the ability of the cell to repair mismatched bases, mutations in the *MSH2* gene lead to the accumulation of other mutations, especially the destabilization of short sequence repeats (often referred to as microsatellites) that occur across the genome (Fishel et al. 1993). When mismatch repair is defective, these short sequence repeats become exceptionally susceptible to replication errors that occur due to slippage of the DNA polymerase (Streisinger et al. 1966). The high microsatellite instability (MSI-H) phenotype, a hallmark of the tumors that occur in most Lynch syndrome families, is characterized by expansions or contractions of these repeats wherever they occur across the genome. This phenotype leads to frameshift mutations within genes, or disruptive modifications to intergenic sequences that regulate gene expression.

In addition to *MSH2*, other Lynch syndrome genes were identified and found to encode other components of the DNA mismatch repair pathway. The protein product of the *MSH2* gene forms an active complex with the protein product of the *MSH6* gene, another homolog of the *E. coli mutS*. Together, they mark genomic sites where base pair mismatches occur, especially those following DNA replication (Marsischky et al. 1996). These marks enable a second complex containing the protein products of the *MLH1* (*E. coli mutL* homolog) and *PMS2* (*S. cerevisiae* postmeiotic segregation increased 2) genes to repair the mismatch. Similar to *MSH2*, the *MSH6*, *MLH1*, and *PMS2* genes can be mutated in the germline of Lynch syndrome families (Papadopoulos et al. 1994; Nicolaides et al. 1994; Miyaki et al. 1997; Akiyama et al. 1997). Of all the germline mutations identified in Lynch syndrome families, approximately 50–60 % are found in *MSH2*, 30–40 % in *MLH1*, 7–10 % in *MSH6*, and <5 % in *PMS2* (Peltomaki 2005). Mutations in any of these genes result in enhanced susceptibility to colorectal and other tumors, all of which exhibit microsatellite instability due to defective mismatch repair.

1.7 Additional Candidate Genes in the DNA Mismatch Repair Pathway

PMS2 was originally identified as a candidate gene for Lynch syndrome along with *PMS1* through a database search for homologs of the yeast *MLH1* gene (Nicolaides et al. 1994). While *PMS2* mutations were subsequently identified in Lynch syndrome families, the importance of *PMS1* mutations remains unclear (Boland and Goel 2010). Although a *PMS1* mutation was originally reported to be the cause of Lynch syndrome in a single patient (Nicolaides et al. 1994), a germline *MSH2* mutation was later identified within the patient's family (Liu et al. 2001). Similarly, mutations in the related *MLH3* gene (a homolog of *E. coli mutL*) have been associated with microsatellite instability (Lipkin et al. 2000) and have been debated as a potential cause for Lynch syndrome (Loukola et al. 2000; Ohmiya et al. 2001). Multiple studies of Lynch syndrome families differ in whether they detect (Wu et al. 2001) or fail to detect (Liu et al. 2003) a correlation between *MLH3* status and familial susceptibility to CRC. Variation in *MLH3* status may make a small contribution to CRC risk that acts in conjunction with mutations in other mismatch repair genes (Liu et al. 2003). Finally, mutations in the *EXO1* gene have been proposed to contribute to Lynch syndrome based on the interaction of the EXO1 protein with other DNA mismatch repair proteins (Wu et al. 2001); however, mutations have not yet been identified in those with Lynch syndrome (Thompson et al. 2004).

1.8 Dominant Inheritance of Lynch Syndrome

In most cases, individuals with Lynch syndrome inherit a mutation in a single DNA mismatch repair allele and develop tumors after the second allele at the same locus has become mutated (Hemminki et al. 1994). This mechanism explains the dominant inheritance of Lynch syndrome in light of Alfred G. Knudson's two-hit hypothesis through an enhanced susceptibility to a second hit (Knudson 1971). The causative "Family G" cancer susceptibility mutation, for example, was ultimately mapped to a splice acceptor site in a single allele of *MSH2* (Yan et al. 2000). There are exceptions to this model, such as families carrying a dominant negative *MLH1* allele (Parsons et al. 1995) or a dominant negative *PMS2* allele (Nicolaides et al. 1998) that lead to mismatch repair defects even in the patients' nonneoplastic cells. Finally, a single allele of *MLH1* can be epigenetically silenced in the germline through hypermethylation in multiple individuals with Lynch syndrome (Gazzoli et al. 2002; Suter et al. 2004). This hemiallelic hypermethylation in the germline is referred to as epimutation of the mismatch repair gene and confers susceptibility to tumors in which the other allele is lost through mutation (Gazzoli et al. 2002). Most examples of MLH1 methylation, however, are found in sporadic colon tumors with high microsatellite instability. An apparently heritable case of epimutation of *MSH2* has also been identified in another Lynch syndrome family (Chan et al. 2006).

1.9 Heterogeneity of Lynch Syndrome in Different Families

As previously described, Lynch syndrome families were formerly characterized by either susceptibility to CRC alone (Type I) or susceptibility to both CRCs and extracolonic tumors (Type II). What is the molecular basis for these observed differences? Part of the explanation lies in differences between the four DNA mismatch repair genes mutated in Lynch syndrome. Studies suggest that *MSH6* mutations, for example, result in a later average onset of disease (Plaschke et al. 2004; Hendriks et al. 2004) and an elevated frequency of tumors at extracolonic sites (Plaschke et al. 2004) relative to Lynch syndrome as a whole. Among unrelated families who carry mutations within the same gene, different mutations also present differently. For example, a particular intronic *MLH1* mutation that silences the affected allele has been reported to result in a relatively low frequency of cancers other than CRC (Jager et al. 1997). Similarly, a mutation leading to hypermethylation of the promoter of a single *MSH2* allele was associated with relatively few extracolonic tumors (Lynch et al. 2011). Clinical differences between various germline mutations could potentially be explained by the fact that complete loss of mismatch repair gene expression removes the potential for dominant negative effects of mutant proteins. Alternatively, missense mutations that lead to the expression of a protein with only partial loss of function may be less clinically severe (Beck et al. 1997).

The heterogeneity of Lynch syndrome is also reflected in its diagnosis. The most widely used standard for identifying an individual affected by Lynch syndrome and subsequent family members is known as the Amsterdam II criteria. To meet this standard, a family must meet all of the following conditions (Vasen et al. 1999):

- At least three relatives must suffer from cancer of one of the following sites: colon/rectum, endometrium, small intestine, ureter, or renal pelvis.
- At least two successive generations must be affected.

- At least one of the family members must be diagnosed with cancer before the age of 50.
- At least one relative must be a first-degree relative of two others.
- FAP must be excluded as the cause in any family member with CRC.
- Pathologic exam must be used to verify tumors.

Another set of specifications used to identify individuals who might have Lynch syndrome is known as the Revised Bethesda Guidelines (Umar et al. 2004). These specifically are useful for recommending when tumors should be tested for microsatellite instability. Unlike the Amsterdam II criteria, only one of the following must be true for an individual with a colorectal tumor:

- CRC is diagnosed in the patient before the age of 50.
- Multiple tumors occur in the same patient either at the same time or at different times, of either colorectal or other Lynch syndrome-associated origin.
- CRC tumor is found in a patient under the age of 60, showing histological evidence of MSI-H phenotype.
- CRC is found in at least one first-degree relative with a Lynch syndrome-associated tumor, with at least one tumor diagnosed before the age of 50.
- CRC is found in at least two first- or second-degree relatives with Lynch syndrome-associated tumors.

Not all Lynch syndrome families meet the Amsterdam II or the Revised Bethesda criteria. An early study found that approximately 82 % of Lynch syndrome families met the original Amsterdam criteria (Peltomaki and Vasen 1997). Only 65 % percent of Amsterdam-positive families and 34 % of Bethesda-positive families carry a germline mutation within the coding region of a mismatch repair gene (Scott et al. 2001). These data reflect the heterogeneity of the syndrome, and particularly highlight the fact that some Lynch syndrome families show susceptibility to tumors, which do not exhibit MSI-H.

1.10 Alternative Lynch Syndrome Inheritance/Familial Colorectal Cancer Type X

The autosomal dominant pattern of Lynch syndrome inheritance suggests the transmission of a single, high-penetrance pathological mutation within each pedigree, yet only 65 % of Lynch syndrome families have an identified germline mutation in a mismatch repair gene (Scott et al. 2001). The remaining 35 % of families, in which the underlying genetic cause remains unclear but who exhibit similar patterns of inheritance, have alternatively been grouped into a syndrome known as familial colorectal cancer type X (FCCTX) (Lindor et al. 2005; Jass 2006). When grouped as a distinct syndrome, in comparison to traditional Lynch syndrome, these individuals exhibit a reduced rate of adenoma to carcinoma progression, a reduced overall risk of cancer with later average age of onset and a greater proportion of distal as opposed to proximal CRCs (Llor et al. 2005; Valle et al. 2007). In general, the tumors in FCCTX individuals may also be less poorly differentiated and less infiltrated with lymphocytes than those occurring in Lynch syndrome (Llor et al. 2005; Valle et al. 2007). Which other genes could harbor the mutations associated with the inherited susceptibility to CRC observed in FCCTX families? An important finding was the identification of a mutation in the *TGFBR2* gene, encoding the type II receptor for the growth-inhibiting TGF-β ligands, as the cause of Lynch syndrome in a family whose tumors did not exhibit high MSI (Lu et al. 1998). As in other Lynch syndrome cases, the wild-type allele of *TGFBR2* was lost in the tumor (Lu et al. 1998). The connection of *TGFBR2* to other families with Lynch syndrome likely relates to the fact that the gene contains at least one microsatellite sequence that is a common site of frameshift mutations in tumors with the MSI-H phenotype (Markowitz et al. 1995; Lu et al. 1995). Defects in the TGF-β signaling pathway attenuate its antiproliferative effect in the colonic epithelium and have been identified in sporadic CRCs and in other types of tumors (Alexandrow and Moses 1995; Markowitz and Roberts 1996). These findings cumulatively suggest that mutation of the *TGFBR2* gene is an important downstream consequence of the MSI-H phenotype through which CRC can be promoted.

A similar study identified mutation of the *BMPR1A* gene in at least one Lynch syndrome family with tumors showing loss of the wild-type allele but not exhibiting the MSI-H phenotype (Nieminen et al. 2011). As discussed in more detail below, the *BMPR1A* gene encodes a receptor for the bone morphogenetic protein family of ligands. Studies have shown that approximately 20 % of gene carriers in definitive Lynch syndrome families form tumors that are microsatellite stable (Iino et al. 2000). While some of these pedigrees harbor *TGFBR2* or *BMPR1A* mutations, other Lynch syndrome genes or mechanisms of tumor formation remain to be discovered.

1.11 *BMPR1A*, *SMAD4*, and Familial Juvenile Polyposis

In addition to FAP and Lynch syndrome, several other recognized familial syndromes lead to the formation of both colorectal polyps and CRC. The first known report of a juvenile polyp was published in the first half of the twentieth century (Diamond 1939). More common in children than in adults, juvenile polyps are hamartomatous, contrasting with adenomatous polyps in that they arise from the underlying tissues rather than from the epithelial lining of the lower intestine. Hamartomas in general consist of multiple tissue components normally found at the site, arranged as a disorganized mass. Sporadically occurring juvenile polyps grow at a rate that is similar to the surrounding tissue, and they are thought to lack malignant potential. Family pedigrees exhibiting a syndrome of multiple juvenile polyposis were first constructed in 1966 (Smilow et al. 1966). Familial juvenile polyposis is an autosomal dominant inherited syndrome characterized by the formation of hamartomatous polyps along the gastrointestinal tract. However, it has been observed that patients with juvenile polyposis syndrome are predisposed to the development of carcinomas along the gastrointestinal tract as well (Brosens et al. 2007). The polyps themselves can exhibit mixed characteristics of hamartomatous and adenomatous polyps (Giardiello et al. 1991), particularly when larger in size, and can show signs of adenomatous epithelial dysplasia (Subramony et al. 1994). The carcinogenic mechanism is not yet clear, but it has been suggested that the microenvironment within the familial juvenile polyp triggers the transformation of adenomatous epithelium that is either incorporated into or adjacent to the polyp (Kinzler and Vogelstein 1998).

Pathogenic mutations for juvenile polyposis have been mapped to two different loci: the SMAD family member 4 (*SMAD4*) gene on chromosome 18q21.1 (Howe et al. 1998) and the bone morphogenic protein receptor type IA (*BMPR1A*) gene on chromosome 10q22.3 (Howe et al. 2001). Between 39 and 60 % of individuals with familial juvenile polyposis inherit a mutation in one of these two genes (Howe et al. 2004; Aretz et al. 2007). The Smad4 protein is a central component of multiple signaling pathways activated by ligands of the transforming growth factor-β (TGF-β) superfamily. *BMPR1A* encodes a transmembrane serine/threonine kinase receptor activated by ligands of this TGF-β superfamily, including bone morphogenic proteins 2 (BMP-2) and 4 (BMP-4) (Natsume et al. 1997; Namiki et al. 1997). In normal cells, upon binding one of these ligands, the BMPR1A (type I) receptor interacts with a type II receptor and becomes phosphorylated on its kinase domain. This modification activates BMPR1A to phosphorylate receptor-regulated Smad proteins such as Smad1, Smad5, or Smad8 (Miyazono 1999). These phosphorylated Smads form activated Smad complexes that always include Smad4, known as the common-mediator Smad (Miyazono 1999). These activated complexes are able to translocate into the nucleus (Miyazono 1999). There, the complexes bind DNA and interact with various transcription factors to regulate gene transcription. In this way, signaling initiated by BMP ligands modulates cellular processes ranging from growth and proliferation to differentiation and apoptosis (Massague et al. 2000).

Most pathogenic mutations in *BMPR1A* produce a truncated protein that is unable to bind ligands and/or phosphorylate target proteins (Zhou et al. 2001). The majority of characterized mutations in *SMAD4* are similarly loss-of-function mutations that compromise the transcriptional effects of BMP signaling to varying degrees (Carr et al. 2012). What is the effect of disabling BMP signaling on tumor initiation or tumor development? BMP signaling can promote the differentiation of CRC cell lines in vitro (Lombardo et al. 2011) and controls cell fate in a variety of tissues (ten Dijke et al. 2003). Paradoxically, BMP-4, in certain contexts, can promote epithelial-to-mesenchymal transition (Theriault et al. 2007), a process that facilitates migration in normal cells and invasion and metastasis in tumor cells (Huber et al. 2005). Similarly, high BMP expression correlates with invasion and metastasis in both melanomas (Rothhammer et al. 2005) and bone cancers (Dai et al. 2005). Contrary to evidence of the positive influence of BMP ligands on tumorigenesis, loss-of-function mutations in *BMPR1A* and *SMAD4* can support both invasion and metastasis in CRC development. It appears that the consequences of BMP signaling vary depending on the specific type of cancer (Thawani et al. 2010). The evidence for CRC in particular will be discussed further in the following section.

Loss of *BMPR1A* and/or *SMAD4* appears to be critical in sporadic CRC development. *BMPR1A* gene expression is suppressed in more than 50 % of primary CRCs (Kim et al. 2008). Deletions within the long arm of chromosome 18, where the *SMAD4* gene is located, have been recognized as one of the most common genetic abnormalities in sporadic CRC, present in more than 70 % of these tumors (Vogelstein et al. 1988). This event occurs relatively late in the process of carcinogenesis, as only 47 % of late adenomas and less than 15 % of early adenomas exhibit these deletions (Vogelstein et al. 1988). *SMAD4* has been identified as one of the genes within this locus whose loss in CRC plays a critical role in tumor development (Miyaki et al. 1999). Deletions at this locus frequently also span the nearby *SMAD2* and *DCC* (Deleted in Colorectal Carcinoma) genes (Fearon et al. 1990). However, both *SMAD2* and *DCC* exhibit lower rates of somatic mutation relative to *SMAD4* and have been studied to a lesser extent as potential tumor suppressor genes (Riggins et al. 1997; Takagi et al. 1998; Mehlen et al. 1998). In addition to deletions within chromosome 18q, biallelic *SMAD4* mutations are found in invasive primary CRCs, especially those with distant metastases, but rarely in adenomas (Miyaki et al. 1999). These findings suggest that loss of *SMAD4* does not function to initiate CRC development but rather contributes to the subsequent stages of tumor progression. This is consistent with the notion that loss of BMP signaling promotes invasion and metastasis.

1.12 *PTEN* and Cowden Syndrome

One of the genes initially proposed to be a site of germline mutations for familial juvenile polyposis was the phosphatase and tensin homolog (*PTEN*) gene on chromosome 10q23.3 (Lynch et al. 1997; Olschwang et al. 1998). The *PTEN* and *BMPR1A* genes are located close to one another on chromosome 10; some affected individuals harbor a germline deletion that includes both genes (Menko et al. 2008). Individuals who inherit *PTEN* mutations are now classified by the designation *PTEN* hamartoma tumor syndrome (PHTS) (Liaw et al. 1997; Marsh et al. 1997). Some of the variants of this syndrome cause phenotypes that overlap with those of familial juvenile polyposis (Eng and Ji 1998). Cowden syndrome, for example, not only exhibits gastrointestinal features that occur early and are difficult to distinguish from familial juvenile polyposis but also features characteristic extraintestinal manifestations that generally present later in life (Eng and Ji 1998).

Cowden syndrome is inherited in an autosomal dominant pattern. Affected individuals develop multiple hamartomas at diverse sites throughout the body and are characterized by enhanced susceptibility to breast, thyroid (Lloyd and Dennis 1963; Brownstein et al. 1978), endometrial (Marsh et al. 1998), renal, and CRCs (Riegert-Johnson et al. 2010). PHTS variants include Cowden, Bannayan–Riley–Ruvalcaba, proteus, and proteus-like syndromes, as well as Lhermitte–Duclos disease, and collectively exhibit a diverse spectrum of phenotypes (Marsh et al. 1998). Bannayan–Riley–Ruvalcaba syndrome is also characterized by the formation of multiple hamartomas in the intestines, as well as various other benign lesions (Cohen 1990).

Whether individuals with Bannayan–Riley–Ruvalcaba syndrome exhibit significant predisposition to the development of CRC is not well characterized. However, it is clear that the broader PHTS is associated with elevated risk of CRC (Tan et al. 2012).

How do germline mutations in the *PTEN* gene cause benign tumor growth and enhance CRC risk? *PTEN* encodes a phosphatase that suppresses tumor formation by controlling the cell cycle. Specifically, the PTEN protein removes phosphate groups from a class of lipids known as phosphoinositides (Maehama and Dixon 1998). This results in inhibition of the Akt/PKB signaling pathway, a mechanism through which growth factors stimulate cell proliferation (Stambolic et al. 1998; Sun et al. 1999). *PTEN* mutations therefore result in prolonged growth factor signaling and acceleration of the cell cycle (Datta et al. 1996). *PTEN* mutations can also reverse cell cycle arrest in G1-phase (Li and Sun 1998) and promote cell survival through the evasion of apoptosis (Kennedy et al. 1997).

PTEN mutations occur in many sporadic cancers; frequently both alleles are inactivated (Sansal and Sellers 2004). In one study, *PTEN* mutations were detected in CRC at a frequency of only 2.2 %, although *PTEN* expression is lost in approximately 35 % of CRCs, most likely via epigenetic mechanisms (Naguib et al. 2011). *PTEN* expression may prove to be a valuable predictive marker, as loss of expression correlates clinically with lack of tumor response to the epidermal growth factor receptor inhibitor cetuximab (Negri et al. 2010). Another study identified *PTEN* mutations in approximately 17 % of CRCs, all of which were found in tumors from patients with either advanced local or metastatic disease (Dicuonzo et al. 2001). Based on these observations, *PTEN* mutations may be acquired relatively late in the development of CRC, often through a microsatellite-instability-mediated mechanism (Dicuonzo et al. 2001).

1.13 *STK11/LKB1* and Peutz–Jeghers Syndrome

Physicians Jan Peutz and Harold Jeghers identified a syndrome that is characterized by gastrointestinal polyps as well as mucocutaneous hyperpigmented lesions of the mouth, hands, and feet (Peutz 1921; Jeghers et al. 1949). Similar to many of the previously described CRC syndromes, Peutz–Jeghers displays an autosomal dominant pattern of inheritance and is associated with increased risk of cancer at multiple sites (Giardiello et al. 1987). In particular, affected individuals exhibit a predisposition to carcinomas at various sites along the gastrointestinal tract, the reproductive organs, lungs, and liver (Boardman et al. 1998). Unlike the adenomatous polyps that characterize FAP, Peutz–Jeghers polyps are hamartomatous lesions, meaning that they are composed primarily of smooth muscle rather than epithelial tissue. They are relatively few in number and occur all along the gastrointestinal tract, but particularly in the small intestine. Also in contrast to adenomatous polyps, these hamartomas bear little malignant potential as precursor lesions in and of themselves.

Pathogenic mutations have been mapped to the gene encoding Serine/Threonine Kinase 11, also known as Liver Kinase B1 (*STK11/LKB1*) (Jenne et al. 1998;

Hemminki et al. 1998). Loss-of-function germline mutations in a single allele of this putative tumor suppressor gene have been identified in the majority of Peutz–Jeghers cases (Amos et al. 1993) but are found only in a small percentage of sporadic CRC (Wang et al. 1998; Avizienyte et al. 1999). While biallelic loss of *APC* efficiently initiates the formation of a benign lesion that possesses long-term malignant potential, biallelic loss of *STK11/LKB1* cannot. Relative to *APC* mutations, *STK11/LKB1* mutations likely occur later in the sequence of carcinoma development and contribute less to the rate of carcinogenesis. Germline mutations inherited by Peutz–Jeghers patients do, however, accelerate many types of carcinogenesis. It is known that the serine/threonine kinase encoded by this gene can phosphorylate and activate a number of other kinases, including the 5′ adenosine monophosphate-activated protein kinase (AMPK), which controls cellular glucose and fat metabolism (Hawley et al. 2003; Woods et al. 2003). STK11/LKB1 deficiency is thought to drive carcinogenesis by causing an inability to suppress growth and arrest cell cycle progression in G1 in the absence of adequate energetic and nutrient resources (Shaw et al. 2004).

1.14 Summary and Conclusions

Through the past 100 years of studying CRC genetics, a number of themes have emerged regarding its development as a heterogeneous disease driven by aberrant genetics and gene expression. First, the order in which sporadic mutations occur in relation to tumor histopathology reflects the fact that certain hits increase the probability of specific subsequent mutations. Mutations that disable mismatch repair and lead to microsatellite instability predispose cells to the acquisition of mutations in the *TGFBR2* gene which in turn inactivate growth inhibition by the TGF-β signaling pathway (Akiyama et al. 1997; Calin et al. 2000). Distinct pathways of CRC progression are shaped or selected by their initial or early mutations. The order in which mutations occur in the sequence of CRC development also reflects the fact that the malignant potential of a tumor is limited or abolished by the premature occurrence of particular mutations earlier than their optimal timing for CRC progression. For example, *KRAS* mutation early in tumor development leads to the formation of a hyperplastic lesion of limited potential, possibly due to the absence of activated Wnt signaling (Vogelstein and Kinzler 2004). Negative regulation of Wnt signaling through proteins, such as APC, has therefore been described as a "gatekeeper" to colorectal transformation. These gatekeepers must be bypassed to enable the rate of cell growth to exceed cell death in the colorectal epithelium (Kinzler and Vogelstein 1996). It is unknown what mechanisms bypass this requirement in the subset of CRCs in which APC and β-catenin remain wild type, but observations from FAP patients with tumors demonstrate that these mutations represent very effective means to initiate adenoma formation. In the future, additional studies correlating CRC genotypes with clinical phenotypes will increasingly guide personalized treatment of patients according to their specific subtype of CRC. As our grasp of the genetic events underlying CRC formation and progression becomes more complete, additional changes in coding and noncoding genes will also provide a foundation for the development of targeted therapeutic interventions.

Many of the most common genetic defects in CRC disrupt regulators of transcription and therefore lead to widespread changes in gene expression. The Wnt and TGF-β superfamily signaling pathways, in particular, are critical in regulating the balance between proliferation and differentiation in normal colorectal crypts and in CRCs (Fevr et al. 2007; Kosinski et al. 2007), reflecting the nature of CRC as a disease of colorectal stem cell origin (Ricci-Vitiani et al. 2009; Abdul Khalek et al. 2010). As a result, the search for diagnostic, prognostic, and predictive biomarkers has expanded beyond the level of coding genes to include noncoding genes, as well as transcript and protein levels. Epigenetic mechanisms dysregulate tumor suppressor genes such as *PTEN* (Naguib et al. 2011) or DNA repair genes such as *MLH1* (Goel et al. 2007). MicroRNA dysregulation is also a new area of great potential in translational CRC research (Nakajima et al. 2006; Slaby et al. 2007). In the future, CRC genetics will integrate with epigenetics and gene regulation, as well as with the growing fields of genomics, transcriptomics, and proteomics. The intersection of these fields will provide a more complete picture of the molecular events underlying the critical steps of tumor formation, progression, invasion, and metastasis and will enable more meaningful stratification of CRCs for both prognosis and treatment.

References

Aaltonen LA et al (1993) Clues to the pathogenesis of familial colorectal cancer. Science 260(5109):812–816

Abdul Khalek FJ, Gallicano GI, Mishra L (2010) Colon cancer stem cells. Gastrointest Cancer Res (suppl 1):S16–S23

Akiyama Y et al (1997) Transforming growth factor beta type II receptor gene mutations in adenomas from hereditary nonpolyposis colorectal cancer. Gastroenterology 112(1):33–39

Alexandrow MG, Moses HL (1995) Transforming growth factor beta and cell cycle regulation. Cancer Res 55(7):1452–1457

Al-Tassan N et al (2002) Inherited variants of MYH associated with somatic G:C→T:A mutations in colorectal tumors. Nat Genet 30(2):227–232

Amos CI et al (1993) Peutz-Jeghers syndrome. In: Pagon RA et al (eds) Gene reviews. University of Washington, Seattle, WA

Aretz S et al (2006) MUTYH-associated polyposis: 70 of 71 patients with biallelic mutations present with an attenuated or atypical phenotype. Int J Cancer 119(4):807–814

Aretz S et al (2007) High proportion of large genomic deletions and a genotype phenotype update in 80 unrelated families with juvenile polyposis syndrome. J Med Genet 44(11):702–709

Avizienyte E et al (1999) LKB1 somatic mutations in sporadic tumors. Am J Pathol 154(3):677–681

Beck NE et al (1997) Genetic testing is important in families with a history suggestive of hereditary non-polyposis colorectal cancer even if the Amsterdam criteria are not fulfilled. Br J Surg 84(2):233–237

Behrens J et al (1998) Functional interaction of an axin homolog, conductin, with beta-catenin, APC, and GSK3beta. Science 280(5363):596–599

Boardman LA et al (1998) Increased risk for cancer in patients with the Peutz-Jeghers syndrome. Ann Intern Med 128(11):896–899

Bodmer WF et al (1987) Localization of the gene for familial adenomatous polyposis on chromosome 5. Nature 328(6131):614–616
Boland CR, Goel A (2010) Microsatellite instability in colorectal cancer. Gastroenterology 138(6):2073–2087 e3
Brosens LA et al (2007) Risk of colorectal cancer in juvenile polyposis. Gut 56(7):965–967
Brownstein MH, Wolf M, Bikowski JB (1978) Cowden's disease: a cutaneous marker of breast cancer. Cancer 41(6):2393–2398
Calin GA et al (2000) Genetic progression in microsatellite instability high (MSI-H) colon cancers correlates with clinico-pathological parameters: a study of the TGRbetaRII, BAX, hMSH3, hMSH6, IGFIIR and BLM genes. Int J Cancer 89(3):230–235
Carr JC et al (2012) Germline mutations in SMAD4 disrupt bone morphogenetic protein signaling. J Surg Res 174(2):211–214
Chan TL et al (2006) Heritable germline epimutation of MSH2 in a family with hereditary nonpolyposis colorectal cancer. Nat Genet 38(10):1178–1183
Cohen MM Jr (1990) Bannayan-Riley-Ruvalcaba syndrome: renaming three formerly recognized syndromes as one etiologic entity. Am J Med Genet 35(2):291–292
Cripps WH (1882) Two cases of disseminated polypus of the rectum. Trans Pathol Soc Lond 33:165–168
Dai J et al (2005) Bone morphogenetic protein-6 promotes osteoblastic prostate cancer bone metastases through a dual mechanism. Cancer Res 65(18):8274–8285
Datta K et al (1996) Akt is a direct target of the phosphatidylinositol 3-kinase. Activation by growth factors, v-src and v-Ha-ras, in Sf9 and mammalian cells. J Biol Chem 271(48):30835–30839
Diamond M (1939) Adenoma of the rectum in children: report of a case in a thirty-month-old girl. Am J Dis Child 57:360–367
Dicuonzo G et al (2001) Colorectal carcinomas and PTEN/MMAC1 gene mutations. Clin Cancer Res 7(12):4049–4053
Dolwani S et al (2003) MYH polyposis: a new autosomal recessive form of familial adenomatous polyposis due to defective base excision repair—reappraisal of genetic risk and family management. Gastroenterology 124(suppl 1):A46–A198
Douglas JA et al (2005) History and molecular genetics of Lynch syndrome in family G: a century later. JAMA 294(17):2195–2202
Dukes C (1930) The hereditary factor in polyposis intestini, or multiple adenomata. Cancer Rev Br 5:241–256
Duncan BK, Miller JH (1980) Mutagenic deamination of cytosine residues in DNA. Nature 287(5782):560–561
Dutrillaux B (1988) [Recent data on the cytogenetics of colorectal adenocarcinoma]. Bull Cancer 75(6):509–516
Eng C, Ji H (1998) Molecular classification of the inherited hamartoma polyposis syndromes: clearing the muddied waters. Am J Hum Genet 62(5):1020–1022
Fearon ER et al (1990) Identification of a chromosome 18q gene that is altered in colorectal cancers. Science 247(4938):49–56
Fevr T et al (2007) Wnt/beta-catenin is essential for intestinal homeostasis and maintenance of intestinal stem cells. Mol Cell Biol 27(21):7551–7559
Fishel R et al (1993) The human mutator gene homolog MSH2 and its association with hereditary nonpolyposis colon cancer. Cell 75(5):1027–1038
Fuchs CS et al (1994) A prospective study of family history and the risk of colorectal cancer. N Engl J Med 331(25):1669–1674
Galiatsatos P, Foulkes WD (2006) Familial adenomatous polyposis. Am J Gastroenterol 101(2):385–398
Gardner E (1951) A genetic and clinical study of intestinal polyposis, a predisposing factor for carcinoma of the colon and rectum. Am J Hum Genet 3:167–176
Gardner EJ, Plenk HP (1952) Hereditary pattern for multiple osteomas in a family group. Am J Hum Genet 4(1):31–36

Gardner EJ, Richards RC (1953) Multiple cutaneous and subcutaneous lesions occurring simultaneously with hereditary polyposis and osteomatosis. Am J Hum Genet 5(2):139–147
Gardner EJ, Stephens FE (1950) Cancer of the lower digestive tract in one family group. Am J Hum Genet 2(1):41–48
Gazzoli I et al (2002) A hereditary nonpolyposis colorectal carcinoma case associated with hypermethylation of the MLH1 gene in normal tissue and loss of heterozygosity of the unmethylated allele in the resulting microsatellite instability-high tumor. Cancer Res 62(14):3925–3928
Giardiello F (1995) Gastrointestinal polyposis syndromes and hereditary nonpolyposis colorectal cancer. In: Rustigi AK (ed) Gastrointestinal cancers: biology, diagnosis, and therapy. Lippincott-Raven, Philadelphia, pp 367–377
Giardiello FM et al (1987) Increased risk of cancer in the Peutz-Jeghers syndrome. N Engl J Med 316(24):1511–1514
Giardiello FM et al (1991) Colorectal neoplasia in juvenile polyposis or juvenile polyps. Arch Dis Child 66(8):971–975
Goel A et al (2007) The CpG island methylator phenotype and chromosomal instability are inversely correlated in sporadic colorectal cancer. Gastroenterology 132(1):127–138
Groden J et al (1991) Identification and characterization of the familial adenomatous polyposis coli gene. Cell 66(3):589–600
Groen EJ et al (2008) Extra-intestinal manifestations of familial adenomatous polyposis. Ann Surg Oncol 15(9):2439–2450
Haggar FA, Boushey RP (2009) Colorectal cancer epidemiology: incidence, mortality, survival, and risk factors. Clin Colon Rectal Surg 22(4):191–197
Half E, Bercovich D, Rozen P (2009) Familial adenomatous polyposis. Orphanet J Rare Dis 4:22
Halford SE et al (2003) Germline mutations but not somatic changes at the MYH locus contribute to the pathogenesis of unselected colorectal cancers. Am J Pathol 162(5):1545–1548
Hamilton SR et al (1995) The molecular basis of Turcot's syndrome. N Engl J Med 332(13):839–847
Hawley SA et al (2003) Complexes between the LKB1 tumor suppressor, STRAD alpha/beta and MO25 alpha/beta are upstream kinases in the AMP-activated protein kinase cascade. J Biol 2(4):28
He TC et al (1998) Identification of c-MYC as a target of the APC pathway. Science 281(5382):1509–1512
Hemminki A et al (1994) Loss of the wild type MLH1 gene is a feature of hereditary nonpolyposis colorectal cancer. Nat Genet 8(4):405–410
Hemminki A et al (1998) A serine/threonine kinase gene defective in Peutz-Jeghers syndrome. Nature 391(6663):184–187
Henderson BR (2000) Nuclear-cytoplasmic shuttling of APC regulates beta-catenin subcellular localization and turnover. Nat Cell Biol 2(9):653–660
Hendriks YM et al (2004) Cancer risk in hereditary nonpolyposis colorectal cancer due to MSH6 mutations: impact on counseling and surveillance. Gastroenterology 127(1):17–25
Herrera L et al (1986) Gardner syndrome in a man with an interstitial deletion of 5q. Am J Med Genet 25(3):473–476
Holliday R (1964) A mechanism for gene conversion in fungi. Genet Res 5:282–304
Howe JR et al (1998) Mutations in the SMAD4/DPC4 gene in juvenile polyposis. Science 280(5366):1086–1088
Howe JR et al (2001) Germline mutations of the gene encoding bone morphogenetic protein receptor 1A in juvenile polyposis. Nat Genet 28(2):184–187
Howe JR et al (2004) The prevalence of MADH4 and BMPR1A mutations in juvenile polyposis and absence of BMPR2, BMPR1B, and ACVR1 mutations. J Med Genet 41(7):484–491
Huber MA, Kraut N, Beug H (2005) Molecular requirements for epithelial-mesenchymal transition during tumor progression. Curr Opin Cell Biol 17(5):548–558
Hulsken J, Birchmeier W, Behrens J (1994) E-cadherin and APC compete for the interaction with beta-catenin and the cytoskeleton. J Cell Biol 127(6 Pt 2):2061–2069
Ichii S et al (1993) Detailed analysis of genetic alterations in colorectal tumors from patients with and without familial adenomatous polyposis (FAP). Oncogene 8(9):2399–2405

Iino H et al (2000) DNA microsatellite instability and mismatch repair protein loss in adenomas presenting in hereditary non-polyposis colorectal cancer. Gut 47(1):37–42
Ionov Y et al (1993) Ubiquitous somatic mutations in simple repeated sequences reveal a new mechanism for colonic carcinogenesis. Nature 363(6429):558–561
Jager AC et al (1997) Reduced frequency of extracolonic cancers in hereditary nonpolyposis colorectal cancer families with monoallelic hMLH1 expression. Am J Hum Genet 61(1):129–138
Jass JR (2006) Hereditary non-polyposis colorectal cancer: the rise and fall of a confusing term. World J Gastroenterol 12(31):4943–4950
Jeghers H, Mc KV, Katz KH (1949) Generalized intestinal polyposis and melanin spots of the oral mucosa, lips and digits; a syndrome of diagnostic significance. N Engl J Med 241(26):1031–1036
Jen J et al (1994) Molecular determinants of dysplasia in colorectal lesions. Cancer Res 54(21):5523–5526
Jenne DE et al (1998) Peutz-Jeghers syndrome is caused by mutations in a novel serine threonine kinase. Nat Genet 18(1):38–43
Jones S et al (2002) Biallelic germline mutations in MYH predispose to multiple colorectal adenoma and somatic G:C→T:A mutations. Hum Mol Genet 11(23):2961–2967
Joslyn G et al (1991) Identification of deletion mutations and three new genes at the familial polyposis locus. Cell 66(3):601–613
Kennedy SG et al (1997) The PI 3-kinase/Akt signaling pathway delivers an anti-apoptotic signal. Genes Dev 11(6):701–713
Kim JC et al (2008) Individual tumorigenesis pathways of sporadic colorectal adenocarcinomas are associated with the biological behavior of tumors. Cancer Sci 99(7):1348–1354
Kinzler KW, Vogelstein B (1996) Lessons from hereditary colorectal cancer. Cell 87(2):159–170
Kinzler KW, Vogelstein B (1998) Landscaping the cancer terrain. Science 280(5366):1036–1037
Kinzler KW et al (1991) Identification of FAP locus genes from chromosome 5q21. Science 253(5020):661–665
Knudson AG Jr (1971) Mutation and cancer: statistical study of retinoblastoma. Proc Natl Acad Sci U S A 68(4):820–823
Korinek V et al (1997) Constitutive transcriptional activation by a beta-catenin-Tcf complex in APC–/– colon carcinoma. Science 275(5307):1784–1787
Kosinski C et al (2007) Gene expression patterns of human colon tops and basal crypts and BMP antagonists as intestinal stem cell niche factors. Proc Natl Acad Sci U S A 104(39):15418–15423
Lanspa SJ et al (1990) Colorectal adenomas in the Lynch syndromes. Results of a colonoscopy screening program. Gastroenterology 98(5 Pt 1):1117–1122
Leppert M et al (1987) The gene for familial polyposis coli maps to the long arm of chromosome 5. Science 238(4832):1411–1413
Levy DB et al (1994) Inactivation of both APC alleles in human and mouse tumors. Cancer Res 54(22):5953–5958
Li DM, Sun H (1998) PTEN/MMAC1/TEP1 suppresses the tumorigenicity and induces G1 cell cycle arrest in human glioblastoma cells. Proc Natl Acad Sci U S A 95(26):15406–15411
Liaw D et al (1997) Germline mutations of the PTEN gene in Cowden disease, an inherited breast and thyroid cancer syndrome. Nat Genet 16(1):64–67
Lindor NM et al (2005) Lower cancer incidence in Amsterdam-I criteria families without mismatch repair deficiency: familial colorectal cancer type X. JAMA 293(16):1979–1985
Lipkin SM et al (2000) MLH3: a DNA mismatch repair gene associated with mammalian microsatellite instability. Nat Genet 24(1):27–35
Liu T et al (2001) The role of hPMS1 and hPMS2 in predisposing to colorectal cancer. Cancer Res 61(21):7798–7802
Liu HX et al (2003) The role of hMLH3 in familial colorectal cancer. Cancer Res 63(8):1894–1899
Llor X et al (2005) Differential features of colorectal cancers fulfilling Amsterdam criteria without involvement of the mutator pathway. Clin Cancer Res 11(20):7304–7310
Lloyd KM II, Dennis M (1963) Cowden's disease. A possible new symptom complex with multiple system involvement. Ann Intern Med 58:136–142

Lombardo Y et al (2011) Bone morphogenetic protein 4 induces differentiation of colorectal cancer stem cells and increases their response to chemotherapy in mice. Gastroenterology 140(1):297–309
Loukola A et al (2000) Germline and somatic mutation analysis of MLH3 in MSI-positive colorectal cancer. Am J Pathol 157(2):347–352
Lu SL et al (1995) Mutations of the transforming growth factor-beta type II receptor gene and genomic instability in hereditary nonpolyposis colorectal cancer. Biochem Biophys Res Commun 216(2):452–457
Lu SL et al (1998) HNPCC associated with germline mutation in the TGF-beta type II receptor gene. Nat Genet 19(1):17–18
Luongo C et al (1994) Loss of Apc+ in intestinal adenomas from Min mice. Cancer Res 54(22):5947–5952
Lynch HT, Krush AJ (1971) Cancer family "G" revisited: 1895–1970. Cancer 27(6):1505–1511
Lynch HT, Lynch J (2000) Lynch syndrome: genetics, natural history, genetic counseling, and prevention. J Clin Oncol 18(21 suppl):19S–31S
Lynch HT et al (1966) Hereditary factors in cancer. Study of two large midwestern kindreds. Arch Intern Med 117(2):206–212
Lynch HT et al (1985) Hereditary nonpolyposis colorectal cancer (Lynch syndromes I and II. I. Clinical description of resource). Cancer 56(4):934–938
Lynch HT et al (1993) Genetics, natural history, tumor spectrum, and pathology of hereditary nonpolyposis colorectal cancer: an updated review. Gastroenterology 104(5):1535–1549
Lynch ED et al (1997) Inherited mutations in PTEN that are associated with breast cancer, cowden disease, and juvenile polyposis. Am J Hum Genet 61(6):1254–1260
Lynch HT et al (2011) Lynch syndrome-associated extracolonic tumors are rare in two extended families with the same EPCAM deletion. Am J Gastroenterol 106(10):1829–1836
Maehama T, Dixon JE (1998) The tumor suppressor, PTEN/MMAC1, dephosphorylates the lipid second messenger, phosphatidylinositol 3,4,5-trisphosphate. J Biol Chem 273(22):13375–13378
Markowitz SD, Roberts AB (1996) Tumor suppressor activity of the TGF-beta pathway in human cancers. Cytokine Growth Factor Rev 7(1):93–102
Markowitz S et al (1995) Inactivation of the type II TGF-beta receptor in colon cancer cells with microsatellite instability. Science 268(5215):1336–1338
Marsh DJ et al (1997) Germline mutations in PTEN are present in Bannayan-Zonana syndrome. Nat Genet 16(4):333–334
Marsh DJ et al (1998) Germline PTEN mutations in Cowden syndrome-like families. J Med Genet 35(11):881–885
Marsischky GT et al (1996) Redundancy of Saccharomyces cerevisiae MSH3 and MSH6 in MSH2-dependent mismatch repair. Genes Dev 10(4):407–420
Massague J, Blain SW, Lo RS (2000) TGFbeta signaling in growth control, cancer, and heritable disorders. Cell 103(2):295–309
Mehlen P et al (1998) The DCC gene product induces apoptosis by a mechanism requiring receptor proteolysis. Nature 395(6704):801–804
Menko FH et al (2008) Variable phenotypes associated with 10q23 microdeletions involving the PTEN and BMPR1A genes. Clin Genet 74(2):145–154
Migliore L et al (2011) Genetics, cytogenetics, and epigenetics of colorectal cancer. J Biomed Biotechnol 2011:792362
Miyaki M et al (1997) Germline mutation of MSH6 as the cause of hereditary nonpolyposis colorectal cancer. Nat Genet 17(3):271–272
Miyaki M et al (1999) Higher frequency of Smad4 gene mutation in human colorectal cancer with distant metastasis. Oncogene 18(20):3098–3103
Miyazono K (1999) Signal transduction by bone morphogenetic protein receptors: functional roles of Smad proteins. Bone 25(1):91–93
Miyoshi Y et al (1992) Somatic mutations of the APC gene in colorectal tumors: mutation cluster region in the APC gene. Hum Mol Genet 1(4):229–233
Modrich P (1991) Mechanisms and biological effects of mismatch repair. Annu Rev Genet 25:229–253

Morin PJ et al (1997) Activation of beta-catenin-Tcf signaling in colon cancer by mutations in beta-catenin or APC. Science 275(5307):1787–1790
Munemitsu S et al (1994) The APC gene product associates with microtubules in vivo and promotes their assembly in vitro. Cancer Res 54(14):3676–3681
Munemitsu S et al (1995) Regulation of intracellular beta-catenin levels by the adenomatous polyposis coli (APC) tumor-suppressor protein. Proc Natl Acad Sci U S A 92(7):3046–3050
Nagase H, Nakamura Y (1993) Mutations of the APC (adenomatous polyposis coli) gene. Hum Mutat 2(6):425–434
Naguib A et al (2011) Alterations in PTEN and PIK3CA in colorectal cancers in the EPIC Norfolk study: associations with clinicopathological and dietary factors. BMC Cancer 11:123
Nakajima G et al (2006) Non-coding MicroRNAs hsa-let-7g and hsa-miR-181b are associated with chemoresponse to S-1 in colon cancer. Cancer Genomics Proteomics 3(5):317–324
Namiki M et al (1997) A kinase domain-truncated type I receptor blocks bone morphogenetic protein-2-induced signal transduction in C2C12 myoblasts. J Biol Chem 272(35):22046–22052
Natsume T et al (1997) Interaction between soluble type I receptor for bone morphogenetic protein and bone morphogenetic protein-4. J Biol Chem 272(17):11535–11540
Negri FV et al (2010) PTEN status in advanced colorectal cancer treated with cetuximab. Br J Cancer 102(1):162–164
Neufeld KL et al (2000) APC-mediated downregulation of beta-catenin activity involves nuclear sequestration and nuclear export. EMBO Rep 1(6):519–523
Nicolaides NC et al (1994) Mutations of two PMS homologues in hereditary nonpolyposis colon cancer. Nature 371(6492):75–80
Nicolaides NC et al (1998) A naturally occurring hPMS2 mutation can confer a dominant negative mutator phenotype. Mol Cell Biol 18(3):1635–1641
Nieminen TT et al (2011) BMPR1A mutations in hereditary nonpolyposis colorectal cancer without mismatch repair deficiency. Gastroenterology 141(1):e23–e26
Nishisho I et al (1991) Mutations of chromosome 5q21 genes in FAP and colorectal cancer patients. Science 253(5020):665–669
Ohmiya N et al (2001) Germline and somatic mutations in hMSH6 and hMSH3 in gastrointestinal cancers of the microsatellite mutator phenotype. Gene 272(1–2):301–313
Ohtsubo T et al (2000) Identification of human MutY homolog (hMYH) as a repair enzyme for 2-hydroxyadenine in DNA and detection of multiple forms of hMYH located in nuclei and mitochondria. Nucleic Acids Res 28(6):1355–1364
Olschwang S et al (1998) PTEN germ-line mutations in juvenile polyposis coli. Nat Genet 18(1):12–14
Papadopoulos N et al (1994) Mutation of a mutL homolog in hereditary colon cancer. Science 263(5153):1625–1629
Parsons R et al (1995) Mismatch repair deficiency in phenotypically normal human cells. Science 268(5211):738–740
Peltomaki P (2005) Lynch syndrome genes. Fam Cancer 4(3):227–232
Peltomaki P, Vasen HF (1997) Mutations predisposing to hereditary nonpolyposis colorectal cancer: database and results of a collaborative study. The International Collaborative Group on Hereditary Nonpolyposis Colorectal Cancer. Gastroenterology 113(4):1146–1158
Peltomaki P et al (1993) Genetic mapping of a locus predisposing to human colorectal cancer. Science 260(5109):810–812
Peutz J (1921) Ned Tijdschr Geneeskd 10:134
Plaschke J et al (2004) Lower incidence of colorectal cancer and later age of disease onset in 27 families with pathogenic MSH6 germline mutations compared with families with MLH1 or MSH2 mutations: the German Hereditary Nonpolyposis Colorectal Cancer Consortium. J Clin Oncol 22(22):4486–4494
Powell SM et al (1992) APC mutations occur early during colorectal tumorigenesis. Nature 359(6392):235–237
Qian J et al (2007) Caspase cleavage of the APC tumor suppressor and release of an amino-terminal domain is required for the transcription-independent function of APC in apoptosis. Oncogene 26(33):4872–4876

Ricci-Vitiani L et al (2009) Colon cancer stem cells. J Mol Med (Berl) 87(11):1097–1104
Riegert-Johnson DL et al (2010) Cancer and Lhermitte-Duclos disease are common in cowden syndrome patients. Hered Cancer Clin Pract 8(1):6
Riggins GJ et al (1997) Frequency of Smad gene mutations in human cancers. Cancer Res 57(13):2578–2580
Rosin-Arbesfeld R, Townsley F, Bienz M (2000) The APC tumour suppressor has a nuclear export function. Nature 406(6799):1009–1012
Rothhammer T et al (2005) Bone morphogenic proteins are overexpressed in malignant melanoma and promote cell invasion and migration. Cancer Res 65(2):448–456
Samowitz WS et al (1999) Beta-catenin mutations are more frequent in small colorectal adenomas than in larger adenomas and invasive carcinomas. Cancer Res 59(7):1442–1444
Sampson JR et al (2003) Autosomal recessive colorectal adenomatous polyposis due to inherited mutations of MYH. Lancet 362(9377):39–41
Sansal I, Sellers WR (2004) The biology and clinical relevance of the PTEN tumor suppressor pathway. J Clin Oncol 22(14):2954–2963
Satoh S et al (2000) AXIN1 mutations in hepatocellular carcinomas, and growth suppression in cancer cells by virus-mediated transfer of AXIN1. Nat Genet 24(3):245–250
Scott RJ et al (2001) Hereditary nonpolyposis colorectal cancer in 95 families: differences and similarities between mutation-positive and mutation-negative kindreds. Am J Hum Genet 68(1):118–127
Shaw RJ et al (2004) The tumor suppressor LKB1 kinase directly activates AMP-activated kinase and regulates apoptosis in response to energy stress. Proc Natl Acad Sci U S A 101(10): 3329–3335
Shtutman M et al (1999) The cyclin D1 gene is a target of the beta-catenin/LEF-1 pathway. Proc Natl Acad Sci U S A 96(10):5522–5527
Sieber OM et al (2003) Multiple colorectal adenomas, classic adenomatous polyposis, and germ-line mutations in MYH. N Engl J Med 348(9):791–799
Slaby O et al (2007) Altered expression of miR-21, miR-31, miR-143 and miR-145 is related to clinicopathologic features of colorectal cancer. Oncology 72(5–6):397–402
Slupska MM et al (1999) Functional expression of hMYH, a human homolog of the Escherichia coli MutY protein. J Bacteriol 181(19):6210–6213
Smilow PC, Pryor CA Jr, Swinton NW (1966) Juvenile polyposis coli. A report of three patients in three generations of one family. Dis Colon Rectum 9(4):248–254
Smith KJ et al (1994) Wild-type but not mutant APC associates with the microtubule cytoskeleton. Cancer Res 54(14):3672–3675
Soker S et al (1998) Neuropilin-1 is expressed by endothelial and tumor cells as an isoform-specific receptor for vascular endothelial growth factor. Cell 92(6):735–745
Solomon E et al (1987) Chromosome 5 allele loss in human colorectal carcinomas. Nature 328(6131):616–619
Spirio L et al (1993) Alleles of the APC gene: an attenuated form of familial polyposis. Cell 75(5):951–957
Stambolic V et al (1998) Negative regulation of PKB/Akt-dependent cell survival by the tumor suppressor PTEN. Cell 95(1):29–39
Steigerwald K et al (2005) The APC tumor suppressor promotes transcription-independent apoptosis in vitro. Mol Cancer Res 3(2):78–89
Streisinger G et al (1966) Frameshift mutations and the genetic code. This paper is dedicated to Professor Theodosius Dobzhansky on the occasion of his 66th birthday. Cold Spring Harb Symp Quant Biol 31:77–84
Stryker SJ et al (1987) Natural history of untreated colonic polyps. Gastroenterology 93(5): 1009–1013
Subramony C et al (1994) Familial juvenile polyposis. Study of a kindred: evolution of polyps and relationship to gastrointestinal carcinoma. Am J Clin Pathol 102(1):91–97
Sun H et al (1999) PTEN modulates cell cycle progression and cell survival by regulating phosphatidylinositol 3,4,5,-trisphosphate and Akt/protein kinase B signaling pathway. Proc Natl Acad Sci U S A 96(11):6199–6204

Suter CM, Martin DI, Ward RL (2004) Germline epimutation of MLH1 in individuals with multiple cancers. Nat Genet 36(5):497–501
Takagi Y et al (1998) Somatic alterations of the SMAD-2 gene in human colorectal cancers. Br J Cancer 78(9):1152–1155
Takao M et al (1999) Differential subcellular localization of human MutY homolog (hMYH) and the functional activity of adenine:8-oxoguanine DNA glycosylase. Nucleic Acids Res 27(18):3638–3644
Tan MH et al (2012) Lifetime cancer risks in individuals with germline PTEN mutations. Clin Cancer Res 18(2):400–407
ten Dijke P et al (2003) Controlling cell fate by bone morphogenetic protein receptors. Mol Cell Endocrinol 211(1–2):105–113
Tetsu O, McCormick F (1999) Beta-catenin regulates expression of cyclin D1 in colon carcinoma cells. Nature 398(6726):422–426
Thawani JP et al (2010) Bone morphogenetic proteins and cancer: review of the literature. Neurosurgery 66(2):233–246; discussion 246
Theriault BL et al (2007) BMP4 induces EMT and Rho GTPase activation in human ovarian cancer cells. Carcinogenesis 28(6):1153–1162
Thompson E et al (2004) Hereditary non-polyposis colorectal cancer and the role of hPMS2 and hEXO1 mutations. Clin Genet 65(3):215–225
Tsuchiya R, Fujisawa N (1999) Historical survey of carcinoma of the pancreas. J Hepatobiliary Pancreat Surg 6(2):165–170
Umar A et al (2004) Revised Bethesda guidelines for hereditary nonpolyposis colorectal cancer (Lynch syndrome) and microsatellite instability. J Natl Cancer Inst 96(4):261–268
Valle L et al (2007) Clinicopathologic and pedigree differences in amsterdam I-positive hereditary nonpolyposis colorectal cancer families according to tumor microsatellite instability status. J Clin Oncol 25(7):781–786
Vasen HF et al (1996) Cancer risk in families with hereditary nonpolyposis colorectal cancer diagnosed by mutation analysis. Gastroenterology 110(4):1020–1027
Vasen HF et al (1999) New clinical criteria for hereditary nonpolyposis colorectal cancer (HNPCC, Lynch syndrome) proposed by the International Collaborative group on HNPCC. Gastroenterology 116(6):1453–1456
Vogelstein B, Kinzler KW (2004) Cancer genes and the pathways they control. Nat Med 10(8):789–799
Vogelstein B et al (1988) Genetic alterations during colorectal-tumor development. N Engl J Med 319(9):525–532
Wall ME, Wani MC (1995) Camptothecin and taxol: discovery to clinic—thirteenth Bruce F. Cain Memorial Award Lecture. Cancer Res 55(4):753–760
Wang ZJ et al (1998) Genetic pathways of colorectal carcinogenesis rarely involve the PTEN and LKB1 genes outside the inherited hamartoma syndromes. Am J Pathol 153(2):363–366
Warthin A (1913) Heredity with reference to carcinoma. Arch Intern Med 12:546–555
Woods A et al (2003) LKB1 is the upstream kinase in the AMP-activated protein kinase cascade. Curr Biol 13(22):2004–2008
Wu Y et al (2001) Germline mutations of EXO1 gene in patients with hereditary nonpolyposis colorectal cancer (HNPCC) and atypical HNPCC forms. Gastroenterology 120(7):1580–1587
Yan H et al (2000) Conversion of diploidy to haploidy. Nature 403(6771):723–724
Zhang T et al (2001) Evidence that APC regulates survivin expression: a possible mechanism contributing to the stem cell origin of colon cancer. Cancer Res 61(24):8664–8667
Zhang H et al (2007) ErbB receptors: from oncogenes to targeted cancer therapies. J Clin Invest 117(8):2051–2058
Zhou XP et al (2001) Germline mutations in BMPR1A/ALK3 cause a subset of cases of juvenile polyposis syndrome and of Cowden and Bannayan-Riley-Ruvalcaba syndromes. Am J Hum Genet 69(4):704–711
Zimmerman M (2003) In: Aufderheide A (ed) The scientific study of mummies. Cambridge University Press: Cambridge

Chapter 2
Molecular Mechanisms of Colorectal Carcinogenesis

Jatin Roper and Kenneth E. Hung

Abstract Colorectal cancer (CRC) presents in three major forms: inherited, sporadic, and familial. Although the mechanisms underlying familial CRC are poorly understood, a large body of evidence suggests that inherited and sporadic CRC are caused by sequential genetic and molecular events. There are three distinct pathways of CRC pathogenesis: the chromosomal instability pathway (CIN), the microsatellite instability pathway (MSI), and the serrated pathway. The majority of CRCs arise from the CIN pathway, which is characterized by defects in chromosomal segregation, telomere stability, and the DNA damage response. Microsatellite instability derives from the loss of DNA mismatch repair and is found in about 15 % of all CRCs, 3 % of which are associated with Lynch syndrome. The serrated pathway, recognized only in the last 15 years, describes the progression of serrated polyps to CRC. The goal of this chapter is to discuss the key genetic and molecular elements of each pathway from a historical perspective and to describe the relevance of this knowledge to the care of patients with CRC.

Keywords Colorectal cancer • Chromosomal instability • Microsatellite instability • Mouse models • Serrated pathway

2.1 Introduction

Colorectal cancer (CRC) continues to be an enormous public health burden. It is the third most common cancer in men and second most common cancer in women worldwide, with nearly 1.2 million new cases yearly, and the third leading cause of

J. Roper, M.D. • K.E. Hung, M.D., Ph.D. (✉)
Division of Gastroenterology, Department of Medicine, Tufts Medical Center,
800 Washington Street, Box #233, Boston, MA 02111, USA
e-mail: khung@tuftsmedicalcenter.org

K.M. Haigis (ed.), *Molecular Pathogenesis of Colorectal Cancer*,
DOI 10.1007/978-1-4614-8412-7_2, © Springer Science+Business Media New York 2013

cancer-related mortality, with approximately 600,000 deaths each year. The 5-year prognosis for patients with newly diagnosed metastatic colon cancer continues to be less than 20 % (Jemal et al. 2011). The underlying causes of CRC are complex and heterogeneous. Both environmental factors and genetic events contribute to CRC risk. Among the environmental risk factors for CRC are diets rich in unsaturated fats and red meat, total energy intake, excessive alcohol consumption, and reduced physical activity. Many studies have examined other exposures for their effects on CRC risk but have yielded ambiguous results (Chan and Giovannucci 2010). In contrast, there has been significant progress in identification of the specific genetic defects underlying the majority of CRCs. To develop effective CRC prevention, diagnosis, and treatment strategies, an understanding of the pathways and molecular events that drive CRC carcinogenesis is essential.

CRC presents in one of three patterns: inherited, familial, and sporadic. Inherited and familial CRC derive, at least in part, from germline mutations. Inherited CRC accounts for 10 % of cases and presents as well-characterized cancer predisposition syndromes including Lynch syndrome and familial adenomatous polyposis (FAP). Familial CRC accounts for 25 % of CRCs and presents without precisely defined Mendelian inheritance patterns or genetic etiology (Pino and Chung 2010). Sporadic CRC derives from somatic mutation, accounts for approximately 70 % of CRCs, and is not associated with family history. This chapter focuses on the genetic and molecular events underlying the three major pathways for sporadic and inherited colorectal carcinogenesis: chromosomal instability (CIN), microsatellite instability (MSI), and the CpG island methylator phenotype (CIMP) pathways.

2.2 CIN: Chromosomal Instability Pathway

2.2.1 The Adenoma–Carcinoma Sequence

By the mid-1970s, several pieces of indirect evidence suggested that colorectal adenocarcinomas may progress from adenomas (1) residual benign adenomatous tissue was found in carcinomas, (2) malignant foci were observed in larger polyps, and (3) there were rare observations of a benign-appearing polyp developing into an invasive carcinoma (Morson 1974). In 1987, Stryker and colleagues reported the natural history of unresected colonic polyps >1 cm in size in 226 patients who declined surgical resection. After 20-year follow-up, they found a 24 % risk of invasive adenocarcinoma at the site of the index polyp, and a 35 % risk of carcinoma at any colonic site (Stryker et al. 1987). Individuals affected by cancer predisposition syndromes, such as FAP, invariably develop CRC by the third or fourth decade of life if their colons are not removed (Lynch and de la Chapelle 2003). The National Polyp Study confirmed the hypothesis that colorectal carcinomas arise from adenomas through showing that polypectomy by colonoscopy reduces the

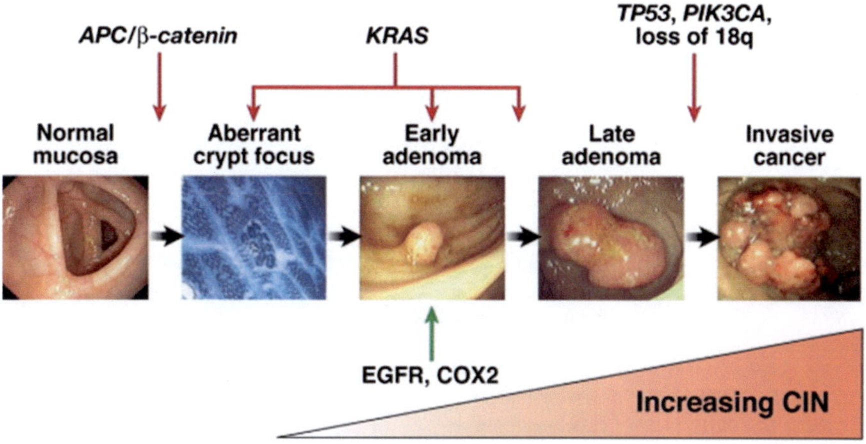

Fig. 2.1 The adenoma–carcinoma sequence. The initial step in colorectal carcinogenesis is thought to be the formation of aberrant crypt foci (ACF). Activation of the Wnt pathway occurs during this step as a result of inactivating mutations in the APC gene. Progression to adenoma and carcinoma is usually mediated by activating mutations in *KRAS* and loss of *TP53* expression, respectively. A subset of advanced adenomas may progress due to mutations in *PIK3CA* and loss of 18q. Reproduced with permission from: Pino MS, Chung DC (2010) The chromosomal instability pathway (CIN) in colon cancer. Gastroenterology 138(6):2059–2072

risk of subsequent CRC (Winawer et al. 1993). In 1990, Fearon and Vogelstein proposed a multistep genetic model of colorectal carcinogenesis in which inactivation of the adenomatous polyposis coli (*APC*) tumor suppressor gene occurs first in normal colonic mucosa, followed by activating mutations in the *KRAS* gene and subsequent additional mutations (e.g., *PIK3CA*, *TP53*, and *TGF*-β pathway genes) (Fig. 2.1 and Table 2.1) (Fearon and Vogelstein 1990; Vogelstein et al. 1988; Fearon 2011). Several key principles of the so-called adenoma–carcinoma sequence have been established (1) multiple genetic hits are required, (2) there are discrete intermediaries in the progression to cancer (Pino and Chung 2010; Haigis et al. 2008), and (3) adenomas arise from aberrant crypt foci in the colonic epithelium (Takayama et al. 1998).

2.2.2 Genomic Instability and Cancer

The mutation rate per nucleotide base pair is far too low (estimated to be approximately 10^{-9} per cell generation) to account for the multiple genetic mutations required for tumorigenesis (Albertini et al. 1990). Therefore, it has been proposed that cancer cells must acquire a "mutator phenotype" that increases the rate of spontaneous mutations (Loeb et al. 2003). 65–70 % of sporadic colorectal cancers exhibit

Table 2.1 Somatic mutations in oncogenes and tumor suppressor genes implicated in colorectal carcinogenesis

Gene	Chromosomal location	Type of mutation	Prevalence (%)	Function of gene product
Oncogenes				
KRAS	12p12	Point mutation (codons 12, 13 of exon 2)	40	Cell proliferation and survival
PIK3CA	3q26	Point mutations (E545K on exon 9, H1047R on exon 20)	15–30	Cell proliferation and survival
CDK8	13q12	Gene amplification	10–15	β-catenin activation
EGFR	7p12	Gene amplification	5–15	Cell proliferation and survival
BRAF	7q34	Point mutations activating kinase activity (most commonly V600E)	5–10	Cell proliferation and survival
CMYC	8q24	Gene amplification	5–10	Cell proliferation and survival
CCNE1	19q12	Gene amplification	5	
NRAS	1p13	Point mutation	<5	Cell proliferation and survival
CTNNB1	3p22	Stabilizing point mutations and in-frame deletions near N terminus	<5	Regulation of Wnt pathway target genes that promote tumor growth and invasion
ERBB2 (*HER2*)	17q21	Gene amplification	<5	Cell proliferation and survival
MYB	6q22-q23	Gene amplification	<5	Stimulates growth of intestinal stem cells
Tumor-suppressor genes				
APC	5q21	Frameshift, point mutation, deletion, allele loss leading to truncated protein	70–80	Inhibition of Wnt signaling
TP53	17q13	Point mutation (missense), allele loss	60–70	Cell cycle arrest, apoptosis and autophagy induction
DCC	18q21	Point mutation	50	Cell surface receptor for netrin-1, triggers tumor cell apoptosis
TGFβRII (*TGFBR2*)	3p22	Frameshift, nonsense	25	Inhibition of cell growth
SMAD4	18q21	Nonsense, missense, allele loss	10–15	Intracellular mediator of the *TGF-β* pathway
PTEN	10q23	Nonsense, deletion	10	Inhibition of PI3K activity
ACVR2A	2q22	Frameshift	10	Cellular growth
SMAD2	18q21	Nonsense, deletion, allele loss	5–10	Intracellular mediator of the *TGF-β* pathway
FBXW7	4q31	Nonsense, missense, deletion	9	Targets oncoproteins for ubiquitin-mediated degradation
SMAD3	15q22	Nonsense, deletion	5	Intracellular mediator of the *TGF-β* pathway
TCF7L2	10q25	Frameshift, nonsense	5	Regulation of the Wnt signaling
BAX	19q13	Frameshift	5	Apoptotic activator
LKB1 (*STK11*)	19p13	Deletion	Rare (limited to PJS)	Regulation of cell polarity

Modified from Fearon ER. Molecular genetics of colorectal cancer. Annu Rev Pathol 2011;6:479–507

an accelerated rate of gains or losses of whole or large portions of chromosomes that result in karyotypic variability between cells. This chromosomal instability (CIN) appears to be a dominant trait (Lengauer et al. 1997, 1998). Consequences of CIN include an imbalance in chromosomal number (aneuploidy), subchromosomal genomic amplifications, and a high frequency of loss of heterozygosity (LOH). There are currently no standardized measures of chromosomal instability; hence, CIN-positive vs. CIN-negative tumors cannot be clearly defined (Pino and Chung 2010). One theory views cancers as clonal in origin, in that they arise from a single, genomically unstable cell, but develop genetic heterogeneity due to CIN. This explains observed heterogeneity within single tumors with regard to DNA content, chromosomal number, gene expression, metabolism, resistance to cytotoxic drugs, and metastatic potential (Duesberg et al. 2004).

2.2.3 Mechanisms Leading to Chromosomal Instability

2.2.3.1 Defects in Chromosomal Segregation

The CIN phenotype can result from defects in pathways that regulate chromosomal segregation. The mitotic or spindle checkpoint ensures proper chromosome segregation by delaying the metaphase-to-anaphase transition until all pairs of duplicated chromatids are properly aligned on the spindle. Genes that encode proteins operating as spindle checkpoint regulators include *mitotic arrest-deficient* (*MAD1L1* and *MAD2L1*), *budding uninhibited by benzimidazoles 1* (*BUB1*), and *kinesin family member 11* (*KIF11*). Mutations in *BUB1* result in abnormal spindle checkpoint and CIN in chromosomally stable cell lines (Bardelli et al. 2001). Cells from dominant-negative mBub1 mutant mice demonstrate escape from apoptosis, continued cell cycle progression, and disrupted spindles (Taylor and McKeon 1997). Kinesin spindle protein, also known as Eg5, is a motor protein responsible for mitotic spindle formation and chromosomal separation during mitosis. Overexpression of *Eg5* in mice leads to spindle defects, CIN, and solid tumor formation (Castillo et al. 2007). Mutations in the hZw10, hZwilch/FLJ10036, and hROD/KNTC genes, which encode kinetochore proteins, have been reported in CRC (Wang et al. 2004).

Chromosomal missegregation due to defects in the mitotic checkpoint may lead to aneuploidy, a concept first proposed by Theodor Boveri in 1902—well before the advent of chromosomal karyotyping (Boveri 2008). The aneuploidy hypothesis proposes a two-step mechanism for tumor initiation. The first step is an event (i.e., a defect in spindle formation) that promotes chromosomal missegregation and aneuploidy. In the second step, aneuploidy destabilizes the genome, gives rise to polyclonal mutations, and results in heterogeneous karyotypes. Aneuploidy therefore stimulates tumorigenesis either by increasing the chances of LOH of a tumor-suppressor gene or by amplifying an oncogene through chromosomal duplication (Duesberg et al. 2004; Castillo et al. 2007).

2.2.3.2 Centromere Dysfunction

Another proposed cause of CIN is abnormal centromere number and function. Centrosomes serve to anchor cytoplasmic microtubules as they are arranged into a mitotic spindle apparatus. Extra centrosomes in cancer cell lines may lead to the formation of multiple spindle poles during mitosis, resulting in unequal distribution of chromosomes and CIN (Ganem et al. 2009). Polo-like kinases (Plk) are serine/threonine kinases, which regulate centrosome duplication. Elevated expression of Plk1 has been observed in 73 % of CRCs and correlate with tumor invasion, lymph node involvement, and stage (Takahashi et al. 2003). The centrosome-associated Aurora A protein is amplified and positively associated with CIN in CRC, but metastatic CRC patients with increased Aurora A gene copy number have longer overall and progression-free survival, particularly in *KRAS* wild-type tumors (Dotan et al. 2002; Herz et al. 2011). The related Aurora B protein regulates chromatid segregation, and its expression is correlated with advanced stages of CRC (Katayama et al. 1999a).

2.2.3.3 Telomere Dysfunction

CIN may also be driven by telomere dysfunction. Telomeres are hexameric DNA repeats (TTAGGG in humans) that protect the ends of eukaryotic chromosomes from fusing and breaking during segregation. A portion of telomeric DNA is lost after each round of DNA replication due the inability of DNA polymerase to completely synthesize the 3′ end of chromosomes. Cells with sufficiently shortened telomeres are targeted for senescence and apoptosis by DNA damage checkpoints. Cells that survive the checkpoint activate telomerase, which elongates telomeres. In mice deficient in the RNA component of telomerase (*Terc* −/−), telomere shortening results in aberrant crypt foci, adenomas, and gastrointestinal tumors (Rudolph et al. 2001; Plentz et al. 2003). 77–90 % of CRCs harbor shorter telomeres, compared to adjacent normal tissue, but increased telomerase activity has also been reported (Engelhardt et al. 1997; Takagi et al. 1999; Katayama et al. 1999b; Nakamura et al. 2000; Gertler et al. 2004; Chadeneau et al. 1995; Tatsumoto et al. 2000). These findings suggest that telomere shortening promotes CIN that initiates carcinogenesis, whereas telomerase activation in established carcinomas leads to immortality of cancer cells.

2.2.3.4 Loss of Heterozygosity

LOH is a key feature of CIN-positive tumors and distinguishes tumors arising from the CIN pathway from tumors arising from the MSI pathway. Approximately 25–30 % of alleles are lost in tumors (Lengauer et al. 1998). Mitotic nondisjunction, recombination between homologous chromosomes, and chromosomal deletion are among the implicated mechanisms. One study found that the majority of losses on chromosome 18 involved the whole chromosome and were caused by mitotic nondisjunction. Losses limited to a part of a chromosome were thought to be due to interchromosomal recombinations and deletions associated with DNA double-strand breaks (Thiagalingam et al. 2001).

2.2.3.5 Deficiencies in DNA Damage Response

Deficiencies in DNA damage response have been linked to human cancer. Inactivating mutations in ataxia telangiectasia mutated (*ATM*) and ataxia telangiectasia and Rad3-related (*ATR*) protein kinases lead to the ataxia telangiectasia and Seckel syndromes, respectively (Khanna and Jackson 2001). Other syndromes linked to impaired DNA damage response include Li–Fraumeni (*TP53* mutations) and hereditary breast–ovarian cancer (*BRCA1* and *BRCA2* mutations). Of these genes, only TP53 has been directly implicated in human colorectal cancer. Haploinsufficiency of histone H2AX, an ATM and ATR substrate, leads to genomic instability and tumor susceptibility in a p53-deficient background, and mouse embryonic fibroblasts derived from ATM- and H2Ax-deficient mice show severe genomic instability (Bassing et al. 2003; Celeste et al. 2003; Zha et al. 2008). Deficiency in *Chk1*, a DNA damage checkpoint protein, causes mitotic defects and disrupts Aurora B during mitosis, resulting in failure of cytokinesis and multinucleation (Peddibhotla et al. 2009).

2.2.4 Genetic Abnormalities Implicated in the Chromosomal Instability Pathway

Recent comprehensive sequencing studies have identified over 80 somatic mutations in exons of colorectal tumors. However, a limited number of these mutations are found in a significant percentage of tumors. Wood et al. predicted that perhaps 15 or fewer of these mutations in any given CRC are critical drivers of tumor initiation, progression, and/or maintenance (Wood et al. 2007). Many of the genes identified by sequencing analysis were already well known to be somatically mutated in CRC (e.g., *APC*, *KRAS*, and *TP53*). Table 2.1 describes data on oncogenes and tumor suppressor genes that are somatically mutated in CRC.

2.2.4.1 APC and the Wnt Pathway

The earliest genetic event in colorectal carcinogenesis is activation of the Wnt pathway, typically via disruption of *APC* on 5q21 (Powell et al. 1992). The APC gene product is an approximately 300-kDa protein with multiple functional domains that regulates differentiation, adhesion, polarity, migration, development, apoptosis, and chromosomal segregation (Fig. 2.2). Restoration of APC protein expression in CRC cells that lack endogenous APC expression promotes apoptosis. In the absence of

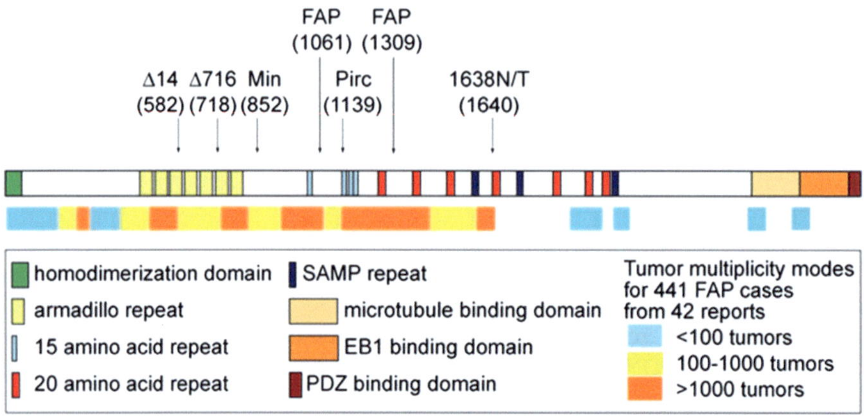

Fig. 2.2 The Wnt signaling pathway. (**a**) In the absence of Wnt ligand, the complex containing APC, glycogen synthase kinase 3β (GSK3β), casein kinase 1α (CK1α), and an Axin scaffold targets cytoplasmic β-catenin for proteasomal degradation. In the nucleus, Wnt target genes are silenced by Groucho. (**b**) In the presence of Wnt ligand, the receptors Frizzled (Fz) and low-density lipoprotein receptor-related protein (LRP) trigger the phosphorylation of the cytoplasmic tail of LRP by GSK3β. Disheveled (Dsh) recruits Axin to the phosphorylated tail of LRP. Phosphorylation of β-catenin does not occur; β-catenin accumulates in the cytoplasm and translocates into the nucleus, where it activates the transcription of multiple target genes by interacting with the TCF family of transcription factors. Reproduced with permission from: Pino MS, Chung DC (2010) The CIN in colon cancer. Gastroenterology 138(6):2059–2072

Wnt ligand signaling, APC binds to the scaffold protein Axin to promote sequential phosphorylation of the N-terminus region of β-catenin by casein kinase 1 and glycogen synthase kinase-3β (GSK3β), thus targeting phosphorylated β-catenin for ubiquitination and subsequent proteasomal degradation. In the setting of CRC in which both *APC* alleles are mutated, loss of APC protein allows for cytoplasmic accumulation of β-catenin, which then complexes with DNA-binding proteins of the TCF/LEF (T-cell factor/lymphoid enhancer family) family, and translocates to the nucleus, where it drives transcription of multiple genes with TCF DNA-binding sites involved in tumor growth and invasion (Fig. 2.3) (Mann et al. 1999).

In sporadic CRC, *APC* mutations are present in microscopic adenomas, 50–60 % of small (<0.5 cm) adenomas and are found at similar frequency in advanced adenomas and carcinomas, indicating that inactivating mutations in *APC* are an early event in colorectal carcinogenesis (Powell et al. 1992; Miyaki et al. 1994; Cottrell et al. 1992). Kinzler and Vogelstein argue that APC is a "gatekeeper gene" which is "responsible for maintaining a constant cell number in renewing cell populations." (Kinzler and Vogelstein 1996) Nearly all somatic mutations lead to premature truncation of the APC protein. Both *APC* alleles are inactivated in adenomas and carcinomas that arise in FAP patients as well as in sporadic disease. While germline-inactivating mutations in *APC* are located throughout the gene, somatic mutations are clustered between codons 1286 and 1513 (Miyoshi et al. 1992).

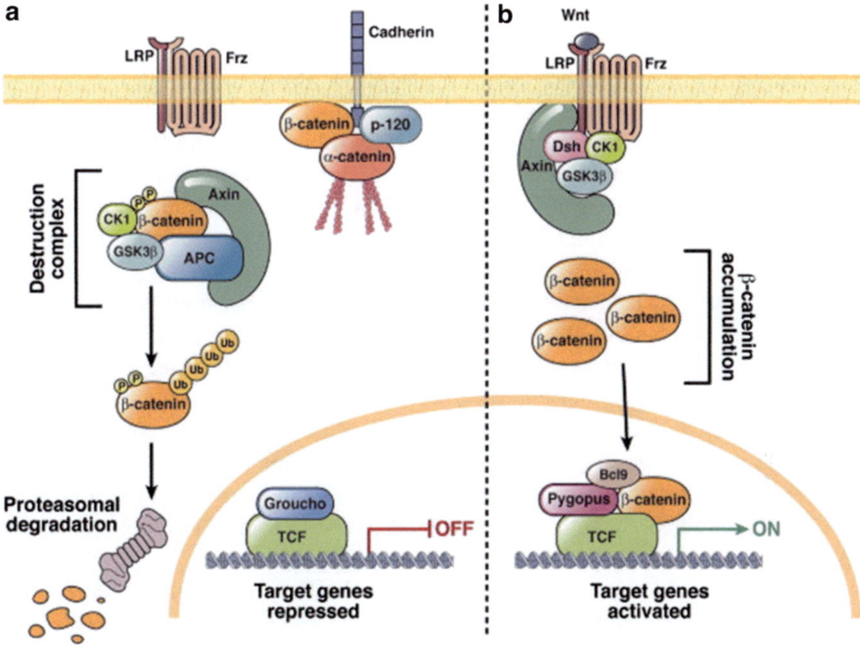

Fig. 2.3 The APC protein. Cartoon of the 2,843 amino acid adenomatous polyposis coli (APC) protein with selected sequence motifs and interaction partners. The N-terminus has a domain which regulates homodimerization. Repeated sequences with homology to *Drosophila* armadillo protein are in the N-terminus third of APC ("armadillo repeat"). Multiple 20-amino acid repeats mediate binding to β-catenin and Axin in the central third of APC. The C-terminal third of APC has a basic region that is involved in microtubule binding and interactions with the protein EB1. *Arrows* indicate orthologous locations of mouse and rat model mutations and the most common FAP mutation sites. Reproduced with permission from William Dove (http://www.mcardle.wisc.edu/dove/Data/Apc.htm)

An alternative mechanism for *APC* gene inactivation may be hypermethylation of the *APC* promoter, which has been reported in 18 % of colorectal adenomas and carcinomas (Esteller et al. 2000).

2.2.4.2 Other Mutations in Wnt Pathway Components

Gain-of-function somatic mutations in β-catenin (*CTNNB1*) that affect key amino acids in β-catenin's N-terminal phosphorylation and ubiquitination motifs have been identified in a subset of CRCs, although they are common in other cancer types. However, these mutations have been found in 50 % of CRC with wild-type *APC*, which underscores the importance of the Wnt pathway in CRC (Sparks et al. 1998). A germline mutation in *AXIN2* was identified in a family with familial CRC and tooth agenesis, which suggests that the mutation may have interfered with the function of Axin in regulating β-catenin (Lammi et al. 2004).

2.2.4.3 KRAS

The RAS family of small G-proteins consists of K-RAS4A, K-RAS4B, H-RAS, and N-RAS, which are molecular switches downstream of growth factor receptors such as the epidermal growth factor receptor (EGFR) (Malumbres and Barbacid 2003). EGFR is affected by somatic mutations (e.g., point mutations or gene amplification) in fewer than 5 % of CRCs. In contrast, the *KRAS* oncogene is mutated in 40 % of CRC. Single nucleotide point mutations in codons 12 and 13 of exon 2, codon 146 in exon 4, and rarely in codon 61 of exon 3, lock the enzyme in the guanosine triphosphate bound (GTP), activated form, which leads to constitutive activation of RAS. A small number of CRCs have *NRAS* mutations at codon 12, 13, or 61. *KRAS* mutations are frequently found in aberrant crypt foci but are not required for adenoma initiation (Pretlow and Pretlow 2005). *KRAS* mutations are demonstrated in 10 % of adenomas smaller than 1 cm and 40–50 % of adenomas >1 cm, suggesting that *KRAS* plays a role in colorectal adenoma progression (Vogelstein et al. 1988). Targeted disruption of mutant *KRAS* alleles in CRC cell lines reduced cell growth, and activating $Kras^{G12D}$ mutation accelerated tumor growth in a mouse model for sporadic CRC (Shirasawa et al. 1993; Hung et al. 2010).

The best characterized effector of *KRAS* is the Raf-mitogen-activated protein kinase (MEK)-extracellular signal-regulated kinase (ERK) pathway. The Raf family includes three serine–threonine kinases (A-RAF, B-RAF, and C-RAF) that phosphorylate MEK1 and MEK2, which then activate ERK1 and ERK2. ERK in turn activates substrates such as JUN and ELK1, transcription factors that regulate genes such as cyclin D1, which is involved in cell cycle control (Pruitt and Der 2001). RAS is linked to nuclear factor-kB (NF-kB), a transcription factor that regulates immune response and cell survival. TBK1 can activate NF-kB by phosphorylating its inhibitor IkB. TBK1 and NF-kB signaling are essential in KRAS-mediated tumors; suppression of TBK1 induced apoptosis specifically in KRAS-transformed cancer cell lines, whereas inhibition of NF-kB blocked RAS-induced formation of lung tumors in mice (Barbie et al. 2009; Meylan et al. 2009).

2.2.4.4 PIK3CA and PTEN

PIK3CA, the gene encoding the catalytic p110α subunit of type I PI3Ks, is somatically mutated in 15–30 % of CRCs, most commonly in exons 9 (E532K, E545K) and 20 (H1047R) (Samuels et al. 2004). These *PIK3CA* mutations are oncogenic in CRC cell lines (Samuels et al. 2005). *PIK3CA* mutations predict reduced progression free survival in response to EGFR-inhibitor therapy (Souglakos et al. 2009). The PTEN protein is a phospholipid phosphatase that mediates dephosphorylation from PIP_3 to PIP_2. Germline mutations in the *PTEN* tumor suppressor gene are found in patients with Cowden syndrome, who demonstrate benign GI tumors but not an increased risk for CRC. However, approximately 10 % of sporadic CRCs exhibit somatic *PTEN* mutations, and loss of PTEN likely enhances PIP3-mediated activation of AKT, which in turn acts on downstream antiapoptotic factors and the mTOR pathway. However, the significance of PTEN mutations in sporadic CRC is still unclear (Chalhoub and Baker 2009).

2.2.4.5 TP53

TP53 is located on chromosome 17p and encodes a transcription factor that is a tumor suppressor and master regulator of hundreds of genes involved in DNA metabolism, apoptosis, autophagy, cell cycle regulation, senescence, angiogenesis, immune response, cell differentiation, motility, and migration. P53 dysfunction is almost universal in human tumors, and loss of p53 function is reported in 4–26 % of adenomas, 50 % of adenomas with foci of carcinoma, and 50–75 % of CRC, which suggests that mutation and LOH of *TP53* plays a major role in the transition from adenoma to carcinoma (Leslie et al. 2002). Selection for *TP53* defects at the adenoma–carcinoma transition may reflect the fact that stresses such as DNA-strand breakage, telomere erosion, and hypoxia may activate apoptotic and cell-cycle arrest pathways in tumor cells with wild-type TP53 function. As such, mutations in *TP53* may facilitate continued growth and invasion in the setting of stresses that might otherwise hinder tumor cell survival at the adenoma–carcinoma transition. Approximately 80 % of *TP53* mutations are missense mutations, which lead to the synthesis of a partially inactive protein. TP53 is induced by oncogenic proteins such as c-Myc, RAS, and adenovirus E1A. TP53 is normally negatively regulated by MDM2, E3-ubiquitin ligase, and MDM4, which target TP53 for ubiquitination, while in stress situations TP53 is allowed function (Levine 1997; Vogelstein et al. 2000).

2.2.4.6 Aneuploidy: 18q Loss

Allelic loss at chromosome 18q has been identified in as many as 70 % of CRCs, particularly at advanced stages. Candidate tumor suppressors located on 18q include *deleted in colorectal carcinoma* (*DCC*), *SMAD2*, *SMAD4*, and *Cables*. *DCC* gene expression is absent or markedly reduced in a majority of advanced colorectal cancers (Fearon et al. 1990; Takagi et al. 1996; Mehlen and Fearon 2004). *DCC* encodes a receptor for netrin-1 and induces apoptosis unless bound to its ligand (Mehlen et al. 1998). However, a *DCC* mutant mouse model did not develop cancer, so doubts were raised about the role of *DCC* in carcinogenesis (Fazeli et al. 1997). A group led by Patric Mehlen recently reported that mice in which the proapoptotic activity of DCC is genetically silenced develop spontaneous intestinal neoplasia and, in an *Apc* mutant background, more invasive adenocarcinoma. Thus, DCC suppresses colorectal tumor formation via induction of tumor cell apoptosis (Castets et al. 2011). *SMAD2* and *SMAD4* mutations have been found in 10 % and 15 % of CRCs, respectively (Takagi et al. 1998). Mutations in *SMAD4* are found in a subset of patients with juvenile polyposis syndrome (JPS), which is characterized by childhood onset of multiple hamartomatous polyps throughout the GI tract and an increased incidence of stomach, small intestinal, colon, and pancreatic cancers (Merg and Howe 2004). Cables protein increases tyrosine phosphorylation of cyclin-dependent kinases (cdk2, cdk3, and cdk5) by nonreceptor tyrosine kinases (Src, Abl, and Wee1). Loss of Cables expressions is found in 60–70 % of sporadic CRC, and loss of Cables in mice potentiates carcinogen-induced colonic tumorigenesis (Park et al. 2007; Kirley et al. 2005).

2.2.4.7 TGF-β Type II Receptor

Inactivating mutations in the TGF-β type II receptor (*TGFβIIR* or *TGFBR2*) are found in approximately 25 % of CRCs, principally in those with MSI (see Sect. 3.5). In addition to MSI-associated tumors, somatic *TGFβIIR* mutations are found in 15 % of MSS tumors. TGF-β-mediated receptor phosphorylation regulates the function of the SMAD2 and SMAD3 proteins (Grady et al. 1999).

2.2.4.8 Aneuploidy: Inactivation of CDC4 and Chromosome 1p Deletion

Chromosome 1p deletions occur at an early stage of colorectal carcinogenesis (Lothe et al. 1995; Bomme et al. 1994; Di Vinci et al. 1996) and are linked to karyotypic evolution during CRC development (Höglund et al. 2002). Introduction of chromosomal band 1p36 into CRC cell lines suppressed tumorigenicity (Tanaka et al. 1993). Interestingly, 76 % of patients with deletions in chromosome 1p in colorectal cancers were reported to harbor similar 1p deletions in distant normal-appearing mucosa (Cianciulli et al. 2004). Chromosome 1p deletions may influence carcinogenesis via loss of genes associated with DNA repair, spindle checkpoint function, apoptosis, miRNAs, the Wnt signaling pathway, tumor suppression, antioxidant functions, and defense against environmental toxins (Roschke et al. 2008; Negrini et al. 2010).

2.2.4.9 CMYC, CCNE1, and FBW7

The role of the *CMYC* gene in human cancer was first identified in the early 1980s, in the setting of chromosomal translocation in lymphoma and gene amplifications in small-cell lung cancer (Eilers and Eisenman 2008). The c-Myc protein is a transcription factor that regulates genes involved in cell-cycle progression and cellular survival. High and moderate copy amplification of the CMYC gene is seen in 10 % and 30 % of CRCs, respectively (Camps et al. 2009; Leary et al. 2008). Expression of *CMYC* is repressed by wild-type APC and activated by β-catenin, and this effect is mediated by TCF-4 binding sites in the *CMYC* promoter. *APC* inactivation may thus in part explain amplifications in *CMYC* expression (He et al. 1998).

High copy amplification of the cyclin E gene (*CCNE1*) is found in 5 % of CRCs, although modest increases are found in 15–20 % of CRCs (Leary et al. 2008; Bondi et al. 2005). More commonly, elevated cyclin E protein expression is due to inactivating mutations in the *FBXW7* gene, the human homologue of yeast gene *Cdc4*. Fbxw7/hCdc4 is a member of the F-box family of proteins, which acts as a substrate recognition component for the SCG ubiquitin ligase complex. Inactivation of Fbxw7/hCdc4 in CRC cells results in a CIN phenotype due to a defect in execution of metaphase (Rajagopalan et al. 2004). Fbxw7/hCdc4 mediates the ubiquitin-dependent proteolysis of several oncoproteins including cyclin E, c-Myc, c-Jun, and Notch (Tan et al. 2008). Somatic mutations that inactivate *FBXW7* are found in 9 % of CRCs (Akhoondi et al. 2007). Low tumor *FBXW7* mRNA expression corresponds to significantly poorer prognosis in CRC patients (Iwatsuki et al. 2010). Together, these data implicate *FBXW7* as a tumor suppressor in CRC.

2.2.4.10 CDK8

The *CKD8* oncogene, located at 13q12, is amplified in approximately 10–15 % of CRCs. CDK8 is a cyclin-dependent kinase that complexes with cyclin C to phosphorylate substrates such as RNA polymerase II and DNA-binding transcription factors. CDK8 kinase activity is necessary for β-catenin activity and for expression of several β-catenin transcriptional targets (Firestein et al. 2008). Overexpression of the *CDK8* gene is associated with increased CRC-related mortality (Firestein and Hahn 2009; Firestein et al. 2010).

2.2.4.11 COX2

Overexpression of cyclooxygenase-2 (COX2) is believed to play a role in CRC tumorigenesis. The *COX2* gene is overexpressed in 43 % of adenomas and 86 % of carcinomas (Eberhart et al. 1994), which is consistent with epidemiologic data for a protective role of aspirin and other nonsteroidal anti-inflammatory drugs (NSAIDs) in CRC (Hahn et al. 2010; Garcia-Albeniz and Chan 2011; Ruder et al. 2011). Direct evidence for a role for COX2 in CRC carcinogenesis came from a study in which the number of small intestinal polyps in $APC^{\Delta 716}$ knockout mice was reduced by 34 % when one copy of *COX2* was knocked out and by 86 % when both alleles were deleted (Oshima et al. 2001). A recent meta-analysis found that aspirin users in four randomized, placebo-controlled trials had a pooled risk ratio of 0.83 (95 % CI, 0.72–0.96) for any adenoma and 0.72 (95 % CI, 0.57–0.90) for advanced adenomas (Cole et al. 2009). Three randomized trials showed that the COX-2 selective inhibitors celecoxib and rofecoxib prevent adenoma recurrence among patients with a history of adenoma (Arber et al. 2006; Bertagnolli et al. 2006; Baron et al. 2006), but enthusiasm for chemoprevention of CRC with COX2 inhibitors was dampened after reports of increased cardiovascular mortality in the COX2 arms of these trials (Bresalier et al. 2005; Curfman et al. 2005).

2.2.4.12 LKB1

LKB1 is a tumor suppressor gene that encodes a ubiquitously expressed and evolutionarily conserved serine–threonine kinase, which in turn regulates and number of downstream kinases. *LKB1* inactivation leads to stimulation of the mammalian target of rapamycin (mTOR) pathway, which promotes cell division and growth. Deletion mutations in *LKB1* are found in a majority of cases of Peutz–Jeghers syndrome (PJS), a rare autosomal dominant syndrome. Loss of function *LKB1* mutations have also been identified in 5–15 % of sporadic nonsmall lung cancers and 5 % of pancreatic cancers and melanomas (Hezel and Bardeesy 2008).

2.2.5 *Timing of CIN*

Is CIN the cause or a consequence of colorectal carcinogenesis? A number of studies have found allelic imbalances in early stages of tumorigenesis; Shih et al. found allelic imbalances of at least one chromosomal arm in over 90 % of adenomas 2 mm in size (Bardi et al. 1997; Stoler et al. 1999; Shih et al. 2001; Cardoso et al. 2006). However, these studies did not ask whether CIN occurred before or after APC inactivation. Nowak et al. used a stochastic mathematical model to conclude that under a variety of conditions, CIN mutation is likely the initiating event or the second event following mutation of one allele of *APC* (Nowak et al. 2002). APC inactivation has been proposed as a potential initiator of CIN. Mouse embryonic stem cells with *APC* mutations, but not wild-type cells, became aneuploid and accumulated chromosomal abnormalities (Fodde et al. 2001; Kaplan et al. 2001), while other studies have found that Wnt signaling might contribute to CIN (Aoki et al. 2007; Hadjihannas et al. 2006). Chromosomal instability has not been conclusively linked to acquisition of key mutations required for colorectal carcinogenesis but is common in the early stages of malignancy and likely increases mutation rate and facilitates CRC progression.

2.2.6 *Clinical Implications of CIN*

Our insights into the genetic basis for CRC have allowed the identification of prognostic molecular markers. Patients with activating *KRAS* and *BRAF* mutations may experience worse overall survival outcomes compared to wild-type patients (Van Cutsem et al. 2011; Ogino et al. 2009a, 2011). Patients with tumor harboring *KRAS* and *PIK3CA* mutations are more likely to develop liver metastases compared to wild-type patients (Li et al. 2011). *TP53* mutation may be associated with greater mortality, but this risk may be limited to patients with metastatic disease (Munro et al. 2005; Russo et al. 2005). There are contradictory reports on whether deletion of chromosome 18q is associated with poor outcomes; individual chromosomal deletions are currently used as molecular markers for CRC prognosis (Zhou et al. 2002; Diep et al. 2003; Ogino et al. 2009b).

Years of research on the molecular mechanisms of CRC are slowly translating into the clinic. Patients with *KRAS* mutant tumors do not appreciably respond to inhibition of the EGFR; use of agents such as Cetuximab is thus limited to patients with *KRAS* wild-type cancer (Karapetis et al. 2008). A recent phase I clinical trial examined treatment of $BRAF^{V600E}$ CRC with Vemurafenib, a specific inhibitor of the $BRAF^{V600E}$ protein and demonstrated mixed results, which suggest the presence of primary resistance mechanisms (Tol et al. 2009). Inhibition of the PI3K and downstream mTOR pathways has shown efficacy in a mouse model for *PIK3CA* wild-type CRC, and phase I clinical trials are planned (Roper et al. 2011). Small molecule inhibitors of Aurora kinase, Plks, and the spindle motor protein Eg5 have shown promise in preclinical studies and have demonstrated safety and antitumor efficacy in phase I human trials (Jani et al. 2010; Schöffski et al. 2012; Infante et al. 2012).

2.3 Microsatellite Instability in Colorectal Cancer

In 1993, Manuel Perucho and colleagues performed PCR amplification of thousands of sequences in colon cancer and matched normal tissue samples using randomly chosen primers. His group found that 12 % of the tumors had bands that were shorter in length. The sequences from these bands contained simple repetitive elements (i.e., microsatellites), primarily in polyadenine (A_n) tracts associated with *Alu* sequences. Further work revealed that tumors with these somatic mutations were associated with distinct clinical characteristics. The tumors were significantly more likely to arise in the proximal colon, less likely to be invasive, less likely to harbor mutations in *KRAS* or *TP53*, more likely to be poorly differentiated, and were found in younger patients (Ionov et al. 1993). Concurrently, the laboratory of Stephen Thibodeau identified deletion mutations in $[CA]_n$ sequences in chromosomes 5q, 15q, 17p, and 18q in colorectal tumors and coined the term *microsatellite instability*. Similar to Perucho's findings, Thibodeau's group reported MSI in 28 % of colorectal tumors and found that 89 % of tumors with MSI were located in the proximal colon and were associated with a better prognosis than MSS tumors (Blake et al. 2001; Thibodeau et al. 1993). Allotyping studies of CRC found that 15 % of CRCs had no apparent LOH; these tumors were later found to harbor MSI (Thibodeau et al. 1993; Vogelstein et al. 1989). Both the Perucho and Thibodeau groups recognized that microsatellite instability represents a unique pathway to CRC development.

2.3.1 DNA MMR System

Further investigations revealed that MSI arises from defects in the DNA mismatch repair (MMR) system, which is one of a number of DNA repair systems. In prokaryotes, the MMR system consists of a family of enzymes encoded by the *mutS* and *mutL* genes that detect DNA replication errors in which the newly synthesized strand has incorporated the wrong nucleotide. These single base-pair mismatches usually result in point mutations. DNA polymerase is more likely to make such errors during replication of long repetitive DNA sequences such as microsatellites. Slippage during replication of a repetitive sequence results in formation of an insertion–deletion loop that can be identified and corrected by the MMR system. If this loop is not repaired a frameshift mutation results, which can produce a truncated, nonfunctional protein. This results in MSI (Boland and Goel 2010). In yeast, MMR is encoded by the genes *Mut S homologue* (*MSH*), *Mut L homologue* (*MLH*), and *postmeiotic segregation-1* (*PMS1*). Homologous copies of these genes are designated *MSH1* to *MSH6*, and *MLH1* through *MLH3*.

2.3.2 Lynch Syndrome

Lynch syndrome (also known as hereditary nonpolyposis CRC or HNPCC), one of the first inherited disease syndromes to be identified, was first described in 1913 by Warthin (1913). Many years later, Henry Lynch and colleagues further characterized kindreds with autosomal dominant patterns of CRC that lacked extensive polyposis. Patients with Lynch syndrome develop CRC at early ages, at a mean age of 40, and also present with extracolonic tumors of the endometrium, stomach, ovary, urinary tract, small intestine, and other sites (Vasen 2005). Without a putative genetic etiology to define the syndrome, the Amsterdam Criteria were developed to facilitate clinical diagnosis and research on families with clustering of CRC. According to these criteria, Lynch syndrome is defined as three CRC cases in a family in which one individual is a first-degree relative of the other two, CRC in at least two generations (in which FAP is excluded), and one affected family member younger than age 50 (Vasen et al. 1991). The Amsterdam II Criteria were developed in 1999 to include the presence of noncolonic tumors (i.e., cancer of the endometrium or small bowel, and transitional cell carcinoma of the ureter or renal pelvis) in the diagnosis (Vasen et al. 1999).

2.3.3 Sporadic MSI

Two of the three initial descriptions of MSI were made in samples from sporadic colon cancers, rather than tumors from patients with familial CRC (Ionov et al. 1993; Blake et al. 2001). Approximately 12–17 % of all colorectal tumors have MSI, whereas only 3 % of CRCs are identified in Lynch syndrome families; thus, most CRCs with MSI are sporadic (Ward et al. 2001; Hampel et al. 2005). Characteristically, sporadic CRC with MSI is associated with (1) absence of significant clustering in families, (2) biallelic methylation of the *MLH1* promoter (Veigl et al. 1998), (3) absence of MLH1 and PMS2 proteins (not MSH2), (4) diploidy (74 %), (5) frequent mutation in *BRAF* (usually V600E) (Carragher et al. 2010), and (6) better prognosis than MSS tumors (Sinicrope et al. 2006). Nevertheless, MSI is associated with poorer survival in metastatic CRC in the context of BRAF mutation (Tran et al. 2011). Patients with sporadic CRC with MSI tend to be older than those with microsatellite stable sporadic CRC, and loss of *MLH1* expression increases with age (Kakar et al. 2003).

2.3.4 Epigenetic Changes in CRC and CpG Island Methylator Phenotype

Unlike colorectal tumors from Lynch syndrome, sporadic CRC with MSI arises via a mechanism involving the CIMP (Toyota et al. 1999). The combination of a cytosine nucleotide followed by a guanine nucleotide (CpG dinucleotide) is relatively uncommon in the human genome. However, pockets of CpG dinucleotides, termed CpG islands, are found in the promoter regions of approximately 50 % of all genes (Bird 1986). The addition of a methyl group to cytosine bases in these CpG regions (i.e., DNA methylation) has been associated with silencing of genes that encode tumor suppressors (e.g., *p16*, *insulin-like growth factor 2*, and *HIC1*), DNA repair genes such as *methylguanine methyltransferase* (*MGMT*) and *MLH1*, and Wnt signaling antagonists known as SFRPs (secreted Frizzled-related proteins), leading to cancer (Jones and Laird 1999; Kim et al. 2010a). Hypermethylation of *MLH1* is the major cause of MSI in sporadic CRC (Kane et al. 1997). Other tumor suppressor genes are also more commonly silenced by methylation in MSI associated, compared to MSS-associated CRC; this lead to the observation that 20–30 % of colorectal cancers are associated with hypermethylation of CpG islands—a phenomenon that was termed CIMP (Benatti et al. 2005; Des Guetz et al. 2010). A subsequent study used a more sensitive method for detecting methylation to develop a more specific classification of CIMP and found the phenotype in 18 % of colorectal tumors (Weisenberger et al. 2006). Although most sporadic MSI-associated tumors have CIMP, half of all tumors with CIMP do not have methylation of *MLH1* or MSI (Samowitz et al. 2005a; Hawkins et al. 2002). Many of these tumors carry BRAF mutations and arise from the serrated pathway (discussed later in this chapter) (Leggett and Whitehall 2010).

In contrast to the specific hypermethylation found in CpG islands, in benign and malignant colorectal tumors there is an overall decrease in total DNA methylation (i.e., hypomethylation) compared to adjacent normal tissue, perhaps leading to activation of oncogenes, though the functional significance of this finding is still unclear (Goelz et al. 1985; Feinberg et al. 1988). DNA hypomethylation of pericentrosomic sequences may impair chromosomal segregation, a theory that would link hypomethylation to the CIN pathway (Ji et al. 1997).

2.3.5 Pathophysiology of Colorectal Carcinogenesis with MSI

In 1995, Markowitz et al. examined the role of transforming growth factor-β (TGF-β) in MSI; TGF-β signaling inhibits proliferation of colonic epithelial cells. They found that *transforming growth factor B* (*TGF*-β) *type II receptor* (*TGFβR2*) was not expressed in CRC cell lines with MSI but was expressed in MSS cell lines. Those cell lines without *TGFβR2* expression did not slow proliferation in response to TGF-β. The group further demonstrated that a single base-pair deletion in a repetitive A_{10} sequence in *TGFβR2* was found in 90 % of 111 MSI-positive colorectal tumor samples, which suggested a model in which repetitive DNA sequences are sensitive to loss of DNA MMR activity, leading to frameshift mutations, premature stop codons, and gene inactivation (Markowitz et al. 1995). An additional tumor suppressor gene in MSI-H CRC is *ACVR2A*, which encodes the activin type II receptor. Both alleles of the *ACVR2A* gene are somatically mutated in a polyadenine repeat tract at exon 10 in approximately 85 % of MSI-H CRCs. The resulting frameshift mutation is associated with loss of the activin type II receptor and poorer prognosis. MSI-H CRC cells in which *ACVR2* or *TGFβR2* function has been restored exhibit slower growth (Jung et al. 2006, 2007). Approximately one-third of MSI-H CRCs harbor mutations in a repeat tract of the *TCF7L2* gene, which encodes the TCF4 protein. TCF4 suppresses DNA transcription of Wnt pathway target genes in the setting of stabilized β-catenin, which may provide an additional pathway for Wnt activation in MSH-H cancer (Cuilliere-Dartigues et al. 2006).

Several other genes affected by MSI have since been identified that encode regulators of cellular proliferation (*GTB1*, *TCG-4*, *WISP3*, *insulin-like growth factor-2 receptor*, *axin-2*, and *CDX2*), cell cycle (*BAX*, *caspase-5*, *RIZ*, *BCL-10*, *PTEN*, *hG4-1*, and *FAS*), and DNA repair (*MBD-4*, *BLM*, *CHK1*, *MLH3*, *RAD50*, *MSH3*, and *MSH6*) (O'Brien et al. 2006). However, it is unclear which of these mutations are of functional significance (as has been determined for *TGFβR2*) and which are simply markers of MSI, because biallelic inactivation of these genes has not been documented in all of the tumors. For instance, a recent retrospective study found no association between *BAX* mutations in MSI-H tumors and patient survival (Shima et al. 2011). Genes associated with MSI in CRC are summarized in Table 2.2. The key steps in the MSI pathway to CRC are outlined in Fig. 2.4.

The discovery of multiple genetic targets of MMR deficiency that differ from the classic Fearon and Vogelstein model indicates that MSI-associated CRC occurs via a different biological pathway than conventional MSS tumors. Tumors in the CIN pathway arise from a combination of genetic mutations and LOH, resulting in biallelic inactivation of APC. Colorectal tumors with MSI, on the other hand, harbor an increased number of point mutations compared to MSS cancers, are more likely to be diploid, and do not exhibit widespread LOH. A vast majority of MSI-associated tumors have normal expression of *APC* but have mutations in β-catenin that prevent binding to the APC protein and degradation, which is functionally equivalent to loss of the APC protein (Miyaki et al. 1999; Mirabelli-Primdahl et al. 1999). Other MSI-associated tumors have neither inactivated *APC* nor mutated β-*catenin* but instead have frameshift mutations in other Wnt pathway factors such as *TCF-4* (Boland and Goel 2010).

Table 2.2 Genes that regulate chromosomal instability

Microsatellite length	Gene
A10	*AIM2*
	CASPACE-5
	MBD-4
	OGT
	SEC63 (also, A9)
	TGFBR2
A9	*BLM*
	CHK1
	GRB-14
	MLH3
	RAD50
	RHAMM
	RIZ (also, A8)
	TCF-4
	WISP3
A8	*ACVR2*
	APAF
	BCL-10
	hG4-1
	MSH3
A6	*PTEN*
T10	*OGT*
T9	*KIAA0971*
	NIADH-UOB
G8	*BAX*
	IGF2R
C9	*SLC23A1*
C8	*MSH6*
G7	*AXIN-2*
	CDX2
T7	*FAS*

Modified from Duval A, Hamelin R. Mutations at coding repeat sequences in mismatch repair-deficient human cancers: toward a new concept of target genes for instability. Cancer Res 2002;62(9):2447–54

2.3.6 MSI and Inflammatory Bowel Disease

CRC risk is increased in inflammatory bowel disease, but the mechanisms are not well established. Inflammation may increase mutagenesis via generation of oxidative stress and free radicals that may promote proliferation of colorectal cells. Although seemingly paradoxical, oxidative stress can inactivate the DNA MMR system and is associated with an increased mutation rate (Lee et al. 2003; Chang et al. 2002; Gasche et al. 2001). MSI has been identified in colorectal cancers of patients with ulcerative colitis; 21 % of 63 colitis-associated tumors and areas of dysplasia had at least 1 of 5 dinucleotide repeat markers mutated (Suzuki et al. 1994). Interestingly, MSI has been found in at least 1 of 7 dinucleotide repeat

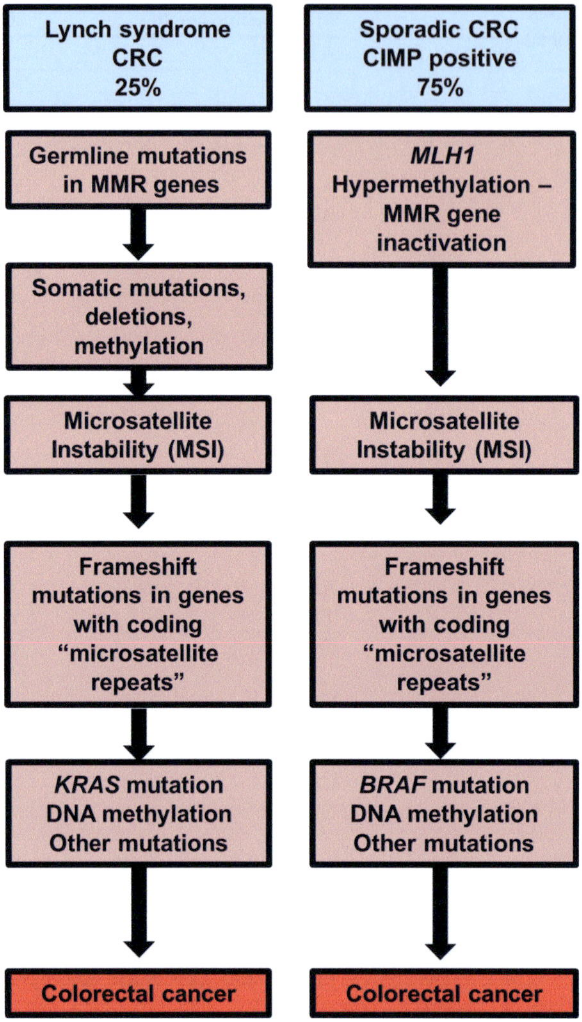

Fig. 2.4 Molecular pathways to MSI-associated colorectal cancer (CRC). Approximately 25 % of MSI-associated CRC arises from the Lynch syndrome, in which inactivating germline mutations in MMR genes are inherited in an autosomal dominant pattern. Additional "second hits" to the wild-type copy of the gene (inherited from the unaffected parent) in the form of somatic mutations, deletions, and methylation lead to MSI. 75 % of MSI-associated CRC is sporadic. These cancers are associated with CIMP and undergo MMR gene inactivation via hypermethylation of the *MLH1* promoter. Both Lynch syndrome and sporadic CIMP positive-associated defects in MMR lead to MSI and rapid accumulation of somatic mutations in genes with coding "microsatellite repeats." Many of these microsatellite repeats may not contribute to carcinogenesis but provide a signature that can be used for identification of MSI. Lynch-associated CRCs often harbor *KRAS* mutations, whereas sporadic CIMP-associated tumors are often *BRAF* mutant. Modified from Boland CR, Goel A (2010) Microsatellite instability in colorectal cancer. Gastroenterology 138(6):2073–2087.e3

markers in 50 % of nonneoplastic tissue in patients with chronic ulcerative colitis, but not in controls with acute infectious colitis (Brentnall et al. 1996).

2.3.7 Clinical Diagnosis of MSI

The definition of MSI was standardized at an international consensus meeting in 1997. The term "MSI" refers to MSI-high, in which >30 % of a defined microsatellite marker panel is mutated. Those CRCs in which at least 1, but <30 %, of the markers are mutated are called MSI-low and have clinical features of MSS tumors (Boland et al. 1998). Another type of MSI has been recognized, called "elevated microsatellite alterations at selected tetranucleotide repeats" (EMAST). EMAST is largely found in noncolonic tumors such as lung, is associated with *TP53* mutations, and is not caused by inactivation of the MMR system (Ahrendt et al. 2000).

MSI testing is used clinically to identify patients with Lynch syndrome, which comprises 2–3 % of all CRCs. MSI identifies MMR-deficient colorectal tumors with 93 % sensitivity, whereas the sensitivity and specificity of immunohistochemical analysis of MLH1 and MSH2 is 92.3 % and 100 %, respectively (Shia 2008). The sensitivity of IHC improves with expression of MSH6 and PMS2 are included in the analysis. Staining of tumors for MMR proteins can be heterogeneous, which may limit sensitivity (Shia 2008; Zhang 2008). MSI-H tumors can also be distinguished from MSS tumors by the presence of tumor-infiltrating cytotoxic lymphocytes on histologic examination, the degree of which independently confers improved survival (Ogino et al. 2009c; Phillips et al. 2004).

2.3.8 MSI and Response to Chemotherapy

The MMR phenotype is associated with resistance to cytotoxic agents in human CRC cell lines such as HCT-116 (Bhattacharyya et al. 1994). Stable restoration of MMR activity in cell lines increases sensitivity to alkylating agents, 6-thioguanine, 5-fluorouracil, and platinum compounds (Mäkinen et al. 2001; Samowitz et al. 2005b; Chan et al. 2002; Wynter et al. 2004; O'Brien et al. 2004; Minoo et al. 2006). With the exception of one study with potential methodological flaws (Elsaleh et al. 2000), multiple studies, including two meta-analyses, have shown no benefit for chemotherapy among patients with MSI-associated colorectal tumors (de Vos tot Nederveen Cappel et al. 2004; Ribic et al. 2003; Storojeva et al. 2005; Benatti et al. 2005; Popat et al. 2005; Lanza et al. 2006; Jover et al. 2006; Kim et al. 2007; Des Guetz et al. 2010). The largest such study, a prospective, multicenter, randomized controlled trial, a threefold increased mortality was found in Stage II CRC patients with MSI-associated tumors compared to without (Ribic et al. 2003). However, MSI is associated with improved response to regimens containing a topoisomerase I inhibitor, irinotecan (Bertagnolli et al. 2009; Fallik et al. 2003).

2.4 The Serrated Pathway in Colorectal Cancer Pathogenesis

Colorectal polyps have traditionally been classified as either hyperplastic or adenomatous, with only the latter progressing into carcinoma. However, beginning in the late 1980s, an increasing number of reports suggested that CRC can arise from hyperplastic polyps in the setting of what is now known as the hyperplastic polyposis syndrome (HPS), in which a large number of hyperplastic polyps are found throughout the colon (these polyps are distinguished from typical hyperplastic polyps, which are small, left sided, and benign) (Samowitz et al. 2006; Ji et al. 2006; Pérez et al. 2010; Shrubsole et al. 2008; Chirieac et al. 2005; Ogino et al. 2006b; Ward et al. 2003; Glazer et al. 2008). These studies identified a 35 % risk of CRC in patients with HPS, as well as increased risk of synchronous cancers (Boparai et al. 2010). Polyps in patients with HPS are characterized by gland serrations, which led pathologists to reexamine the malignant potential of other polyps with histologic serrations. Data from screening colonoscopy cohorts have demonstrated that serrated polyps are strongly associated with the development of synchronous and metachronous advanced adenoma and CRC (Li et al. 2009; Schreiner et al. 2010).

2.4.1 *Classification of Serrated Polyps*

Serrated polyps are characterized by a "sawtooth" pattern, or serrations, in the colonic crypts. In 1990, Longacre and Fenoglio-Presiser proposed the term "serrated adenoma" for polyps exhibiting features of both adenomatous and hyperplastic polyps (Longacre and Fenoglio-Preiser 1990). In 1996, Torlakovic and Snover first showed that polyps in HPS have serrated features similar to serrated adenomas, though with less atypia, and were more likely to be sessile than standard hyperplastic polyps (Torlakovic and Snover 1996). Further detailed work identified a subset of serrated polyps with abnormal proliferation, crypt distortion, and dilation that were typically sessile and found on the right side of the colon. These polyps were distinguished from traditional serrated adenomas (TSAs), which more closely resembled conventional adenomas (Torlakovic et al. 2003). These findings eventually led to a proposal for a new nomenclature for serrated polyps in 2005 (Snover et al. 2005).

The use of the term "adenoma" to describe sessile lesions has been controversial because conventional adenomas are dysplastic, whereas SSAs lack cytological dysplasia, though they manifest disordered proliferation and crypt architecture. Robert Odze and colleagues have thus opted for the term "sessile serrated polyp" in a recent pathology textbook (Hornick and Odze 2009), whereas a recent European publication chose the term "sessile serrated lesion." (Lambert et al. 2009) As the term SSA has grown in research and clinical practice, we will use it in this chapter.

2.4.1.1 Hyperplastic Polyp

Hyperplastic polyps (HPs) have a narrow crypt base lined with proliferative cells and serrations in the upper third of the gland. HPs have been subdivided into goblet cell-rich type, microvesicular type (which are precursers to SSAs), and the rare mucin-poor variant. Overall, HPs are highly prevalent sessile lesions that are commonly located in the distal colon and rectum (Tedesco et al. 1982; Imperiale et al. 2002). Endoscopically, HPs are identified by their smooth, symmetrical, and pale appearance. Microvesicular type HPs are precursor lesions to SSAs and, like SSAs, harbor *BRAFV*^{600E} mutations. Goblet cell HPs, on the other hand, often contain *KRAS* mutations (43 % in one study), which are mutually exclusive of *BRAF* mutations (O'Brien et al. 2006). Large goblet cell HPs may progress into *KRAS* mutant dysplastic serrated polyps (Boparai et al. 2008).

2.4.1.2 Sessile Serrated Polyp

Sessile serrated adenomas (SSAs) are characterized by crypt architectural alterations that reflect disordered growth. These include serration of the epithelium, often at the base of the crypts; dilation of the base of the crypts; and T- or L-shaped crypts. SSAs may contain areas of cytologic dysplasia and adenocarcinoma; tumors with neoplastic progression tend to lose serrated features (Fujita et al. 2011). SSAs likely evolve from preexisting microvesicular type HPs (Spring et al. 2006). Endoscopically, SSAs are usually larger than 5 mm, flat or sessile (height one half or less than width), and often mucous covered (Jaramillo et al. 2005). They are generally larger than HPs and located in the proximal colon. The surface is often smooth, and the edges are poorly defined and irregular. These features make SSAs difficult to detect endoscopically (Higuchi et al. 2005).

2.4.1.3 Dysplastic Serrated Polyps

Dysplastic serrated polyps contain gland serrations and cytologic dysplasia. There are two categories of dysplastic serrated polyp (1) SSA with dysplasia, which exhibits SSA morphologic characteristics contiguous to an area of conventional dysplasia and (2) TSA, which has not only serrations but also dysplastic epithelial cells and ectopic crypts with bases not adjacent to the muscularis mucosa, in contrast to SSAs in which new crypts are generally anchored to the muscularis mucosa. TSAs differ from SSAs in that they are typically distally located, polypoid, contain tubulovillous architecture, and marked cytoplasmic eosinophilia (O'Brien 2007; Torlakovic et al. 2008). TSAs not only are frequently *KRAS* mutant (which heralds an aggressive phenotype) but may also be *KRAS/BRAF* wild-type or *BRAF* mutant (Kim et al. 2010b).

2.4.2 Epidemiology

It is estimated that up to 20 % of CRCs arise from the serrated pathway, or nearly 30,000 cases annually (Jass 2007). One study reported a prevalence of 29 % HPs, 9 % SSAs, 1.7 % mixed polyps, and 0.7 % TSAs from a cohort of colonoscopy-resected specimens (Spring et al. 2006). Dysplastic serrated polyps are much less common than conventional polyps or HPs, representing 1–2 % of all polyps (Higuchi et al. 2005; Jass et al. 2006).

The molecular basis of the serration of the crypt epithelium has not been determined, though it has been proposed that serrations occur due to cell crowding or because of failure of apoptosis or anoikis (Tateyama et al. 2002). Crypt serration is strongly associated with the presence of BRAF mutation; hyperplastic polyps with KRAS rather than BRAF mutation have less, or absence of, gland serration.

2.4.3 Serrated Polyps, MSI, CIMP, and BRAF

Tumors associated with HPS have a higher than expected incidence of MSI (Leggett and Whitehall 2010; Jeevaratnam et al. 1996; Rashid et al. 2000; Jass et al. 2000). Serrated polyps from colectomy specimens were more likely to have MSI than MSS, and another study found that MSI was more common in serrated adenomas than in control tumors (37.5 % vs. 11 %, respectively) (Hawkins and Ward 2001; Mäkinen et al. 2001). O'Brien et al. found MSI only in the areas of advanced SSAs with carcinoma, which suggests that MSI develops late in the serrated pathway. Epigenetic silencing of *MLH1* is the underlying cause of MSI in serrated lesions and is an important driver of the progression to invasive cancer (O'Brien et al. 2006). A large proportion serrated cancers are MSS and frequently have *TP53* mutation, which may explain their more aggressive phenotype and poorer prognosis than MSI-associated tumors [hazard rate ratio (HRR), 2.97; 95 % CI, 2.05–4.32] (Samowitz et al. 2005b).

CIMP is commonly observed in both HPs and in proximal SSA (Chan et al. 2002; Wynter et al. 2004; O'Brien et al. 2004). Yang et al. detected CIMP in microvesicular HP (47 %), SSA (75 %), and TSA (80 %). Using a narrower definition of methylation ($\geq$4/5 markers), CIMP was detected in 11 % of MVHP, compared to 40 % of SSA[284]. CIMP has even been detected in histologically normal colonic mucosa of HPS patients (Minoo et al. 2006). Higher CIMP levels (four or more markers positive) were more frequently found in SSAs (with or without carcinoma) than in conventional adenomas or carcinomas (O'Brien et al. 2006). Together, these data indicate that methylation of specific CIMP loci may facilitate the transition from microvesicular HP to SSA.

In a systematic genome-wide screen for genes affecting cell proliferation and death, activating mutations in *BRAF* were identified in a high proportion of melanomas and in a small fraction of other cancers including colon. BRAF is a serine/

threonine kinase that is part of the mitogen-activated protein kinase (MAPK) cell signaling pathway; mutations in *BRAF* result in constitutive activation of the MAPK pathway and transcription of genes promoting cell growth and proliferation (Davies et al. 2002). Rajagopalan et al. sequenced *BRAF* and *KRAS* mutations in colorectal tumors and found that (1) 10 % of tumors harbored somatic mutations in *BRAF* and (2) no tumors exhibited mutations in both *BRAF* and *KRAS* (Rajagopalan et al. 2002). Another group confirmed these findings (Yuen et al. 2002). Chan et al. examined *BRAF* and *KRAS* mutations in a series of serrated polyps and found *BRAF* mutations in 36 % of HPs and 100 % of SSAs. The $BRAF^{V600E}$ substitution is the most common *BRAF* mutation in human cancers including serrated CRCs (Davies et al. 2002). Using current histologic definitions, 70–76 % of MVHPs and 75–83 % of SSAs have $BRAF^{V600E}$ mutations. *BRAF* and *KRAS* mutations are mutually exclusive (O'Brien et al. 2006). The $BRAF^{V600E}$ mutation was found in 5 % of a cohort of MSS tumors and 52 % of MSI-associated tumors (Samowitz et al. 2005b). However, histological reviews have confirmed that *BRAF* is almost never mutated in conventional adenomas or in Lynch syndrome, highlighting the association of *BRAF* mutation with the serrated pathway rather than MSI (O'Brien et al. 2006; Kambara et al. 2004; Wang et al. 2003). Mutation of *BRAF* strongly correlates with CIMP (Weisenberger et al. 2006). These findings support the role of CIMP and the MAPK pathway via activating mutation in *BRAF* or *KRAS* in the serrated adenoma pathway.

2.4.4 Initiation and Progression of the Serrated Pathway

Activation of BRAF in normal melanocyte epithelium and in mouse gastrointestinal epithelium results in an initial burst of proliferation followed by cell senescence (Carragher et al. 2010; Campisi 2005). Methylation-induced silencing of *p16INK4a* is an early event in the serrated pathway and may be sufficient to allow colorectal cells (and possible microvesicular HPs) to escape BRAF-induced senescence (Chen et al. 2005). In melanocytes, activated BRAF is sufficient for synthesis and secretion of insulin-like growth factor binding protein 7 (IGFBP7) which in turn inhibits MAPK signaling and induces senescence and apoptosis (Wajapeyee et al. 2008). The large columnar vacuolated cells of the upper crypts of the microvesicular HP and SSA are a manifestation of cell senescence (Minoo and Jass 2006). Silencing via methylation of *IGFBP7* in *BRAF*-mutant, CIMP-positive CRC cells permits unrestrained cell proliferation and progression to SSA by enabling escape from p53-induced senescence (Suzuki et al. 2010). Therefore, the additive tumorigenic effects of mutated *BRAF* and *CIMP* may result from silencing of tumor suppressor genes such as *p16INK4a* and *IGFBP7* via hypermethylation.

The Wnt signaling pathway is another major regulator of cellular proliferation in CRC. In the absence of APC protein, β-catenin accumulates in the cell nucleus instead of undergoing degradation. Three studies found positive nuclear β-catenin immunostaining in 0–50 % of HPs and 38–67 % of SSAs, 36 % of TSAs, and 100 % of tubular adenomas (TA). 29 % of SSAs without dysplasia and all SSAs with dysplasia displayed aberrant β-catenin staining. Nuclear β-catenin was identified only in the setting of *BRAF*^{V600E} mutation (Wu et al. 2008; Yachida et al. 2009; Sandmeier et al. 2009). A recent histological study found that nuclear β-catenin staining in SSA was limited to dysplastic areas of the polyps, and histologically these dysplastic areas lost serrated features and become more tubulovillous (Fujita et al. 2011). Unlike conventional adenomas, however, *APC* mutation is found in only a minority (19 %) of serrated polyps, and β-catenin gain-of-function mutation in *CTNNB1* has not been identified in serrated polyps (Yachida et al. 2009; de Vogel et al. 2009). A mouse model for *BRAF*^{V600E} CRC demonstrated that expression of *BRAF*^{V600E} in intestinal crypts was sufficient for β-catenin nuclear localization via MAPK-dependent, Akt-independent phosphorylation of Gsk3β (Carragher et al. 2010). However, this mechanism of Wnt activation has not been confirmed in the human serrated pathway. These findings suggest that activation of the Wnt signaling pathway follows *BRAF* mutation and plays an important role in the progression (but not initiation) of the serrated pathway. The molecular steps in the initiation and progression of sessile serrated adenomas are summarized in Fig. 2.5.

2.4.5 An Alternate Serrated Pathway

Recognition of the heterogeneity of serrated polyps has led to the hypothesis that there are two parallel serrated pathways to colorectal carcinogenesis: one driven by *BRAF* mutation and the other driven by *KRAS* mutation (O'Brien et al. 2006; O'Brien 2007; Yang et al. 2004). The *BRAF* pathway has been discussed in detail above. *KRAS* mutant serrated carcinomas have relatively low levels of CIMP, but it is possible that rather than being a true CIMP-low group, these cancers are methylated at different loci (Weisenberger et al. 2006). Silencing of the DNA repair gene *MGMT* by promoter hypermethylation has been associated with *KRAS* mutation and CIMP-low status (Ogino et al. 2006a, 2007; Whitehall et al. 2001). However, no specific panel of markers has been validated to study this pathway. No precursor lesion to *KRAS* mutant serrated carcinomas has been identified, though it has been proposed that large goblet cell HPs, tubulovillous adenomas, and/or serrated polyps with dysplasia may be relevant to the "alternate pathway" (Boparai et al. 2008; Jass et al. 2006).

2.4.6 Risk Factors for Serrated CRC

Susceptibility to serrated neoplasia may be associated with a genetic predisposition to hypermethylation of gene promoters. Rare families with multiple members affected by HPS have been described (Jeevaratnam et al. 1996; Rashid et al. 2000; Chow et al. 2006). Most cases of CIMP-high, *BRAF* mutant serrated polyps appear to be sporadic, although a few families with high incidences of CRC and serrated polyps have been identified (Des Guetz et al. 2010)[306,307]. However, residents of

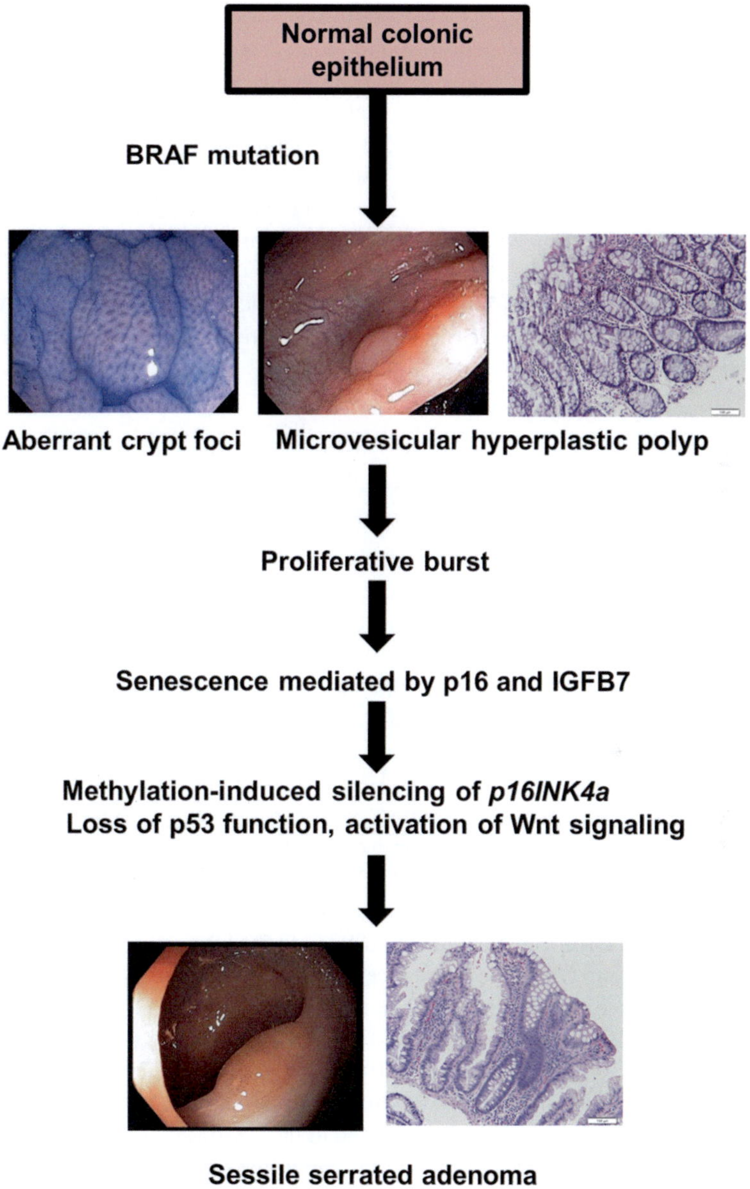

Fig. 2.5 The sessile serrated pathway. Activation of BRAF induces the formation of ACF with serrated features and microvesicular hyperplastic polyp (MVHP). Further cell proliferation is controlled by cell senescence, which is mediated by *p16INK4a* expression and IGFBP7 secretion. Methylation-induced silencing of *p16INK4a* or loss p53 function allows early polyps to escape the senescent state and develop into sessile serrated adenomas. Endoscopic images courtesy of Moises Guelrud, Tufts Medical Center; pathology images courtesy of Barbara Weinstein, Tufts Medical Center. Modified from Leggett B, Whitehall V (2010) Role of the serrated pathway in CRC pathogenesis. Gastroenterology 138(6):2088–2100

Melbourne, Australia of Anglo-Celtic origin were found to have a higher risk of CIMP and *BRAF* mutant CRC compared to those of southern European origin, and serrated polyps were more frequent in Caucasians compared to Hispanics and African Americans (English et al. 2008; Wallace et al. 2009). Cigarette smoking has been strongly associated with CIMP and BRAF mutation and is a stronger risk factor for HPs than adenomatous polyps in multiple studies, although one report found no association between smoking and HPs (Samowitz et al. 2006; Ji et al. 2006; Pérez et al. 2010; Shrubsole et al. 2008). Aspirin is protective against serrated polyps, as with conventional polyps (Wallace et al. 2009). A study on risk factors for CRC found that obesity, smoking, dietary fat, caloric intake, and red meat intake were associated with increased risk for distal, but not proximal, serrated polyps (Wallace et al. 2009).

2.4.7 Clinical Characteristics of Serrated CRC

The presence or absence of BRAF mutation does not affect the excellent prognosis of MSI-associated CRC (Samowitz et al. 2005b). Cancers that arise via the serrated pathway, whether with or without MSI, tend to be proximal, mucinous, occur in older individuals, and present at more advanced stage (Samowitz et al. 2005a; Hawkins et al. 2002; Chirieac et al. 2005; Ogino et al. 2006b). In the context of MSS, increased DNA methylation and *BRAF* mutation is associated with worse prognosis (Weisenberger et al. 2006; Ward et al. 2003). Serrated polyps are strongly associated with synchronous advanced neoplasia (defined as invasive carcinoma, tubular adenoma 1 cm, or adenoma with any villous histology or high-grade dysplasia), particularly proximal CRCs, in large colonoscopy cohort studies (Glazer et al. 2008; Hiraoka et al. 2010; Li et al. 2009; Schreiner et al. 2010).

HPS is an uncommon condition characterized by multiple and/or large HPs. Several reports of CRC in patients with HPS led to the hypothesis that serrated polyps may develop into CRC (Jeevaratnam et al. 1996). The incidence of CRC in HPS is estimated to be 40–50 % (Buchanan et al. 2010; Leggett et al. 2001; Rubio et al. 2006). Type I HPS is defined multiple (five or more), large, proximally located SSAs. There is a high frequency of CIMP and mutated BRAF. Type II HPS, a more heterogeneous condition, describes the finding of numerous (≥30) small HPS distributed throughout the colon, and is believed to have a lower risk of CRC than type I HPS (Ferrández et al. 2004). Although the syndrome has no proven genetic basis, there are reports of familial HPS and ethnic associations in population studies (Young and Jass 2006; Young et al. 2007).

2.4.8 Detection and Surveillance of Serrated Polyps

Detection of serrated polyps via currently available screening modalities may be difficult. Serrated polyps are less likely to bleed, and hence may not be detected by fecal occult blood testing (East et al. 2008). CT Colonography may be less likely to detect flat or sessile lesions, though this has not been studied. Colonoscopy performs relatively poorly in the detection of serrated polyps, which may partly explain findings that morality rates from left-sided CRC, but not right-sided CRC, have decreased in recent years (Baxter et al. 2009; Brenner et al. 2010). This may be due to poor colonic prep on the right side of the colon and/or poor visualization of flat, mucous-covered lesions. Randomized trials have demonstrated that chromoendoscopy improves detection of serrated polyps by twofold. The importance of detection and removal of serrated polyps is highlighted by the findings that interval cancers (found despite appropriate screening or surveillance colonoscopy) were four times as likely to be associated with MSI (Sawhney et al. 2006) and CIMP, and more likely to be proximal and mucinous, which are all features suggestive of *BRAF* mutation (Leggett et al. 1997; Farrar et al. 2006; Bressler et al. 2004; Arain et al. 2010).

2.4.9 Models of the Serrated Pathway

Isogenic $BRAF^{V600E}$ human CRC cell lines (VACO432 and RKO) have been developed in which either the endogenous wild-type or mutant allele has been inactivated through targeted homologous recombination (Yun et al. 2009). Carragher et al. published a Cre–lox-regulated knockin mouse in which $Braf^{V600E}$ is expressed from the endogenous *Braf* gene in the proliferative cells of the intestinal crypts. They showed that intestinal $Braf^{V600E}$ is not only sufficient for formation of hyperplastic crypts via activation of the MAPK and Wnt pathways but also induces cell senescence, and that inactivation of *p16INK4a* through DNA methylation is necessary for tumor progression. However, polyps in this model are adenomas, not carcinomas, and are confined to the small bowel (Carragher et al. 2010). Kenneth Hung and colleagues developed a novel genetically engineered mouse model in which mice with a conditional Apc allele were crossed with those with a latent $Braf^{V600E}$ allele. They showed that combination treatment with BRAF and PI3K/mTOR inhibitors was required to induce apoptosis and tumor regression. This model offers several advantages for preclinical drug testing (1) solitary tumors develop rapidly along a reproducible time line in the colon; (2) tumors can be continuously monitored throughout drug treatment via colonoscopy; and (3) tumors recapitulate the serrated pathway seen in humans, including HPs, SSAs, SSAs with dysplasia, and SSAs with congruent invasive adenocarcinoma (Coffee et al., manuscript under review).

2.5 Conclusions

CRC continues to be a significant public health burden. Whereas there have been significant advances in the development of targeted therapies, the 5-year prognosis for metastatic CRC still continues to be less than 10 %. However, our increased understanding of the molecular events underlying CRC carcinogenesis will enable the development of new targeted therapies and the identification of clinical biomarkers that will inform their effective usage. This is an exciting time for cancer medicine and we believe that the field is poised to make significant therapeutic breakthroughs.

References

Ahrendt SA, Decker PA, Doffek K et al (2000) Microsatellite instability at selected tetranucleotide repeats is associated with p53 mutations in non-small cell lung cancer. Cancer Res 60(9):2488–2491

Akhoondi S, Sun D, von der Lehr N et al (2007) FBXW7/hCDC4 is a general tumor suppressor in human cancer. Cancer Res 67(19):9006–9012

Albertini RJ, Nicklas JA, O'Neill JP, Robison SH (1990) In vivo somatic mutations in humans: measurement and analysis. Annu Rev Genet 24:305–326

Aoki K, Aoki M, Sugai M et al (2007) Chromosomal instability by β-catenin/TCF transcription in APC or β-catenin mutant cells. Oncogene 26(24):3511–3520

Arain MA, Sawhney M, Sheikh S et al (2010) CIMP status of interval colon cancers: another piece to the puzzle. Am J Gastroenterol 105(5):1189–1195

Arber N, Eagle CJ, Spicak J et al (2006) Celecoxib for the prevention of colorectal adenomatous polyps. N Engl J Med 355(9):885–895

Barbie DA, Tamayo P, Boehm JS et al (2009) Systematic RNA interference reveals that oncogenic KRAS-driven cancers require TBK1. Nature 462(7269):108–112

Bardelli A, Cahill DP, Lederer G et al (2001) Carcinogen-specific induction of genetic instability. Proc Natl Acad Sci U S A 98(10):5770–5775

Bardi G, Parada LA, Bomme L et al (1997) Cytogenetic comparisons of synchronous carcinomas and polyps in patients with colorectal cancer. Br J Cancer 76(6):765–769

Baron JA, Sandler RS, Bresalier RS et al (2006) A randomized trial of rofecoxib for the chemoprevention of colorectal adenomas. Gastroenterology 131(6):1674–1682

Bassing CH, Suh H, Ferguson DO et al (2003) Histone H2AX: a dosage-dependent suppressor of oncogenic translocations and tumors. Cell 114(3):359–370

Baxter NN, Goldwasser MA, Paszat LF, Saskin R, Urbach DR, Rabeneck L (2009) Association of colonoscopy and death from colorectal cancer. Ann Intern Med 150(1):1–8

Benatti P, Gafà R, Barana D et al (2005) Microsatellite instability and colorectal cancer prognosis. Clin Cancer Res 11(23):8332–8340

Bertagnolli MM, Eagle CJ, Zauber AG et al (2006) Celecoxib for the prevention of sporadic colorectal adenomas. N Engl J Med 355(9):873–884

Bertagnolli MM, Niedzwiecki D, Compton CC et al (2009) Microsatellite instability predicts improved response to adjuvant therapy with irinotecan, fluorouracil, and leucovorin in stage III colon cancer: Cancer and Leukemia Group B Protocol 89803. J Clin Oncol 27(11):1814–1821

Bhattacharyya NP, Skandalis A, Ganesh A, Groden J, Meuth M (1994) Mutator phenotypes in human colorectal carcinoma cell lines. Proc Natl Acad Sci U S A 91(14):6319–6323

Bird AP (1986) CpG-rich islands and the function of DNA methylation. Nature 321(6067):209–213

Blake C, Tsao J-L, Wu A, Shibata D (2001) Stepwise deletions of PolyA sequences in mismatch repair-deficient colorectal cancers. Am J Pathol 158(5):1867–1870

Boland CR, Goel A (2010) Microsatellite instability in colorectal cancer. Gastroenterology 138(6):2073–2087.e3

Boland CR, Thibodeau SN, Hamilton SR et al (1998) A National Cancer Institute workshop on microsatellite instability for cancer detection and familial predisposition: development of international criteria for the determination of microsatellite instability in colorectal cancer. Cancer Res 58(22):5248–5257

Bomme L, Bardi G, Pandis N, Fenger C, Kronborg O, Heim S (1994) Clonal karyotypic abnormalities in colorectal adenomas: clues to the early genetic events in the adenoma-carcinoma sequence. Genes Chromosomes Cancer 10(3):190–196

Bondi J, Husdal A, Bukholm G, Nesland JM, Bakka A, Bukholm IRK (2005) Expression and gene amplification of primary (A, B1, D1, D3, and E) and secondary (C and H) cyclins in colon adenocarcinomas and correlation with patient outcome. J Clin Pathol 58(5):509–514
Boparai KS, Dekker E, Van Eeden S et al (2008) Hyperplastic polyps and sessile serrated adenomas as a phenotypic expression of MYH-associated polyposis. Gastroenterology 135(6):2014–2018
Boparai KS, Mathus-Vliegen EMH, Koornstra JJ et al (2010) Increased colorectal cancer risk during follow-up in patients with hyperplastic polyposis syndrome: a multicentre cohort study. Gut 59(8):1094–1100
Boveri T (2008) Concerning the origin of malignant tumours by Theodor Boveri. Translated and annotated by Henry Harris. J Cell Sci 121(Suppl 1):1–84
Brenner H, Hoffmeister M, Arndt V, Stegmaier C, Altenhofen L, Haug U (2010) Protection from right- and left-sided colorectal neoplasms after colonoscopy: population-based study. J Natl Cancer Inst 102(2):89–95
Brentnall TA, Crispin DA, Bronner MP et al (1996) Microsatellite instability in nonneoplastic mucosa from patients with chronic ulcerative colitis. Cancer Res 56(6):1237–1240
Bresalier RS, Sandler RS, Quan H et al (2005) Cardiovascular events associated with rofecoxib in a colorectal adenoma chemoprevention trial. N Engl J Med 352(11):1092–1102
Bressler B, Paszat LF, Vinden C, Li C, He J, Rabeneck L (2004) Colonoscopic miss rates for right-sided colon cancer: a population-based analysis. Gastroenterology 127(2):452–456
Buchanan DD, Sweet K, Drini M et al (2010) Phenotypic diversity in patients with multiple serrated polyps: a genetics clinic study. Int J Colorectal Dis 25(6):703–712
Campisi J (2005) Suppressing cancer: the importance of being senescent. Science 309(5736):886–887
Camps J, Nguyen QT, Padilla-Nash HM et al (2009) Integrative genomics reveals mechanisms of copy number alterations responsible for transcriptional deregulation in colorectal cancer. Genes Chromosomes Cancer 48(11):1002–1017
Cardoso J, Molenaar L, De Menezes RX et al (2006) Chromosomal instability in MYH- and APC-mutant adenomatous polyps. Cancer Res 66(5):2514–2519
Carragher LAS, Snell KR, Giblett SM et al (2010) V600EBraf induces gastrointestinal crypt senescence and promotes tumour progression through enhanced CpG methylation of p16INK4a. EMBO Mol Med 2(11):458–471
Castets M, Broutier L, Molin Y et al (2011) DCC constrains tumour progression via its dependence receptor activity. Nature 482(7386):534–537, Available from: http://www.ncbi.nlm.nih.gov/pubmed/22158121
Castillo A, Morse HC III, Godfrey VL, Naeem R, Justice MJ (2007) Overexpression of Eg5 causes genomic instability and tumor formation in mice. Cancer Res 67(21):10138–10147
Celeste A, Difilippantonio S, Difilippantonio MJ et al (2003) H2AX haploinsufficiency modifies genomic stability and tumor susceptibility. Cell 114(3):371–383
Chadeneau C, Hay K, Hirte HW, Gallinger S, Bacchetti S (1995) Telomerase activity associated with acquisition of malignancy in human colorectal cancer. Cancer Res 55(12):2533–2536
Chalhoub N, Baker SJ (2009) PTEN and the PI3-kinase pathway in cancer. Annu Rev Pathol 4:127–150
Chan AT, Giovannucci EL (2010) Primary prevention of colorectal cancer. Gastroenterology 138(6):2029–2043.e10
Chan AO-O, Issa J-PJ, Morris JS, Hamilton SR, Rashid A (2002) Concordant CpG island methylation in hyperplastic polyposis. Am J Pathol 160(2):529–536
Chang CL, Marra G, Chauhan DP et al (2002) Oxidative stress inactivates the human DNA mismatch repair system. Am J Physiol Cell Physiol 283(1):C148–C154
Chen Z, Trotman LC, Shaffer D et al (2005) Crucial role of p53-dependent cellular senescence in suppression of Pten-deficient tumorigenesis. Nature 436(7051):725–730
Chirieac LR, Shen L, Catalano PJ, Issa J-P, Hamilton SR (2005) Phenotype of microsatellite-stable colorectal carcinomas with CpG island methylation. Am J Surg Pathol 29(4):429–436
Chow E, Lipton L, Lynch E et al (2006) Hyperplastic polyposis syndrome: phenotypic presentations and the role of MBD4 and MYH. Gastroenterology 131(1):30–39

Cianciulli A, Cosimelli M, Marzano R et al (2004) Genetic and pathologic significance of 1p, 17p, and 18q aneusomy and the ERBB2 gene in colorectal cancer and related normal colonic mucosa. Cancer Genet Cytogenet 151(1):52–59
Cole BF, Logan RF, Halabi S et al (2009) Aspirin for the chemoprevention of colorectal adenomas: meta-analysis of the randomized trials. J Natl Cancer Inst 101(4):256–266
Cottrell S, Bicknell D, Kaklamanis L, Bodmer WF (1992) Molecular analysis of APC mutations in familial adenomatous polyposis and sporadic colon carcinomas. Lancet 340(8820):626–630
Cuilliere-Dartigues P, El-Bchiri J, Krimi A et al (2006) TCF-4 isoforms absent in TCF-4 mutated MSI-H colorectal cancer cells colocalize with nuclear CtBP and repress TCF-4-mediated transcription. Oncogene 25(32):4441–4448
Curfman GD, Morrissey S, Drazen JM (2005) Expression of concern: Bombardier et al., "Comparison of upper gastrointestinal toxicity of rofecoxib and naproxen in patients with rheumatoid arthritis", N Engl J Med 2000;343:1520–8. N Engl J Med 353(26):2813–2814
Davies H, Bignell GR, Cox C et al (2002) Mutations of the BRAF gene in human cancer. Nature 417(6892):949–954
de Vogel S, Weijenberg MP, Herman JG et al (2009) MGMT and MLH1 promoter methylation versus APC, KRAS and BRAF gene mutations in colorectal cancer: indications for distinct pathways and sequence of events. Ann Oncol 20(7):1216–1222
de Vos tot Nederveen Cappel WH, Meulenbeld HJ, Kleibeuker JH et al (2004) Survival after adjuvant 5-FU treatment for stage III colon cancer in hereditary nonpolyposis colorectal cancer. Int J Cancer 109(3):468–471
Des Guetz G, Lecaille C, Mariani P et al (2010) Prognostic impact of microsatellite instability in colorectal cancer patients treated with adjuvant FOLFOX. Anticancer Res 30(10):4297–4301
Di Vinci A, Infusini E, Peveri C, Risio M, Rossini FP, Giaretti W (1996) Deletions at chromosome 1p by fluorescence in situ hybridization are an early event in human colorectal tumorigenesis. Gastroenterology 111(1):102–107
Diep CB, Thorstensen L, Meling GI, Skovlund E, Rognum TO, Lothe RA (2003) Genetic tumor markers with prognostic impact in Dukes' stages B and C colorectal cancer patients. J Clin Oncol 21(5):820–829
Dotan E, Meropol NJ, Zhu F et al (2012) Relationship of increased aurora kinase A gene copy number, prognosis and response to chemotherapy in patients with metastatic colorectal cancer. Br J Cancer 106(4):748–755, Available from: http://www.ncbi.nlm.nih.gov/pubmed/22240781
Duesberg P, Fabarius A, Hehlmann R (2004) Aneuploidy, the primary cause of the multilateral genomic instability of neoplastic and preneoplastic cells. IUBMB Life 56(2):65–81
East JE, Saunders BP, Jass JR (2008) Sporadic and syndromic hyperplastic polyps and serrated adenomas of the colon: classification, molecular genetics, natural history, and clinical management. Gastroenterol Clin North Am 37(1):25–46, v
Eberhart CE, Coffey RJ, Radhika A, Giardiello FM, Ferrenbach S, Dubois RN (1994) Up-regulation of cyclooxygenase 2 gene expression in human colorectal adenomas and adenocarcinomas. Gastroenterology 107(4):1183–1188
Eilers M, Eisenman RN (2008) Myc's broad reach. Genes Dev 22(20):2755–2766
Elsaleh H, Joseph D, Grieu F, Zeps N, Spry N, Iacopetta B (2000) Association of tumour site and sex with survival benefit from adjuvant chemotherapy in colorectal cancer. Lancet 355(9217):1745–1750
Engelhardt M, Drullinsky P, Guillem J, Moore MAS (1997) Telomerase and telomere length in the development and progression of premalignant lesions to colorectal cancer. Clin Cancer Res 3(11):1931–1941
English DR, Young JP, Simpson JA et al (2008) Ethnicity and risk for colorectal cancers showing somatic BRAF V600E mutation or CpG island methylator phenotype. Cancer Epidemiol Biomarkers Prev 17(7):1774–1780
Esteller M, Sparks A, Toyota M et al (2000) Analysis of adenomatous polyposis coli promoter hypermethylation in human cancer. Cancer Res 60(16):4366–4371
Fallik D, Borrini F, Boige V et al (2003) Microsatellite instability is a predictive factor of the tumor response to irinotecan in patients with advanced colorectal cancer. Cancer Res 63(18):5738–5744

Farrar WD, Sawhney MS, Nelson DB, Lederle FA, Bond JH (2006) Colorectal cancers found after a complete colonoscopy. Clin Gastroenterol Hepatol 4(10):1259–1264

Fazeli A, Dickinson SL, Hermiston ML et al (1997) Phenotype of mice lacking functional deleted in colorectal cancer (Dec) gene. Nature 386(6627):796–804

Fearon ER (2011) Molecular genetics of colorectal cancer. Annu Rev Pathol 6:479–507

Fearon ER, Vogelstein B (1990) A genetic model for colorectal tumorigenesis. Cell 61(5):759–767

Fearon ER, Cho KR, Nigro JM et al (1990) Identification of a chromosome 18q gene that is altered in colorectal cancers. Science 247(4938):49–56

Feinberg AP, Gehrke CW, Kuo KC, Ehrlich M (1988) Reduced genomic 5-methylcytosine content in human colonic neoplasia. Cancer Res 48(5):1159–1161

Ferrández A, Samowitz W, DiSario JA, Burt RW (2004) Phenotypic characteristics and risk of cancer development in hyperplastic polyposis: case series and literature review. Am J Gastroenterol 99(10):2012–2018

Firestein R, Hahn WC (2009) Revving the throttle on an oncogene: CDK8 takes the driver seat. Cancer Res 69(20):7899–7901

Firestein R, Bass AJ, Kim SY et al (2008) CDK8 is a colorectal cancer oncogene that regulates [bgr]-catenin activity. Nature 455(7212):547–551

Firestein R, Shima K, Nosho K et al (2010) CDK8 expression in 470 colorectal cancers in relation to β-catenin activation, other molecular alterations and patient survival. Int J Cancer 126(12):2863–2873

Fodde R, Kuipers J, Rosenberg C et al (2001) Mutations in the APC tumour suppressor gene cause chromosomal instability. Nat Cell Biol 3(4):433–438

Fujita K, Yamamoto H, Matsumoto T et al (2011) Sessile serrated adenoma with early neoplastic progression: a clinicopathologic and molecular study. Am J Surg Pathol 35(2):295–304

Ganem NJ, Godinho SA, Pellman D (2009) A mechanism linking extra centrosomes to chromosomal instability. Nature 460(7252):278–282

Garcia-Albeniz X, Chan AT (2011) Aspirin for the prevention of colorectal cancer. Best Pract Res Clin Gastroenterol 25(4–5):461–472

Gasche C, Chang CL, Rhees J, Goel A, Boland CR (2001) Oxidative stress increases frameshift mutations in human colorectal cancer cells. Cancer Res 61(20):7444–7448

Gertler R, Rosenberg R, Stricker D et al (2004) Telomere length and human telomerase reverse transcriptase expression as markers for progression and prognosis of colorectal carcinoma. J Clin Oncol 22(10):1807–1814

Glazer E, Golla V, Forman R, Zhu H, Levi G, Bodenheimer HC Jr (2008) Serrated adenoma is a risk factor for subsequent adenomatous polyps. Dig Dis Sci 53(8):2204–2207

Goelz SE, Vogelstein B, Hamilton SR, Feinberg AP (1985) Hypomethylation of DNA from benign and malignant human colon neoplasms. Science 228(4696):187–190

Grady WM, Myeroff LL, Swinler SE et al (1999) Mutational inactivation of transforming growth factor beta receptor type II in microsatellite stable colon cancers. Cancer Res 59(2):320–324

Hadjihannas MV, Brückner M, Jerchow B, Birchmeier W, Dietmaier W, Behrens J (2006) Aberrant Wnt/β-catenin signaling can induce chromosomal instability in colon cancer. Proc Natl Acad Sci U S A 103(28):10747–10752

Hahn E, Kraus S, Arber N (2010) Role of cyclooxygenase-2 in pathogenesis and prevention of colorectal cancer. Dig Dis 28(4–5):585–589

Haigis KM, Kendall KR, Wang Y et al (2008) Differential effects of oncogenic K-Ras and N-Ras on proliferation, differentiation and tumor progression in the colon. Nat Genet 40(5):600–608

Hampel H, Frankel WL, Martin E et al (2005) Screening for the Lynch syndrome (hereditary nonpolyposis colorectal cancer). N Engl J Med 352(18):1851–1860

Hawkins NJ, Ward RL (2001) Sporadic colorectal cancers with microsatellite instability and their possible origin in hyperplastic polyps and serrated adenomas. J Natl Cancer Inst 93(17):1307–1313

Hawkins N, Norrie M, Cheong K et al (2002) CpG island methylation in sporadic colorectal cancers and its relationship to microsatellite instability. Gastroenterology 122(5):1376–1387

He T-C, Sparks AB, Rago C et al (1998) Identification of c-MYC as a target of the APC pathway. Science 281(5382):1509–1512

Herz C, Schlürmann F, Batarello D et al (2011) Occurrence of Aurora A positive multipolar mitoses in distinct molecular classes of colorectal carcinomas and effect of Aurora A inhibition. Mol Carcinog 51(9):696–710, Available from: http://www.ncbi.nlm.nih.gov/pubmed/21739483

Hezel AF, Bardeesy N (2008) LKB1; linking cell structure and tumor suppression. Oncogene 27(55):6908–6919

Higuchi T, Sugihara K, Jass JR (2005) Demographic and pathological characteristics of serrated polyps of colorectum. Histopathology 47(1):32–40

Hiraoka S, Kato J, Fujiki S, et al (2010) The presence of large serrated polyps increases risk for colorectal cancer. Gastroenterology 139(5):1503–1510, 1510.e1–3

Höglund M, Gisselsson D, Hansen GB, Säll T, Mitelman F, Nilbert M (2002) Dissecting karyotypic patterns in colorectal tumors: two distinct but overlapping pathways in the adenoma-carcinoma transition. Cancer Res 62(20):5939–5946

Hornick JL, Odze RD (2009) Polyps of the large intestine, 2nd edn. Elsevier, Philadelphia

Hung KE, Maricevich MA, Richard LG et al (2010) Development of a mouse model for sporadic and metastatic colon tumors and its use in assessing drug treatment. Proc Natl Acad Sci U S A 107(4):1565–1570

Imperiale TF, Wagner DR, Lin CY, Larkin GN, Rogge JD, Ransohoff DF (2002) Results of screening colonoscopy among persons 40 to 49 years of age. N Engl J Med 346(23):1781–1785

Infante JR, Kurzrock R, Spratlin J et al (2012) A phase I study to assess the safety, tolerability, and pharmacokinetics of AZD4877, an intravenous Eg5 inhibitor in patients with advanced solid tumors. Cancer Chemother Pharmacol 69(1):165–172

Ionov Y, Peinado MA, Malkhosyan S, Shibata D, Perucho M (1993) Ubiquitous somatic mutations in simple repeated sequences reveal a new mechanism for colonic carcinogenesis. Nature 363(6429):558–561

Iwatsuki M, Mimori K, Ishii H et al (2010) Loss of FBXW7, a cell cycle regulating gene, in colorectal cancer: clinical significance. Int J Cancer 126(8):1828–1837

Jani JP, Arcari J, Bernardo V et al (2010) PF-03814735, an orally bioavailable small molecule aurora kinase inhibitor for cancer therapy. Mol Cancer Ther 9(4):883–894

Jaramillo E, Tamura S, Mitomi H (2005) Endoscopic appearance of serrated adenomas in the colon. Endoscopy 37(3):254–260

Jass JR (2007) Classification of colorectal cancer based on correlation of clinical, morphological and molecular features. Histopathology 50(1):113–130

Jass JR, Iino H, Ruszkiewicz A et al (2000) Neoplastic progression occurs through mutator pathways in hyperplastic polyposis of the colorectum. Gut 47(1):43–49

Jass JR, Baker K, Zlobec I et al (2006) Advanced colorectal polyps with the molecular and morphological features of serrated polyps and adenomas: concept of a "fusion" pathway to colorectal cancer. Histopathology 49(2):121–131

Jeevaratnam P, Cottier DS, Browett PJ, Van De Water NS, Pokos V, Jass JR (1996) Familial giant hyperplastic polyposis predisposing to colorectal cancer: a new hereditary bowel cancer syndrome. J Pathol 179(1):20–25

Jemal A, Bray F, Center MM, Ferlay J, Ward E, Forman D (2011) Global cancer statistics. CA Cancer J Clin 61(2):69–90

Ji W, Hernandez R, Zhang XY et al (1997) DNA demethylation and pericentromeric rearrangements of chromosome 1. Mutat Res 379(1):33–41

Ji B-T, Weissfeld JL, Chow W-H, Huang W-Y, Schoen RE, Hayes RB (2006) Tobacco smoking and colorectal hyperplastic and adenomatous polyps. Cancer Epidemiol Biomarkers Prev 15(5):897–901

Jones PA, Laird PW (1999) Cancer epigenetics comes of age. Nat Genet 21(2):163–167

Jover R, Zapater P, Castells A et al (2006) Mismatch repair status in the prediction of benefit from adjuvant fluorouracil chemotherapy in colorectal cancer. Gut 55(6):848–855

Jung B, Smith EJ, Doctolero RT et al (2006) Influence of target gene mutations on survival, stage and histology in sporadic microsatellite unstable colon cancers. Int J Cancer 118(10): 2509–2513

Jung BH, Beck SE, Cabral J et al (2007) Activin type 2 receptor restoration in MSI-H colon cancer suppresses growth and enhances migration with activin. Gastroenterology 132(2):633–644
Kakar S, Burgart LJ, Thibodeau SN et al (2003) Frequency of loss of hMLH1 expression in colorectal carcinoma increases with advancing age. Cancer 97(6):1421–1427
Kambara T, Simms LA, Whitehall VLJ et al (2004) BRAF mutation is associated with DNA methylation in serrated polyps and cancers of the colorectum. Gut 53(8):1137–1144
Kane MF, Loda M, Gaida GM et al (1997) Methylation of the hMLH1 promoter correlates with lack of expression of hMLH1 in sporadic colon tumors and mismatch repair-defective human tumor cell lines. Cancer Res 57(5):808–811
Kaplan KB, Burds AA, Swedlow JR, Bekir SS, Sorger PK, Näthke IS (2001) A role for the adenomatous polyposis coli protein in chromosome segregation. Nat Cell Biol 3(4):429–432
Karapetis CS, Khambata-Ford S, Jonker DJ et al (2008) K-ras mutations and benefit from cetuximab in advanced colorectal cancer. N Engl J Med 359(17):1757–1765
Katayama H, Ota T, Jisaki F et al (1999a) Mitotic kinase expression and colorectal cancer progression. J Natl Cancer Inst 91(13):1160–1162
Katayama S, Shiota G, Oshimura M, Kawasaki H (1999b) Clinical usefulness of telomerase activity and telomere length in the preoperative diagnosis of gastric and colorectal cancer. J Cancer Res Clin Oncol 125(7):405–410
Khanna KK, Jackson SP (2001) DNA double-strand breaks: signaling, repair and the cancer connection. Nat Genet 27(3):247–254
Kim GP, Colangelo LH, Wieand HS et al (2007) Prognostic and predictive roles of high-degree microsatellite instability in colon cancer: a National Cancer Institute-National Surgical Adjuvant Breast and Bowel Project Collaborative Study. J Clin Oncol 25(7):767–772
Kim MS, Lee J, Sidransky D (2010a) DNA methylation markers in colorectal cancer. Cancer Metastasis Rev 29(1):181–206
Kim K-M, Lee EJ, Kim Y-H, Chang DK, Odze RD (2010b) KRAS mutations in traditional serrated adenomas from Korea herald an aggressive phenotype. Am J Surg Pathol 34(5):667–675
Kinzler KW, Vogelstein B (1996) Lessons from hereditary colorectal cancer. Cell 87(2):159–170
Kirley SD, D'Apuzzo M, Lauwers GY, Graeme-Cook F, Chung DC, Zukerberg LR (2005) The cables gene on chromosome 18Q regulates colon cancer progression in vivo. Cancer Biol Ther 4(8):861–863
Lambert R, Kudo SE, Vieth M et al (2009) Pragmatic classification of superficial neoplastic colorectal lesions. Gastrointest Endosc 70(6):1182–1199
Lammi L, Arte S, Somer M et al (2004) Mutations in AXIN2 cause familial tooth agenesis and predispose to colorectal cancer. Am J Hum Genet 74(5):1043–1050
Lanza G, Gafà R, Santini A, Maestri I, Guerzoni L, Cavazzini L (2006) Immunohistochemical test for MLH1 and MSH2 expression predicts clinical outcome in stage II and III colorectal cancer patients. J Clin Oncol 24(15):2359–2367
Leary RJ, Lin JC, Cummins J et al (2008) Integrated analysis of homozygous deletions, focal amplifications, and sequence alterations in breast and colorectal cancers. Proc Natl Acad Sci 105(42):16224–16229
Lee S-H, Chang DK, Goel A et al (2003) Microsatellite instability and suppressed DNA repair enzyme expression in rheumatoid arthritis. J Immunol 170(4):2214–2220
Leggett B, Whitehall V (2010) Role of the serrated pathway in colorectal cancer pathogenesis. Gastroenterology 138(6):2088–2100
Leggett BA, Cornwell M, Thomas LR et al (1997) Characteristics of metachronous colorectal carcinoma occurring despite colonoscopic surveillance. Dis Colon Rectum 40(5):603–608
Leggett BA, Devereaux B, Biden K, Searle J, Young J, Jass J (2001) Hyperplastic polyposis: association with colorectal cancer. Am J Surg Pathol 25(2):177–184
Lengauer C, Kinzler KW, Vogelstein B (1997) Genetic instability in colorectal cancers. Nature 386(6625):623–627
Lengauer C, Kinzler KW, Vogelstein B (1998) Genetic instabilities in human cancers. Nature 396(6712):643–649
Leslie A, Carey FA, Pratt NR, Steele RJC (2002) The colorectal adenoma-carcinoma sequence. Br J Surg 89(7):845–860

Levine AJ (1997) p53, the cellular gatekeeper for growth and division. Cell 88(3):323–331
Li D, Jin C, McCulloch C et al (2009) Association of large serrated polyps with synchronous advanced colorectal neoplasia. Am J Gastroenterol 104(3):695–702
Li H-T, Lu Y-Y, An Y-X, Wang X, Zhao Q-C (2011) KRAS, BRAF and PIK3CA mutations in human colorectal cancer: relationship with metastatic colorectal cancer. Oncol Rep 25(6):1691–1697, Available from: http://www.ncbi.nlm.nih.gov/pubmed/21424126
Loeb LA, Loeb KR, Anderson JP (2003) Multiple mutations and cancer. Proc Natl Acad Sci U S A 100(3):776–781
Longacre TA, Fenoglio-Preiser CM (1990) Mixed hyperplastic adenomatous polyps/serrated adenomas. A distinct form of colorectal neoplasia. Am J Surg Pathol 14(6):524–537
Lothe RA, Andersen SN, Hofstad B et al (1995) Deletion of 1p loci and microsatellite instability in colorectal polyps. Genes Chromosomes Cancer 14(3):182–188
Lynch HT, de la Chapelle A (2003) Hereditary colorectal cancer. N Engl J Med 348(10):919–932
Mäkinen MJ, George SM, Jernvall P, Mäkelä J, Vihko P, Karttunen TJ (2001) Colorectal carcinoma associated with serrated adenoma—prevalence, histological features, and prognosis. J Pathol 193(3):286–294
Malumbres M, Barbacid M (2003) RAS oncogenes: the first 30 years. Nat Rev Cancer 3(6):459–465
Mann B, Gelos M, Siedow A et al (1999) Target genes of β-catenin-T cell-factor/lymphoid-enhancer-factor signaling in human colorectal carcinomas. Proc Natl Acad Sci U S A 96(4):1603–1608
Markowitz S, Wang J, Myeroff L et al (1995) Inactivation of the type II TGF-beta receptor in colon cancer cells with microsatellite instability. Science 268(5215):1336–1338
Mehlen P, Fearon ER (2004) Role of the dependence receptor DCC in colorectal cancer pathogenesis. J Clin Oncol 22(16):3420–3428
Mehlen P, Rabizadeh S, Snipas SJ, Assa-Munt N, Salvesen GS, Bredesen DE (1998) The DCC gene product induces apoptosis by a mechanism requiring receptor proteolysis. Nature 395(6704):801–804
Merg A, Howe JR (2004) Genetic conditions associated with intestinal juvenile polyps. Am J Med Genet C Semin Med Genet 129C(1):44–55
Meylan E, Dooley AL, Feldser DM et al (2009) Requirement for NF-B signalling in a mouse model of lung adenocarcinoma. Nature 462(7269):104–107
Minoo P, Jass JR (2006) Senescence and serration: a new twist to an old tale. J Pathol 210(2):137–140
Minoo P, Baker K, Goswami R et al (2006) Extensive DNA methylation in normal colorectal mucosa in hyperplastic polyposis. Gut 55(10):1467–1474
Mirabelli-Primdahl L, Gryfe R, Kim H et al (1999) Beta-catenin mutations are specific for colorectal carcinomas with microsatellite instability but occur in endometrial carcinomas irrespective of mutator pathway. Cancer Res 59(14):3346–3351
Miyaki M, Konishi M, Kikuchi-Yanoshita R et al (1994) Characteristics of somatic mutation of the adenomatous polyposis coli gene in colorectal tumors. Cancer Res 54(11):3011–3020
Miyaki M, Iijima T, Kimura J et al (1999) Frequent mutation of beta-catenin and APC genes in primary colorectal tumors from patients with hereditary nonpolyposis colorectal cancer. Cancer Res 59(18):4506–4509
Miyoshi Y, Nagase H, Ando H et al (1992) Somatic mutations of the APC gene in colorectal tumors: mutation cluster region in the APC gene. Hum Mol Genet 1(4):229–233
Morson B (1974) President's address. The polyp-cancer sequence in the large bowel. Proc R Soc Med 67(6 Pt 1):451–457
Munro AJ, Lain S, Lane DP (2005) P53 abnormalities and outcomes in colorectal cancer: a systematic review. Br J Cancer 92(3):434–444
Nakamura K-I, Furugori E, Esaki Y et al (2000) Correlation of telomere lengths in normal and cancers tissue in the large bowel. Cancer Lett 158(2):179–184
Negrini S, Gorgoulis VG, Halazonetis TD (2010) Genomic instability—an evolving hallmark of cancer. Nat Rev Mol Cell Biol 11(3):220–228

Nowak MA, Komarova NL, Sengupta A et al (2002) The role of chromosomal instability in tumor initiation. Proc Natl Acad Sci U S A 99(25):16226–16231

O'Brien MJ (2007) Hyperplastic and serrated polyps of the colorectum. Gastroenterol Clin North Am 36(4):947–968, viii

O'Brien MJ, Yang S, Clebanoff JL et al (2004) Hyperplastic (serrated) polyps of the colorectum: relationship of CpG island methylator phenotype and K-ras mutation to location and histologic subtype. Am J Surg Pathol 28(4):423–434

O'Brien MJ, Yang S, Mack C et al (2006) Comparison of microsatellite instability, CpG island methylation phenotype, BRAF and KRAS status in serrated polyps and traditional adenomas indicates separate pathways to distinct colorectal carcinoma end points. Am J Surg Pathol 30(12):1491–1501

Ogino S, Kawasaki T, Kirkner GJ, Loda M, Fuchs CS (2006a) CpG island methylator phenotype-low (CIMP-low) in colorectal cancer: possible associations with male sex and KRAS mutations. J Mol Diagn 8(5):582–588

Ogino S, Odze RD, Kawasaki T et al (2006b) Correlation of pathologic features with CpG island methylator phenotype (CIMP) by quantitative DNA methylation analysis in colorectal carcinoma. Am J Surg Pathol 30(9):1175–1183

Ogino S, Kawasaki T, Kirkner GJ, Suemoto Y, Meyerhardt JA, Fuchs CS (2007) Molecular correlates with MGMT promoter methylation and silencing support CpG island methylator phenotype-low (CIMP-low) in colorectal cancer. Gut 56(11):1564–1571

Ogino S, Nosho K, Kirkner GJ et al (2009a) CpG island methylator phenotype, microsatellite instability, BRAF mutation and clinical outcome in colon cancer. Gut 58(1):90–96

Ogino S, Nosho K, Irahara N et al (2009b) Prognostic significance and molecular associations of 18q loss of heterozygosity: a cohort study of microsatellite stable colorectal cancers. J Clin Oncol 27(27):4591–4598

Ogino S, Nosho K, Irahara N et al (2009c) Lymphocytic reaction to colorectal cancer is associated with longer survival, independent of lymph node count, microsatellite instability, and CpG island methylator phenotype. Clin Cancer Res 15(20):6412–6420

Ogino S, Shima K, Meyerhardt J et al (2011) Predictive and prognostic roles of BRAF mutation in stage III colon cancer: results from intergroup trial CALGB 89803. Clin Cancer Res 18(3):890–900, Available from: http://www.ncbi.nlm.nih.gov/pubmed/22147942

Oshima M, Murai N, Kargman S et al (2001) Chemoprevention of intestinal polyposis in the ApcΔ716 mouse by rofecoxib, a specific cyclooxygenase-2 inhibitor. Cancer Res 61(4):1733–1740

Park DY, Sakamoto H, Kirley SD et al (2007) The cables gene on chromosome 18q is silenced by promoter hypermethylation and allelic loss in human colorectal cancer. Am J Pathol 171(5):1509–1519

Peddibhotla S, Lam MH, Gonzalez-Rimbau M, Rosen JM (2009) The DNA-damage effector checkpoint kinase 1 is essential for chromosome segregation and cytokinesis. Proc Natl Acad Sci U S A 106(13):5159–5164

Pérez R, Lasa A, Toro DH, Cruz-Correa M, Martinez-Souss J (2010) Relationship between sporadic hyperplastic polyps and colorectal neoplasia in Hispanic veterans. P R Health Sci J 29(4):372–376

Phillips SM, Banerjea A, Feakins R, Li SR, Bustin SA, Dorudi S (2004) Tumour-infiltrating lymphocytes in colorectal cancer with microsatellite instability are activated and cytotoxic. Br J Surg 91(4):469–475

Pino MS, Chung DC (2010) The chromosomal instability pathway in colon cancer. Gastroenterology 138(6):2059–2072

Plentz RR, Wiemann SU, Flemming P et al (2003) Telomere shortening of epithelial cells characterises the adenoma-carcinoma transition of human colorectal cancer. Gut 52(9):1304–1307

Popat S, Hubner R, Houlston RS (2005) Systematic review of microsatellite instability and colorectal cancer prognosis. J Clin Oncol 23(3):609–618

Powell SM, Zilz N, Beazer-Barclay Y et al (1992) APC mutations occur early during colorectal tumorigenesis. Nature 359(6392):235–237

Pretlow TP, Pretlow TG (2005) Mutant KRAS in aberrant crypt foci (ACF): initiation of colorectal cancer? Biochim Biophys Acta 1756(2):83–96

Pruitt K, Der CJ (2001) Ras and Rho regulation of the cell cycle and oncogenesis. Cancer Lett 171(1):1–10

Rajagopalan H, Bardelli A, Lengauer C, Kinzler KW, Vogelstein B, Velculescu VE (2002) Tumorigenesis: RAF/RAS oncogenes and mismatch-repair status. Nature 418(6901):934

Rajagopalan H, Jallepalli PV, Rago C et al (2004) Inactivation of hCDC4 can cause chromosomal instability. Nature 428(6978):77–81

Rashid A, Houlihan PS, Booker S, Petersen GM, Giardiello FM, Hamilton SR (2000) Phenotypic and molecular characteristics of hyperplastic polyposis. Gastroenterology 119(2):323–332

Ribic CM, Sargent DJ, Moore MJ et al (2003) Tumor microsatellite-instability status as a predictor of benefit from fluorouracil-based adjuvant chemotherapy for colon cancer. N Engl J Med 349(3):247–257

Roper J, Richardson MP, Wang WV et al (2011) The dual PI3K/mTOR inhibitor NVP-BEZ235 induces tumor regression in a genetically engineered mouse model of PIK3CA wild-type colorectal cancer. PLoS One 6(9):e25132

Roschke AV, Glebov OK, Lababidi S, Gehlhaus KS, Weinstein JN, Kirsch IR (2008) Chromosomal instability is associated with higher expression of genes implicated in epithelial-mesenchymal transition, cancer invasiveness, and metastasis and with lower expression of genes involved in cell cycle checkpoints, DNA repair, and chromatin maintenance. Neoplasia 10(11):1222–1230

Rubio CA, Stemme S, Jaramillo E, Lindblom A (2006) Hyperplastic polyposis coli syndrome and colorectal carcinoma. Endoscopy 38(3):266–270

Ruder EH, Laiyemo AO, Graubard BI, Hollenbeck AR, Schatzkin A, Cross AJ (2011) Non-steroidal anti-inflammatory drugs and colorectal cancer risk in a large, prospective cohort. Am J Gastroenterol 106(7):1340–1350

Rudolph KL, Millard M, Bosenberg MW, DePinho RA (2001) Telomere dysfunction and evolution of intestinal carcinoma in mice and humans. Nat Genet 28(2):155–159

Russo A, Bazan V, Iacopetta B, Kerr D, Soussi T, Gebbia N (2005) The TP53 colorectal cancer international collaborative study on the prognostic and predictive significance of p53 mutation: influence of tumor site, type of mutation, and adjuvant treatment. J Clin Oncol 23(30):7518–7528

Samowitz WS, Albertsen H, Herrick J et al (2005a) Evaluation of a large, population-based sample supports a CpG island methylator phenotype in colon cancer. Gastroenterology 129(3):837–845

Samowitz WS, Sweeney C, Herrick J et al (2005b) Poor survival associated with the BRAF V600E mutation in microsatellite-stable colon cancers. Cancer Res 65(14):6063–6069

Samowitz WS, Albertsen H, Sweeney C et al (2006) Association of smoking, CpG island methylator phenotype, and V600E BRAF mutations in colon cancer. J Natl Cancer Inst 98(23):1731–1738

Samuels Y, Wang Z, Bardelli A et al (2004) High frequency of mutations of the PIK3CA gene in human cancers. Science 304(5670):554

Samuels Y, Diaz LA, Schmidt-Kittler O et al (2005) Mutant PIK3CA promotes cell growth and invasion of human cancer cells. Cancer Cell 7(6):561–573

Sandmeier D, Benhattar J, Martin P, Bouzourene H (2009) Serrated polyps of the large intestine: a molecular study comparing sessile serrated adenomas and hyperplastic polyps. Histopathology 55(2):206–213

Sawhney MS, Farrar WD, Gudiseva S et al (2006) Microsatellite instability in interval colon cancers. Gastroenterology 131(6):1700–1705

Schöffski P, Awada A, Dumez H et al (2012) A phase I, dose-escalation study of the novel Polo-like kinase inhibitor volasertib (BI 6727) in patients with advanced solid tumours. Eur J Cancer 48(2):179–186

Schreiner MA, Weiss DG, Lieberman DA (2010) Proximal and large hyperplastic and nondysplastic serrated polyps detected by colonoscopy are associated with neoplasia. Gastroenterology 139(5):1497–1502

Shia J (2008) Immunohistochemistry versus microsatellite instability testing for screening colorectal cancer patients at risk for hereditary nonpolyposis colorectal cancer syndrome. J Mol Diagn 10(4):293–300

Shih I-M, Zhou W, Goodman SN, Lengauer C, Kinzler KW, Vogelstein B (2001) Evidence that genetic instability occurs at an early stage of colorectal tumorigenesis. Cancer Res 61(3):818–822

Shima K, Morikawa T, Yamauchi M et al (2011) TGFBR2 and BAX mononucleotide tract mutations, microsatellite instability, and prognosis in 1072 colorectal cancers. PLoS One 6(9):e25062

Shirasawa S, Furuse M, Yokoyama N, Sasazuki T (1993) Altered growth of human colon cancer cell lines disrupted at activated Ki-ras. Science 260(5104):85–88

Shrubsole MJ, Wu H, Ness RM, Shyr Y, Smalley WE, Zheng W (2008) Alcohol drinking, cigarette smoking, and risk of colorectal adenomatous and hyperplastic polyps. Am J Epidemiol 167(9):1050–1058

Sinicrope FA, Rego RL, Halling KC et al (2006) Prognostic impact of microsatellite instability and DNA ploidy in human colon carcinoma patients. Gastroenterology 131(3):729–737

Snover DC, Jass JR, Fenoglio-Preiser C, Batts KP (2005) Serrated polyps of the large intestine: a morphologic and molecular review of an evolving concept. Am J Clin Pathol 124(3):380–391

Souglakos J, Philips J, Wang R et al (2009) Prognostic and predictive value of common mutations for treatment response and survival in patients with metastatic colorectal cancer. Br J Cancer 101(3):465–472

Sparks AB, Morin PJ, Vogelstein B, Kinzler KW (1998) Mutational analysis of the APC/β-catenin/Tcf pathway in colorectal cancer. Cancer Res 58(6):1130–1134

Spring KJ, Zhao ZZ, Karamatic R et al (2006) High prevalence of sessile serrated adenomas with BRAF mutations: a prospective study of patients undergoing colonoscopy. Gastroenterology 131(5):1400–1407

Stoler DL, Chen N, Basik M et al (1999) The onset and extent of genomic instability in sporadic colorectal tumor progression. Proc Natl Acad Sci U S A 96(26):15121–15126

Storojeva I, Boulay J-L, Heinimann K et al (2005) Prognostic and predictive relevance of microsatellite instability in colorectal cancer. Oncol Rep 14(1):241–249

Stryker SJ, Wolff BG, Culp CE, Libbe SD, Ilstrup DM, MacCarty RL (1987) Natural history of untreated colonic polyps. Gastroenterology 93(5):1009–1013

Suzuki H, Harpaz N, Tarmin L et al (1994) Microsatellite instability in ulcerative colitis-associated colorectal dysplasias and cancers. Cancer Res 54(18):4841–4844

Suzuki H, Igarashi S, Nojima M et al (2010) IGFBP7 is a p53-responsive gene specifically silenced in colorectal cancer with CpG island methylator phenotype. Carcinogenesis 31(3):342–349

Takagi Y, Kohmura H, Futamura M et al (1996) Somatic alterations of the DPC4 gene in human colorectal cancers in vivo. Gastroenterology 111(5):1369–1372

Takagi Y, Koumura H, Futamura M et al (1998) Somatic alterations of the SMAD-2 gene in human colorectal cancers. Br J Cancer 78(9):1152–1155

Takagi S, Kinouchi Y, Hiwatashi N et al (1999) Telomere shortening and the clinicopathologic characteristics of human colorectal carcinomas. Cancer 86(8):1431–1436

Takahashi T, Sano B, Nagata T et al (2003) Polo-like kinase 1 (PLK1) is overexpressed in primary colorectal cancers. Cancer Sci 94(2):148–152

Takayama T, Katsuki S, Takahashi Y et al (1998) Aberrant crypt foci of the colon as precursors of adenoma and cancer. N Engl J Med 339(18):1277–1284

Tan Y, Sangfelt O, Spruck C (2008) The Fbxw7/hCdc4 tumor suppressor in human cancer. Cancer Lett 271(1):1–12

Tanaka K, Yanoshita R, Konishi M et al (1993) Suppression of tumourigenicity in human colon carcinoma cells by introduction of normal chromosome 1p36 region. Oncogene 8(8):2253–2258

Tateyama H, Li W, Takahashi E, Miura Y, Sugiura H, Eimoto T (2002) Apoptosis index and apoptosis-related antigen expression in serrated adenoma of the colorectum: the saw-toothed structure may be related to inhibition of apoptosis. Am J Surg Pathol 26(2):249–256

Tatsumoto N, Hiyama E, Murakami Y et al (2000) High telomerase activity is an independent prognostic indicator of poor outcome in colorectal cancer. Clin Cancer Res 6(7):2696–2701

Taylor SS, McKeon F (1997) Kinetochore localization of murine Bub1 is required for normal mitotic timing and checkpoint response to spindle damage. Cell 89(5):727–735
Tedesco FJ, Hendrix JC, Pickens CA, Brady PG, Mills LR (1982) Diminutive polyps: histopathology, spatial distribution, and clinical significance. Gastrointest Endosc 28(1):1–5
Thiagalingam S, Laken S, Willson JKV et al (2001) Mechanisms underlying losses of heterozygosity in human colorectal cancers. Proc Natl Acad Sci U S A 98(5):2698–2702
Thibodeau SN, Bren G, Schaid D (1993) Microsatellite instability in cancer of the proximal colon. Science 260(5109):816–819
Tol J, Nagtegaal ID, Punt CJA (2009) BRAF mutation in metastatic colorectal cancer. N Engl J Med 361(1):98–99
Torlakovic E, Snover DC (1996) Serrated adenomatous polyposis in humans. Gastroenterology 110(3):748–755
Torlakovic E, Skovlund E, Snover DC, Torlakovic G, Nesland JM (2003) Morphologic reappraisal of serrated colorectal polyps. Am J Surg Pathol 27(1):65–81
Torlakovic EE, Gomez JD, Driman DK et al (2008) Sessile serrated adenoma (SSA) vs. traditional serrated adenoma (TSA). Am J Surg Pathol 32(1):21–29
Toyota M, Ahuja N, Ohe-Toyota M, Herman JG, Baylin SB, Issa JP (1999) CpG island methylator phenotype in colorectal cancer. Proc Natl Acad Sci U S A 96(15):8681–8686
Tran B, Kopetz S, Tie J et al (2011) Impact of BRAF mutation and microsatellite instability on the pattern of metastatic spread and prognosis in metastatic colorectal cancer. Cancer 117(20):4623–4632, Available from: http://www.ncbi.nlm.nih.gov/pubmed/21456008
Van Cutsem E, Köhne C-H, Láng I et al (2011) Cetuximab plus irinotecan, fluorouracil, and leucovorin as first-line treatment for metastatic colorectal cancer: updated analysis of overall survival according to tumor KRAS and BRAF mutation status. J Clin Oncol 29(15):2011–2019
Vasen HFA (2005) Clinical description of the Lynch syndrome [hereditary nonpolyposis colorectal cancer (HNPCC)]. Fam Cancer 4(3):219–225
Vasen HFA, Mecklin J-P, Meera Khan P, Lynch HT (1991) The International Collaborative Group on Hereditary Non-Polyposis Colorectal Cancer (ICG-HNPCC). Dis Colon Rectum 34(5):424–425
Vasen HFA, Watson P, Mecklin J-P, Lynch HT (1999) New clinical criteria for hereditary nonpolyposis colorectal cancer (HNPCC, Lynch syndrome) proposed by the International Collaborative Group on HNPCC. Gastroenterology 116(6):1453–1456
Veigl ML, Kasturi L, Olechnowicz J et al (1998) Biallelic inactivation of hMLH1 by epigenetic gene silencing, a novel mechanism causing human MSI cancers. Proc Natl Acad Sci U S A 95(15):8698–8702
Vogelstein B, Fearon ER, Hamilton SR et al (1988) Genetic alterations during colorectal-tumor development. N Engl J Med 319(9):525–532
Vogelstein B, Fearon E, Kern S et al (1989) Allelotype of colorectal carcinomas. Science 244(4901):207–211
Vogelstein B, Lane D, Levine AJ (2000) Surfing the p53 network. Nature 408(6810):307–310
Wajapeyee N, Serra RW, Zhu X, Mahalingam M, Green MR (2008) Oncogenic BRAF induces senescence and apoptosis through pathways mediated by the secreted protein IGFBP7. Cell 132(3):363–374
Wallace K, Grau MV, Ahnen D et al (2009) The association of lifestyle and dietary factors with the risk for serrated polyps of the colorectum. Cancer Epidemiol Biomarkers Prev 18(8):2310–2317
Wang L, Cunningham JM, Winters JL et al (2003) BRAF mutations in colon cancer are not likely attributable to defective DNA mismatch repair. Cancer Res 63(17):5209–5212
Wang Z, Cummins JM, Shen D et al (2004) Three classes of genes mutated in colorectal cancers with chromosomal instability. Cancer Res 64(9):2998–3001
Ward R, Meagher A, Tomlinson I et al (2001) Microsatellite instability and the clinicopathological features of sporadic colorectal cancer. Gut 48(6):821–829
Ward RL, Cheong K, Ku S-L, Meagher A, O'Connor T, Hawkins NJ (2003) Adverse prognostic effect of methylation in colorectal cancer is reversed by microsatellite instability. J Clin Oncol 21(20):3729–3736

Warthin A (1913) Heredity with reference to carcinoma—as shown by the study of the cases examined in the pathological laboratory of the University of Michigan, 1895–1913. Arch Intern Med 12(5):546–555

Weisenberger DJ, Siegmund KD, Campan M et al (2006) CpG island methylator phenotype underlies sporadic microsatellite instability and is tightly associated with BRAF mutation in colorectal cancer. Nat Genet 38(7):787–793

Whitehall VL, Walsh MD, Young J, Leggett BA, Jass JR (2001) Methylation of O-6-methylguanine DNA methyltransferase characterizes a subset of colorectal cancer with low-level DNA microsatellite instability. Cancer Res 61(3):827–830

Winawer SJ, Zauber AG, Ho MN et al (1993) Prevention of colorectal cancer by colonoscopic polypectomy. N Engl J Med 329(27):1977–1981

Wood LD, Parsons DW, Jones S et al (2007) The genomic landscapes of human breast and colorectal cancers. Science 318(5853):1108–1113

Wu JM, Montgomery EA, Iacobuzio-Donahue CA (2008) Frequent beta-catenin nuclear labeling in sessile serrated polyps of the colorectum with neoplastic potential. Am J Clin Pathol 129(3):416–423

Wynter CVA, Walsh MD, Higuchi T, Leggett BA, Young J, Jass JR (2004) Methylation patterns define two types of hyperplastic polyp associated with colorectal cancer. Gut 53(4):573–580

Yachida S, Mudali S, Martin SA, Montgomery EA, Iacobuzio-Donahue CA (2009) Beta-catenin nuclear labeling is a common feature of sessile serrated adenomas and correlates with early neoplastic progression after BRAF activation. Am J Surg Pathol 33(12):1823–1832

Yang S, Farraye FA, Mack C, Posnik O, O'Brien MJ (2004) BRAF and KRAS Mutations in hyperplastic polyps and serrated adenomas of the colorectum: relationship to histology and CpG island methylation status. Am J Surg Pathol 28(11):1452–1459

Young J, Jass JR (2006) The case for a genetic predisposition to serrated neoplasia in the colorectum: hypothesis and review of the literature. Cancer Epidemiol Biomarkers Prev 15(10): 1778–1784

Young J, Jenkins M, Parry S et al (2007) Serrated pathway colorectal cancer in the population: genetic consideration. Gut 56(10):1453–1459

Yuen ST, Davies H, Chan TL et al (2002) Similarity of the phenotypic patterns associated with BRAF and KRAS mutations in colorectal neoplasia. Cancer Res 62(22):6451–6455

Yun J, Rago C, Cheong I et al (2009) Glucose deprivation contributes to the development of KRAS pathway mutations in tumor cells. Science 325(5947):1555–1559

Zha S, Sekiguchi J, Brush JW, Bassing CH, Alt FW (2008) Complementary functions of ATM and H2AX in development and suppression of genomic instability. Proc Natl Acad Sci U S A 105(27):9302–9306

Zhang L (2008) Immunohistochemistry versus microsatellite instability testing for screening colorectal cancer patients at risk for hereditary nonpolyposis colorectal cancer syndrome. Part II. The utility of microsatellite instability testing. J Mol Diagn 10(4):301–307

Zhou W, Goodman SN, Galizia G et al (2002) Counting alleles to predict recurrence of early-stage colorectal cancers. Lancet 359(9302):219–225

Chapter 3
The Association Between Inflammation and Colorectal Cancer

Maria José Oliveira and Sérgia Velho

Abstract Inflammation plays an important role in the development and progression of many forms of cancer, including colorectal cancer (CRC). Several lines of evidence support a major role for an inflammatory background in CRC: (1) inflammatory bowel disease patients have a higher risk of developing CRC; (2) regular use of non-steroidal anti-inflammatory drugs (e.g., aspirin) prevents CRC development in cancer-free patients and promotes regression of established cancers; and (3) even CRCs that do not develop under inflammatory conditions are strongly infiltrated with multiple types of pro-tumorigenic immune cells. In this chapter, the association between chronic inflammation and CRC is reviewed and the major molecular mechanisms leading to tumor development are summarized. In addition, the contributions of distinct immune cell populations for cancer progression are discussed. Finally, the implications of these associations for cancer prevention and treatment are highlighted.

Keywords Colorectal cancer • Colitis-associated cancer • Inflammatory bowel disease • Inflammation • Immune system • Immunotherapy • STAT3 • NFkB • IL-6 • TNF-α

M.J. Oliveira (✉)
INEB—Institute for Biomedical Engineering, University of Porto, Rua do Campo Alegre, 823, Porto 4150-180, Portugal

Department of Pathology and Oncology, Faculty of Medicine, University of Porto, Alameda Professor Hernâni Monteiro, Porto 4200-319, Portugal
e-mail: mariajo@ineb.up.pt

S. Velho
Molecular Pathology Unit, Center for Cancer Research, and Center for Systems Biology, Massachusetts General Hospital, Harvard Medical School, 149 13th Street, 7.372, Charlestown, MA 02129, USA

Institute of Molecular Pathology and Immunology of the University of Porto, Porto, Portugal
e-mail: svelho@partners.org; sergiavelho@ipatimup.pt

K.M. Haigis (ed.), *Molecular Pathogenesis of Colorectal Cancer*,
DOI 10.1007/978-1-4614-8412-7_3,

3.1 The Link Between Inflammation and Cancer

Inflammation is a natural bodily defense raised in response to injury caused by external stimuli, such as pollutants, irritants, radiation, viruses, bacteria, pathogens, or cellular damage. An initial response, known as acute inflammation, occurs when injured tissues transiently recruit, from the blood system, plasma enriched in soluble factors (vascular stage) and leukocytes, such as eosinophils, monocytes and neutrophils (cellular stage). At the site of injury, phagocytic leukocytes engulf and digest the external aggressors, causing necrosis and pro-inflammatory mediators. A cascade of signaling mediators (histamine, cytokines, chemokines, and proteases) propagates and consolidates the inflammatory response, which, once resolved, leads to tissue healing and repair (Coussens and Werb 2002). Nevertheless, if injury is sustained, a prolonged unresolved condition with a distinct pattern of cellular and molecular mediators, known as chronic inflammation, takes place. Several diseases, such as heart disease, autoimmunity, arthritis, Alzheimer's and cancer have been associated with chronic inflammation. The evidence that inflammation and cancer are tightly linked comes from the nineteenth century, when Rudolf Virchow reported that inflammatory cells were present in tumor biopsy specimens and that tumors often developed in the setting of chronic inflammation (Balkwill and Mantovani 2001). Since then, several chronic infections have been associated with cancer development, as is the case for *Helicobacter pylori* and gastric carcinoma (Peek and Blaser 2002) the Human papillomavirus (HPV) and cervical cancer (Woodman et al. 2007; Moody and Laimins 2010), the Epstein–Barr virus and lymphoma (Henle and Henle 1973; Young and Murray 2003) and *Schistosoma haemotobium* and urinary bladder cancer (Botelho et al. 2011; Gelfand et al. 1967).

Over the past decades, efforts have been made to understand the connection between inflammation and cancer. Currently, it is accepted that inflammation plays dual and opposing roles in carcinogenesis (Rizzo et al. 2011): because it promotes the eradication of nascent tumor cells, it protects the organism from cancer, but since it establishes microenvironmental conditions sustaining tumor cell activities, it aids tumor development and selects malignant cells that escape immune system recognition (Schreiber et al. 2011). Concurrently, tumor cells release cytokines and chemokines and can sustain the inflammatory response, modulate the activity of inflammatory cells, and promote cellular phenotypes associated with transformation (e.g., proliferation, migration, and invasion). Three major immune hallmarks have been defined for the successful progression of cancer: (1) the ability to survive in a chronically-inflamed microenvironment; (2) the capacity to evade immune surveillance; (3) the ability to induce immune suppression (Cavallo et al. 2011).

The inflammatory microenvironment is known to create favorable conditions that foster the different steps of cancer development (initiation, promotion, and progression) by inducing genotoxic stress and by enhancing cell proliferation, survival, migration, angiogenesis, invasion, and metastasis (Schreiber et al. 2011; Greten et al. 2004). The role that inflammation plays in the different steps of cancer development results from a complex cross talk between inflammatory cells and cancer cells. This cross talk is ensured through two pathways: the extrinsic pathway through which inflammation induces changes at the cancer cells and the intrinsic pathway by which cancer cells modulate the inflammatory response (Mantovani et al. 2008). The extrinsic pathway involves the continuous release of reactive oxygen species (ROS) and reactive nitrogen intermediates (RNI) by inflammatory cells, namely neutrophils, macrophages and dendritic cells, reducing DNA repair or causing DNA damage and epigenetic changes (Grady and Carethers 2008), and predisposing to the acquisition of genomic instability and dysplasia. In addition, inflammatory mediators, such as cytokines and chemokines, secreted by inflammatory cells may induce, in a paracrine manner, the activation of cancer-related signaling pathways (Mantovani et al. 2008; Grivennikov et al. 2010). In the intrinsic pathway, the activation of certain oncogenes (e.g., RET, RAS, MYC, and B-RAF) in tumor cells can modulate the inflammatory reaction by inducing the expression of an inflammatory transcriptome (Mantovani et al. 2008; Borrello et al. 2008; Sumimoto et al. 2006; Guerra et al. 2007; Sparmann and Bar-Sagi 2004). Both the extrinsic and the intrinsic pathways converge upon the activation of several transcription factors, such as the nuclear factor-kb (NF-kB), signal transducer and activator of transcription 3 (STAT3), hypoxia-inducible factor 1α (HiF1α), and SMAD-family members in both tumor and inflammatory cells. These transcription factors coordinate (1) the production of pro-inflammatory chemokines and cytokines, which in turn increase the production of more inflammatory mediators and contribute for the generation/maintenance of a cancer-related inflammatory microenvironment; (2) the expression of several anti-apoptotic proteins (e.g., Bcl-2, Bcl-XL among others); (3) the expression of TP53; (4) the expression of cell cycle-related proteins; (5) the expression of c-MYC; and (6) the production of other molecules involved in the control of angiogenesis, tumor growth and invasion (Mantovani et al. 2008; Grivennikov et al. 2010).

Chronic inflammation has been associated with distinct forms of cancer, including colorectal cancer (CRC). This association was supported by studies revealing higher CRC incidence in individuals with inflammatory bowel disease (IBD) and reduced incidence upon treatment with common anti-inflammatory drugs, such as aspirin (Munkholm 2003). In addition, the presence of chronic intestinal inflammation is considered, together with hereditary CRC syndromes, a high risk factor for the development of CRC (Xie and Itzkowitz 2008). In this chapter the association between chronic inflammation and CRC will be reviewed and the major molecular mechanisms summarized. The contribution of distinct immune cell populations for cancer progression will be discussed. Finally, the implications of this association for cancer treatment will be highlighted as well.

3.2 The Link Between Inflammation and Colorectal Cancer: Insights from Inflammatory Bowel Disease

IBD, in the form of ulcerative colitis (UC) or Crohn's disease (CD), is characterized by an abnormal immune reaction developed in response to antigens of commensal intestinal bacteria, resulting in chronic inflammation of the gastrointestinal tract.

There is now a general consensus that multiple factors contribute to the development of the disease, for example intestinal microenvironmental changes, alterations to the commensal flora (i.e., dysbiosis), disturbances in the innate adaptive immune responses, and genetic variations increasing susceptibility (Triantafillidis et al. 2011; Schirbel and Fiocchi 2010). Polymorphisms found in several genes, including *NOD2/CARD15*, *DLG5*, *SLC22A4*, *SLC22A5*, *ABCB1/MDR1*, *ATG16L1*, and *IL23R*, have been associated with increased susceptibility to IBD (Wirtz and Neurath 2007; Cummings et al. 2007a, b).

The association between the presence of IBD and an increased risk for CRC was first described by Rosenberg and Crohn in 1925. In fact, this inflammatory condition represents a paradigm for the development of a type of inflammation-driven CRC, known as colitis-associated cancer (CAC). Epidemiologic data shows that, although patients with IBD represent only a small fraction (1–2 %) of all CRC patients (Kraus and Arber 2009), the presence of this inflammatory disorder is considered, together with hereditary syndromes such as familial adenomatous polyposis (FAP) and hereditary non-polyposis colorectal cancer (HNPCC), a high-risk factor for the development of CRC (Xie and Itzkowitz 2008). Progression to CAC is related to the duration of the inflammatory disease and the extent and severity of inflammation (Gupta et al. 2007; Rutter et al. 2004). In general, UC patients have a 2.75 higher overall incidence rate ratio of developing CRC than the general population (Rizzo et al. 2011; Bernstein et al. 2001). A large meta-analysis study estimated that the risk of CRC in UC patients is 2 % after 10 years, 8 % after 20 years and 18 % after 30 years of disease (Eaden 2004). Nevertheless, prospective data obtained from a surveillance program estimated that the likelihood of UC patients developing CRC is 2.5 % at 20 years, 7.6 % at 30 years, and 10.8 % at 40 years (Rutter et al. 2006). The risk of CRC in CD patients is less documented. A meta-analysis study estimated that 2.9 % of the CD patients will develop CRC after 10 years of disease, 5.6 % after 20 years, and 8.3 % after 30 years (Canavan et al. 2006). Nevertheless, in contrast to UC, which exclusively affects the mucosal lining of the colon and rectum, Crohn's patients can develop inflammation in any part of the gastrointestinal tract (although showing predominance in the terminal ileum and colon). When inflammation occurs exclusively in the colon, CD patients have a 5.6-fold increased risk of developing CRC. The risk decreases to 3.2 if the disease locates at the ileocolonic region, whereas, the risk for patients with only ileum disease is not different from the general population (Rizzo et al. 2011; Ekbom et al. 1990). Some studies also showed an association between the age at onset of the IBD and CRC development. An early onset of IBD often correlates with the presence of widespread neoplasia, whereas localized tumors are frequently associated with late-onset (Brackmann et al. 2009; Delaunoit et al. 2006). Other studies, however, have failed to find a correlation between the age at onset and the development of CRC in IBD patients (Rutter et al. 2006). Moreover, other factors such as the presence of sclerosing cholangitis, a chronic liver disease caused by progressive inflammation (Kornfeld et al. 1997; Shetty et al. 1999; Torres et al. 2011; Soetikno et al. 2002), and family history of sporadic CRC (SCRC) (Askling et al. 2001; Nuako et al. 1998) further increase the risk associated with the development CRC in IBD patients (Rutter et al. 2004).

3.2.1 Molecular Mechanisms Underlying Colitis-Associated Cancer Development

Most of our knowledge relating inflammation to CAC derives from studies performed in experimental mouse models. The conventional model to study the mechanisms underlying the relationship between IBD and CAC is the azoxymethane (AOM)-dextran sodium sulfate (DSS) colitis model. DSS is directly toxic to colonic epithelial cells and therefore affects the integrity of the mucosal barrier, leading to the activation of tissue resident macrophages by exposing them to commensal bacteria. Oral administration of DSS solution is widely employed to recapitulate human UC, because it can cause acute inflammatory reaction and ulceration in the entire colon similar to that observed in UC patients (Popivanova et al. 2008; Okayasu et al. 1990). In susceptible mice strains, the administration of DSS for several cycles (e.g., 7 days DSS, 14 days water) results in chronic colitis and, if combined with a single initial dose of the genotoxic colon carcinogen AOM, inflammation-associated CRC (Tanaka et al. 2003; Wirtz et al. 2007; Neufert et al. 2007). In addition, genetically modified mouse models targeting components of the innate and adaptive immune system have also been developed (Wirtz and Neurath 2007), facilitating our understanding of the cellular drivers and molecular mechanisms underlying CAC. The use of murine models has revealed that CAC arises through a cross talk established between inflammatory/immune cells, present in the tumor microenvironment, and epithelial cells, which promotes the acquisition of genomic alterations and potentiates growth and survival of cancer cells through the secretion of inflammatory mediators and growth factors.

3.2.1.1 Inflammation Induces Genetic and Epigenetic Alterations in Epithelial Cells

As mentioned previously, cancer cells need to accumulate genomic alterations in order to overcome the negative pressure imposed by the immune system, to escape its surveillance, and to increase their oncogenic potential. Several lines of evidence reveal that inflammatory cells may contribute to the acquisition of DNA alterations driving initial neoplastic transformation. Macrophages and neutrophils secrete reactive oxygen (ROS) and nitrogen (RNS) species, which are thought to promote DNA damage by inducing base oxidation, deamination, and alkylation (Meira et al. 2008; Hussain et al. 2003) (Fig. 3.1a). Increased levels of ROS and RNS and expression of enzymes responsible to catalyze DNA base alterations have been reported in the inflamed colonic epithelium of IBD patients (for a complete review see Wiseman and Halliwell 1996). In accordance with the role of ROS and RNS in inducing DNA alterations under inflammatory conditions, it was demonstrated that the capacity of colonic epithelial cells to engage an efficient DNA repair response after chronic inflammation is important for the suppression of inflammation-induced tumorigenesis (Meira et al. 2008). In addition, macrophages also secrete macrophage

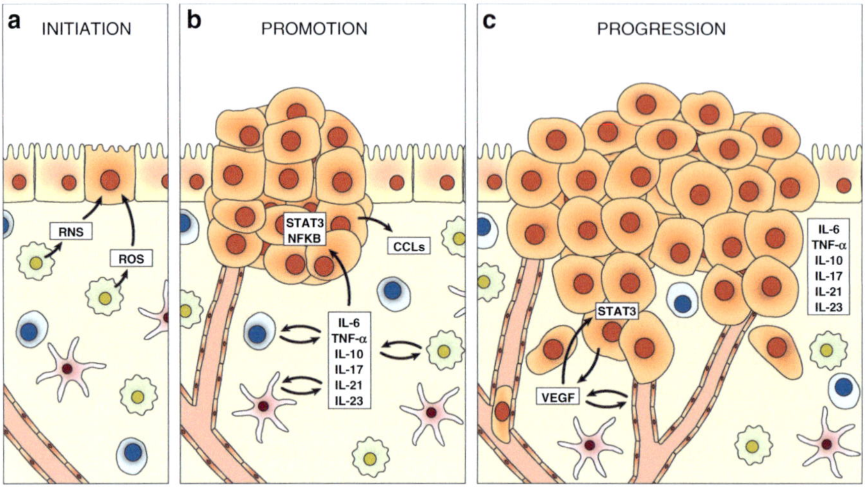

Fig. 3.1 Schematic representation of the molecular and cellular contributors in colitis-associated cancer development. (**a**) Inflammatory cells secrete ROS and RNS that induce genetic and epigenetic alterations in epithelial cells, initiating neoplastic transformation. (**b**) Tumor cell outgrowth is promoted by a complex interaction between immune cells that secrete pro- and anti-inflammatory mediators that will promote proliferation and survival of tumor cells by activating STAT3- and NFkB-dependent signaling. Tumor cells secrete chemokines (CCLs) that will recruit more inflammatory cells thus contributing to sustain an inflammatory reaction (**c**) Secretion of VEGF may contribute to tumor progression by inducing angiogenesis and enabling migration of both tumor and inflammatory cells

inhibitory factor (MIF), which was shown to suppress the transcription of the tumor suppressor *TP53* gene, inhibiting the DNA-damage response and, perhaps, leading to an increase in genomic instability (Fingerle-Rowson et al. 2003).

In addition to genetic alterations, inflammation is also associated with the presence of a methylator phenotype, leading to epigenetic silencing of genes that are important for preventing tumor development (McCabe et al. 2009; Issa et al. 2001). In vivo modeling of CAC also suggests that inflammation-induced DNA methylation occurs in the early phases of the disease, affecting noncancerous mucosa (Hartnett and Egan 2012). In fact, 60 % of genes hypermethylated in CRCs are reported to exhibit aberrant methylation in inflamed noncancerous tissues, suggesting that inflammation creates a signature of aberrant DNA methylation similar to what is found in cancers (Hahn et al. 2008). Some authors speculate that in early phases of tumor development, methylation is important to silence tumor specific antigens, therefore, keeping tumor cells hidden from the immune system in an equilibrium phase (Hosking 2012).

The mechanisms by which inflammation promotes changes in DNA methylation are not fully characterized. Methylation has been reported to occur after DNA damage induced by ROS, as halogenated pyrimidines, one form of ROS-induced damage, mimic 5-methylcytosine and stimulate DNA methyltransferase 1 (DNMT1)-mediated CpG methylation (McCabe et al. 2009; O'Hagan et al. 2011;

Katsurano et al. 2012). In addition, in vitro studies using CRC cell lines and the analyses of CAC from mouse models of colitis have shown that the expression of inflammatory cytokines, such as IL-6, IL-1β, tumor necrosis factor alpha (TNF-α), and interferon-γ, is associated with increased expression and stability of enzymes involved in DNA methylation, such as DNMT1 and DNMT3b (Hartnett and Egan 2012; Katsurano et al. 2012; Foran et al. 2010; Kominsky et al. 2011). Increased expression of DNMT1 was also observed in human CAC (Foran et al. 2010). Furthermore, over-expression of enzymes central to all cellular methylation mechanisms, S-adenosylmethionine synthetase and S-adenosylhomocysteine hydrolase, was found to occur in a DSS mouse model (Kominsky et al. 2011). Recent data using in vitro cell culture and a mouse model of colitis also showed that oxidative damage and inflammation increase the recruitment of silencing complexes containing DNMTs to the promoter CpG islands of genes, some of them previously shown to undergo inflammation and tumor-specific DNA methylation in models of intestinal inflammation and human cancers (O'Hagan et al. 2011).

3.2.1.2 Inflammation Contributes to CAC Promotion and Progression

The pro-tumorigenic properties of the tumor-associated inflammatory reaction are also linked to the capacity of inflammation-related factors to stimulate proliferation, survival, angiogenesis, and migration of epithelial cells, allowing the expansion of tumor initiating cells (Grivennikov et al. 2009). The effect exerted by some of these factors on the modulation of cancer cell-related activities will be discussed in detail in this section.

NFκB

In the DSS mouse model of chronic colitis, NFκB signaling is induced in activated macrophages, stimulating the production and secretion of pro-inflammatory cytokines by these cells. Secreted cytokines then activate NFκB signaling in intestinal epithelial cells (IECs), promoting the expression of survival molecules (Karin and Greten 2005) (Fig. 3.1b). In accordance with these observations, enterocyte-specific ablation of IKK-β, an activator of NFκB, was shown to decrease tumor incidence drastically in response to AOM-DSS treatment, without affecting the size and composition of tumors or the induction of oncogenic mutations. This finding indicates that the IKK-β-dependent NFκB-activated pathway operates during early tumor promotion. In addition, deletion of IKK-β in enterocytes enhanced the loss of intestinal barrier function induced by DSS and caused more inflammation, suggesting that the tumor-promoting function of NFκB in enterocytes is associated with its ability to suppress apoptosis of pre-neoplastic progenitors (Greten et al. 2004; Karin and Greten 2005). On the other hand, deletion of IKK-β in myeloid cells (dendritic cells and macrophages) resulted in a significant decrease in tumor number and size, but without affecting apoptosis of epithelial cells. The difference between enterocyte and myeloid-specific ablation of NFκB was mainly associated with a decrease in the expression of pro-inflammatory cytokines by myeloid cells that may serve as tumor growth factors (Greten et al. 2004).

IL-6/STAT3

Several studies using mouse models of chronic colitis have highlighted the role of the pro-inflammatory cytokine interleukin (IL)-6 in the development of CAC. It has been suggested that IL-6 contributes to increase tumor burden and multiplicity in the early stages of CAC development, as well to maintain tumor growth at late stages of the disease by stimulating proliferation and survival of neoplastic IECs. Additionally, IL-6 also helps to perpetuate inflammation by influencing the continuous recruitment and activation of inflammatory cells to sites of inflammation (Grivennikov et al. 2009; Becker et al. 2004; Bollrath et al. 2009). Myeloid cells and T lymphocytes present in the lamina propria or infiltrating the tumor tissue were described as the main sources of IL-6 secretion to the microenvironment during CAC development (Grivennikov et al. 2009; Becker et al. 2004, 2005; Matsumoto et al. 2010) (Fig. 3.1b). Grivennikov and colleagues (Grivennikov et al. 2009) reported that IL-6 expression by myeloid cells is driven by an NFκB-dependent mechanism, whereas, expression of IL-6 by T-lymphocytes was reported to occur during CAC progression and seems to be controlled by TGF-β signaling (Becker et al. 2004). Upon secretion into the microenvironment, IL-6 stimulates proliferation and survival of IECs through the activation of STAT3 (Greten et al. 2004; Grivennikov et al. 2009; Becker et al. 2004; Bollrath et al. 2009) (Fig. 3.1b). The activation of STAT3 was shown to occur downstream IL-6 binding to its receptors (gp130 and IL-6 receptor) expressed in the surface of epithelial cells or due to IL-6 trans-signaling induced by macrophage-derived IL-6/soluble IL-6R (Matsumoto et al. 2010; Grivennikov and Karin 2011). Consistent with the role of IL-6 in CAC development, a reduction in tumor number and in tumor size was observed in IL-6 null mice treated with DSS, along with inhibition of progression from adenoma to carcinoma. These data show that IL-6 signaling plays an important role during early stages of CAC (Grivennikov et al. 2009). In accordance with the role of IL-6/STAT3 pathway activation, deletion of STAT3 in IECs reduces cell proliferation, increases apoptosis and colitis, and reduces the number and size of tumors. These data demonstrate a critical role for epithelial STAT3 activation in inflammation-induced tumor formation and growth (Grivennikov et al. 2009; Becker et al. 2004; Bollrath et al. 2009). Curiously, the ablation of STAT3 in IECs had a stronger effect than ablation of IL-6, suggesting that other factors may contribute to induce STAT3 activation. In this vein, other cytokines, such as IL-11, IL-22 and IL-23, were also reported to induce STAT3 activation in IECs in mouse models of colitis (Grivennikov et al. 2009; Bollrath et al. 2009; Sugimoto et al. 2008; Pickert et al. 2009; Grivennikov and Karin 2010).

In vivo studies also showed that activation of STAT3 in T cells plays a pathogenic role in chronic colitis by inducing prolonged survival of pro-inflammatory T cells and disruption of immune tolerance (Sugimoto 2008; Atreya et al. 2000; Takeda et al. 1998). In addition, IL-6 was shown to induce polarization of T lymphocytes towards more pro-tumorigenic subtypes, such as T helper 17 (Th17), while inhibiting the differentiation of the suppressor T regulatory cells (Tregs), in this way contributing to enhancement of the inflammatory reaction (Grivennikov and Karin 2011; Dominitzki et al. 2007). Mice lacking STAT3 specifically in macrophages and neutrophils showed abnormal activation of these cells and impaired expression of IL-10 signaling. These mice developed chronic enterocolitis and showed enhanced T cell polarization towards Th1 cell activity, as is the case in IL-10-deficient mice (Kuhn et al. 1993; Berg et al. 1996; Takeda et al. 1999). These findings demonstrate that STAT3 activation in myeloid cells is essential for anti-inflammatory reactions mediated by IL-10 (Takeda et al. 1999). Latter studies confirmed that tumor-associated macrophages (TAMs) released IL-6, which in turn induced STAT3-mediated IL10 production in tumor cells, favoring immunesuppression and tumor progression (Herbeuval et al. 2004). Altogether, these results point to a role for IL-6 in promoting a strong inflammatory response, contributing to the continuous release of pro-tumorigenic factors through T-cells recruitment/activation, which can be counter-balanced by the induction of an immune-suppressive phenotype in myeloid cells, sustaining a pro-oncogenic microenvironment.

Consistent with the in vitro and in vivo studies aforementioned, higher levels of IL-6, an increase in active STAT3, and lower levels of SOCS3 (a negative regulator of STAT3 activation), were reported in the blood and inflamed mucosa, as well as in dysplasias and cancers, of IBD patients in comparison to patients with inactive UC and controls (Li et al. 2010), demonstrating the importance of this signaling pathway in the human context. In addition, constitutive activation of STAT3 was reported in immune cells (mainly macrophages and T lymphocytes) present in actively inflamed areas of both CD and UC patients (Wick et al. 2012; Lovato et al. 2003; Musso et al. 2005).

TNF-α

TNF-α is another NFκB-regulated pro-inflammatory cytokine playing a central role in the initiation and progression of CAC (Bollrath and Greten 2009). Similar to IL-6, TNF-α is also involved in the control of inflammatory cells recruitment and known to induce survival of epithelial cells (Fig. 3.1b). In addition, TNF-α and IL-6 were reported to cross-regulate each other, contributing to the enhancement of chronic inflammation and intestinal tumorigenesis (Grivennikov et al. 2009).

TNF-α expression levels were shown to increase after AOM/DSS treatment and this was associated with an increase in the number of infiltrating leukocytes expressing its major receptor (TNFR1), in the lamina propria and submucosal regions of the colon. During CAC development TNF-α is mostly produced by macrophages and T lymphocytes (Fig. 3.1b), although epithelial cells might also express this cytokine. The tumor-promoting properties of TNF-α are probably linked to its role as an inducer of NFκB signaling in epithelial cells, therefore stimulating survival (Grivennikov and Karin 2011). In accordance with this view, two studies have addressed the role of TNF-α in activating NFκB in epithelial cells. In one study, an increase in the NFκB signaling and simultaneous up-regulation of TNFR2 were observed during the progression from normal mucosa to CAC in DSS-treated mice (Onizawa et al. 2009). In this case, up-regulation of TNFR2 in colon cancer cells was mediated primarily by STAT3 activity upon IL-6 and TNF-α stimulation (Hamilton et al. 2011). Over-expression of this receptor was also reported to occur in epithelial cells from IBD patients (Mizoguchi et al. 2002). In the other study, NFκB activation was shown to occur downstream of the TNFR1-RAF1 signaling cascade (Edelblum et al. 2008). In conclusion, TNF-α stimulates NFκB activation in epithelial cells through both of its receptors.

TNF-α levels are increased in the mucosa of IBD patients (Kollias 2004; Roberts-Thomson et al. 2011) and its contribution to CAC development makes it a valuable molecule for targeted therapies. Consistent with this notion, AOM/DSS treated mice lacking TNFR1 or treated with Etanercept, a specific antagonist of TNF-α, showed reduced mucosal damage, reduced infiltration of macrophages and neutrophils, and attenuated tumor formation (Popivanova et al. 2008). The same inhibitory effect on tumor growth was observed when TNF-α was inhibited using monoclonal antibodies during late stages of CAC development (Grivennikov et al. 2009; Onizawa et al. 2009). In addition, TNF-α inhibition in AOM/DSS-treated mice was also shown to reduce angiogenesis, possibly due to inhibition of leucocytes recruitment and consequent inhibition of ciclooxigenase 2 (COX-2) expression (Popivanova et al. 2008; Goel et al. 2011). The beneficial effects of anti-TNF-α targeted therapies using monoclonal antibodies have been extensively shown in CD and UC patients that are refractory to conventional therapies, such as aminosalicylates, corticosteroids, or immunosuppressors (Triantafillidis et al. 2011). From the observations made in experimental mouse models of colitis, TNF-α inhibition seems to be anti-tumorigenic, but the effect of TNF-α inhibitors in CAC from human patients remains to be elucidated.

IL-21

Interleukin (IL)-21, a T-cell-derived cytokine, is over-produced in IBD (Fina et al. 2008) and its role in CAC development has been recently demonstrated. After AOM/DSS treatment, IL-21 knockout mice showed reduced mucosal damage, reduced infiltration of T cells, and diminished production of IL-6 and IL-17A. Absence of IL-21 reduced STAT3 activation in epithelial and stromal cells and resulted in the development of fewer and smaller tumors compared with wild-type mice (Stolfi et al. 2011). In addition, IL-21 induces the polarization of T cells towards Th17-mediated chronic intestinal inflammation, characterized by high levels of IL-17A, and reduces the production of interferon-γ (IFN-γ), which exerts anti-tumor activity by enhancing the capacity of cytotoxic CD8 T cells. Thus IL-21 supports chronic inflammation and reduces tumor immune surveillance, promoting a tumor-supportive microenvironment in the colon (Danese et al. 2011; Jauch et al. 2011).

Chemokine Expression

Chemokines are important components of cancer-related inflammation where they play a key role in orchestrating the recruitment and positioning of leukocytes (Bonecchi et al. 2011). NFκB activation in IECs was shown to induce the expression of several chemokines involved in the recruitment of more myeloid cells to sites of inflammation (Bollrath and Greten 2009; Eckmann et al. 2008). In the DSS model of colitis, mice lacking CCR5 do not develop colitis (Goel et al. 2011; Andres et al. 2000). In the AOM-DSS model of inflammation-induced colon carcinogenesis, CCR2 knockout mice exhibit less macrophage infiltration and lower tumor numbers, indicating that CCL2, its ligand, is a crucial mediator of the initiation and promotion of CAC and that targeting CCR2 may be useful in treating CAC (Goel et al. 2011; Popivanova et al. 2009). The use of CCL2 antagonists inhibited COX-2 expression, reduced angiogenesis, and decreased the number and size of colon tumors in mice (Popivanova et al. 2009).

Angiogenesis

Angiogenesis is a fundamental process underlying tumor growth and progression, since it allows the diffusion of oxygen and nutrients to tumor cells and provides a conduit through which cancer cells can metastasize (Keith and Simon 2008). An increase in the microvessel density has also been shown to play an important role in the pathogenesis of IBD and CAC (Danese et al. 2006). Vascular endothelial growth factor (VEGF), a strong pro-angiogenic factor involved in the induction of endothelial cell proliferation, migration, survival and permeability, was shown to be up-regulated in the inflamed mucosa of IBD (Danese et al. 2006; Tsiolakidou et al. 2008; Alkim et al. 2012; Scaldaferri et al. 2009) and in CAC patients (Fig. 3.1c) (Waldner et al. 2010). Similarly, the expression of VEGF receptor 2 was found to be up-regulated in the mucosa of IBD patients, mainly localized in endothelial cells, however others have also shown it in epithelial cells from inflamed mucosa and CAC (Scaldaferri et al. 2009; Waldner et al. 2010; Frysz-Naglak et al. 2011). Together, the abovementioned observations highlight the importance of the microvasculature and of the associated angiogenic factors in colitis and CAC.

3.3 Colitis-Associated Colorectal Cancer Versus Sporadic Colorectal Cancer

CAC is an example of the extrinsic pathway of inflammation-induced cancer in which components of the immune system are the main drivers of carcinogenesis. In sporadic colorectal cancer (SCC), inflammation also plays a role, but these tumors are unlikely to be initiated by inflammation because most tumor immune cells are presumably recruited after the tumor is formed (to be discussed in more detail in Sects. 3.4 and 3.5 of this chapter) (Terzic et al. 2010). Because the etiologic factors underlying cancer development in these two subsets of malignancies are different, it is reasonable to ask if both types of cancers share the same clinical, pathologic, and molecular features, or if they are two distinct entities. In the next subsections, we will highlight the similarities and differences between these two subtypes of CRC.

3.3.1 Clinical and Pathologic Features of CAC Versus SCC

Carcinomas arising in CD and UC patients have strikingly similar clinicopathologic features, namely age at onset, tumor location, and histology, suggestive of an inflammation-related signature of carcinogenesis (Choi and Zelig 1994; Svrcek et al. 2007). Some studies have, however, shown that CRC arising in CD patients appears at more advanced stages (Kiran et al. 2010) and are associated with worse overall survival (Ouaissi et al. 2011) when compared with CRC from UC patients. Nevertheless, compared with sporadic colorectal carcinoma (SCRC), CRC arising in patients with IBD has several distinguishing clinical features. In general, CAC affects individuals at a younger age (Itzkowitz and Yio 2004) and CAC presents more often with mucinous or signet ring cell histology. Due to the presence of widespread inflammation, CRC arising in IBD patients tend to be macroscopically heterogeneous and poorly delimited, irregular, and frequently multifocal, suggesting a broader field effect of mucosal inflammation (Brackmann et al. 2009; Delaunoit et al. 2006). CAC is frequently anaplastic, broadly infiltrating, and rapidly growing, progressing to invasive adenocarcinoma from flat and non-polypoid dysplasia more frequently than SCRC. In accordance, it was reported that CAC patients show a poorer survival rate when compared to SCRC in the background population (Brackmann et al. 2009). In contrast, other studies failed to identify a difference in the survival rates for IBD-associated and SCRC (Delaunoit et al. 2006; Kiran et al. 2010).

3.3.2 Molecular Alterations of CAC Versus SCRC

The molecular alterations accompanying pathogenesis are well described for SCRC. Regarding the molecular landscape, SCRC can be primarily divided in two major groups according to the type of genetic instability they exhibit. The majority (85 %) of SCRC are characterized by the presence of chromosomal instability (CIN), which consists of great losses and gains of chromosomal material. The remaining 15 % of the tumors are characterized by the accumulation of numerous mutations throughout the genome due to inactivation of mismatch repair (MMR) genes, creating a phenotype known as microsatellite instability (MSI) (Ionov et al. 1993). It is accepted that CIN and MSI SCRCs develop through different pathways of neoplastic transformation, which are associated with specific molecular alterations. CIN

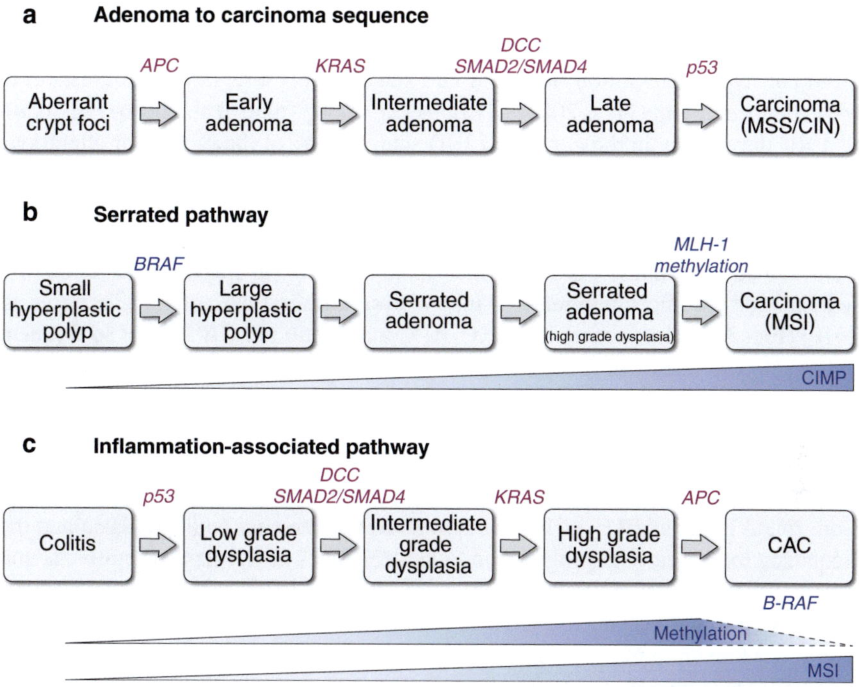

Fig. 3.2 Pathways for colorectal cancer development. (**a**) In the adenoma-carcinoma sequence, inactivating mutations of *APC* gene are the onset of adenoma formation. Acquisition of further genetic alterations, namely *KRAS* oncogenic activation and *TP53* inactivating mutations, will then function as driving forces towards progression from adenoma to carcinoma; (**b**) *BRAF* mutation is a marker for the serrated polyp pathway that has its origin in hyperplastic polyps and a potential end point as a MSI carcinoma. CIMP develops early in this sequence and MSI develops late due to *MLH1* silencing by promoter methylation; (**c**) CAC development frequently occurs through the adenoma-carcinoma sequence although a serrated pathway has also been described. In both cases, the sequence of molecular events driving tumor development differs from the one described for non-inflammation-associated CRCs. In CAC development, *TP53* mutations are the initiating event in the adenoma-carcinoma sequence and loss of *APC* occurs prior to malignant transformation. In the serrated pathway of CAC development, *BRAF* mutations occur in a late stage of tumor development. In its turn, the role of CpG island methylation is still controversial. It seems to be important only for the progression of pre-malignant lesions and not so relevant in an established cancer

tumors follow the adenoma—carcinoma pathway described by Fearon and Vogelstein in (1990). In this pathway, inactivation of the adenomatous polyposis coli (*APC*) gene is considered to be the trigger to adenoma formation and it occurs in about 80 % of SCCs. Progression to more advanced stages and the eventual formation of adenocarcinoma proceeds with the acquisition of alterations in other fundamental genes such as *KRAS*, *deleted in CRC* (*DCC*), *SMAD2*, *SMAD4*, and, lastly, *TP53* (Fig. 3.2a). On the other hand, a serrated neoplasia pathway, which is frequently associated with the early occurrence of *BRAF* mutations and the CpG island methylator phenotype, a surrogate marker of widespread methylation at the promoter regions of cancer-associated genes, is mainly associated with the development of sporadic MSI tumors. In this pathway, loss of function of MLH1 due to promoter hypermethylation occurs at late stages and results in the progression to MSI adenocarcinoma (Jass 2005; O'Brien et al. 2006; Velho et al. 2008) (Fig. 3.2b).

CRC developing in the context of IBD shares many of the molecular alterations found in both CIN and MSI SCRCs, although the type of precursor lesion and the sequence of molecular events leading to neoplastic transformation differ from the one described for SCRC (Xie and Itzkowitz 2008). Unlike SCRC that arises from well-defined adenomas, CAC may also arise from flat dysplasia areas through a sequence of chronic inflammation, injury, dysplasia and carcinoma (Terzic et al. 2010) (Fig. 3.2c). In contrast to SCRC, *APC* inactivation in CAC is not so frequent (14–33 %) and occurs in the late stages of the inflammation-associated dysplasia—carcinoma pathway (Umetani et al. 1999; Sepulveda and Aisner 2010). Conversely, *TP53* mutations, which are often found in late stages of the adenoma—carcinoma sequence, occur early in the development of CAC and are often found in non-dysplastic mucosa (Xie and Itzkowitz 2008; Kraus and Arber 2009). *TP53* mutations were found in about 19 % of biopsies from IBD patients without dysplasia and the frequency increased with progression to higher grades of dysplasia (Sepulveda and Aisner 2010). In CAC, *TP53* mutations were described to occur in frequencies above 50 % (Harpaz and Polydorides 2010; Aust et al. 2005; Sanchez et al. 2011). Other alterations known to play a role in the adenoma-carcinoma sequence, such as loss of *DCC*, *SMAD2* and *SMAD4*, were also found to occur at early phases of CAC development (Harpaz and Polydorides 2010). Another important gene in the development and progression of SCRC is *KRAS*. Mutations in this gene also play a role in the development of CAC and are found in high-grade dysplasia (Umetani et al. 1999). *KRAS* mutations, which are very frequent in SCRC (approximately 35 %) (Lau and Haigis 2009; Oliveira et al. 2004), occur in only about 20 % of IBD-related cancers (Umetani et al. 1999; Aust et al. 2005; Lyda et al. 2000; Holzmann et al. 1998).

Alterations similar to the mutator pathway were also described to occur during the pathogenesis of CAC, and MSI is also considered one of the mechanisms accounting for neoplastic progression in IBD patients (Umetani et al. 1999) (Fig. 3.2c). MSI was early found in inflamed mucosa of IBD patients without signs of dysplasia (Tahara et al. 2005). In addition, a high incidence of MSI was described in UC patients with long-standing severe inflammation, probably reflecting genomic instability caused by repeated inflammatory stress (Ishitsuka et al. 2001). The majority of the studies in which the frequency of MSI was analyzed reported that approximately 15 % of CAC display the mutator phenotype (Svrcek et al. 2007; Umetani et al. 1999; Schulmann et al. 2005), although others described higher frequencies (Tahara et al. 2005). Compared with sporadic MSI CRCs, MSI CAC patients presented with a younger age at diagnosis, and there was neither female nor right-sided predominance as it is characteristic of MSI SCC (Svrcek et al. 2007; Schulmann et al. 2005). Furthermore, there is some disagreement about the MMR defects underlying MSI in CAC. Methylation of *MLH1* promoter region, which is the foremost mechanism causing MSI in SCRC, was described to occur in IBD-related neoplasias (Aust et al. 2005; Schulmann et al. 2005; Fleisher et al. 2000), although most of these studies addressed methylation in a region of the promoter that is not related with silencing of the gene (Svrcek et al. 2007). Indeed, only a small proximal region of the *MLH1* promoter (C region) has been demonstrated to harbor a methylation status that correlates invariably with the loss of gene expression (Svrcek et al. 2007; Deng et al. 1999; Capel et al. 2007). When analyzing a large series of IBD-related neoplasias, Svrcek and colleagues (Kiran et al. 2010) found that, unlike sporadic MSI CRCs, methylation at the C-region of *MLH1* promoter occurs in a low frequency and, instead, MSI IBD neoplasias presented heterogeneous MMR defects involving *MLH1*, *MSH2*, *MSH6*, or *PMS2* genes.

Other molecular markers of the MSI pathway, such as *BRAF* mutations and CIMP, were also described to occur during CAC development. *BRAF* V600E hotspot mutations were described in 33.3 % of MSI IBD-associated neoplasias, a frequency comparable to the one found in MSI SCRC (Kiran et al. 2010), although, they are not considered an initiating event in CAC development (Aust et al. 2005) (Fig. 3.2c). The presence of CIMP is also a common feature of SCRC, in particular in MSI tumors, but in CAC its role is still controversial. Due to the aforementioned effect of inflammation in the induction of DNA methylation, it would be expected that CAC exhibited similar, if not higher levels, of DNA methylation than SCC. Instead, lower levels of CIMP and lower levels of methylation of age-associated genes have been reported to occur in CAC in comparison to SCRC (Sanchez et al. 2011; Konishi et al. 2007; Olaru et al. 2012). As described previously, the presence of methylation is a common feature of colitis-associated dysplastic lesions, however, the methylator phenotype does not seem to play a major role during CAC progression. In order to explain this discrepancy, some authors proposed that the presence of colitis-associated methylation creates a field defect resulting in premature aging of epithelial cells, therefore, increasing the risk of malignancy. Because, in CAC, methylation seems to play a minor role, genetic alterations are thought to be the main drivers of immune escape, leading to a more aggressive clinical course than epigenetic changes (Issa et al. 2001; Konishi et al. 2007).

3.4 The Role of Inflammation in Other Forms of Colorectal Cancer

CRC can be classified as: (1) inherited, including non-polyposis colorectal cancer (HNPCC) and familial adenomatous polyposis coli (FAP); (2) SCRC; and (3) inflammation-driven CRC (CAC) (Xie and Itzkowitz 2008). Although most CRCs are unlikely to be initiated by inflammation, they recruit and activate distinct immune cells, creating an inflammatory microenvironment. The cytokines and chemokines released by immune cells may then promote a pro-inflammatory response, counteracting tumor growth and survival, or an anti-inflammatory response, sustaining tumor cell activities (Terzic et al. 2010; Mantovani et al. 2004) (Fig. 3.3). Early studies profiling immune population distribution within CRC indicated that, at peritumoral regions, the inflammatory infiltrate consisted of 47 % lymphocytes, 19 % plasma cells, 15 % macrophages/monocytes, 5 % granulated mast cells, and 15 % polymorphonucleated (PMN) cells. Necrotic areas of the tumors were abundant in PMN and

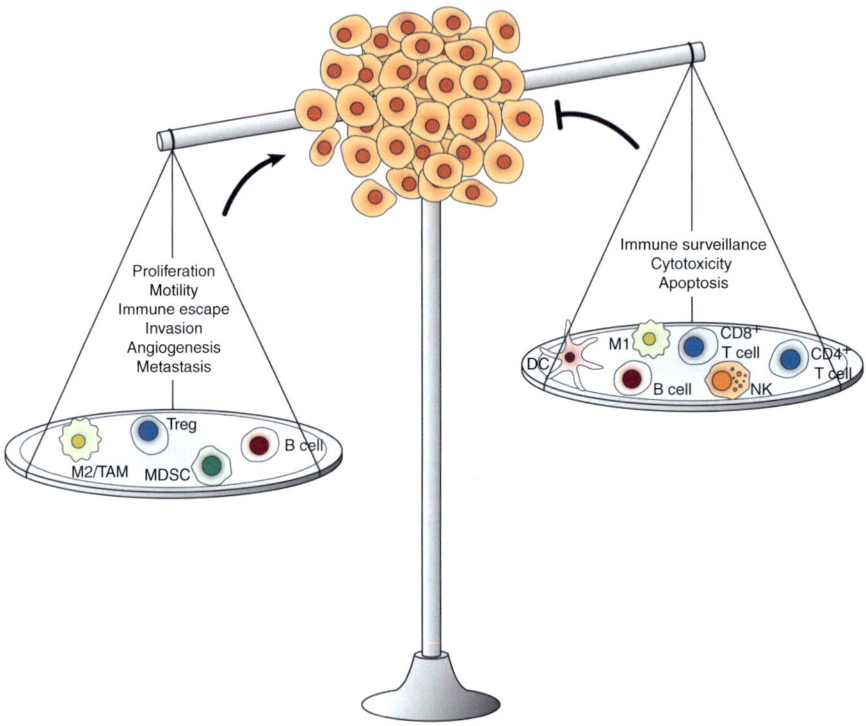

Fig. 3.3 Fine-tune regulation of the immune response in sporadic colorectal cancer. A strong inflammatory component is often present in the stroma of CRCs that do not develop in response to a pre-existent chronic inflammatory condition. The stroma of these cancers may be populated by pro-tumorigenic immune cells (M2/TAM, B cell, Treg, MDSC) that promote immune evasion, proliferation, motility, invasion, and metastization, or by anti-tumorigenic immune cells (Dc, $CD4^+$ T cell, $CD8^+$ T cell, B cell, NK e M1/TAM) that stimulate immune surveillance, cytotoxicity, and apoptosis. Inflammatory modulators (antigens, cytokines, and chemokines) produced by tumors cells shift the balance towards a more pro-tumorigenic or a more anti-tumorigenic immune response

macrophages (Svennevig et al. 1982). The role of innate immune cells, such as macrophages, natural killer (NK) and dendritic (DC) cells or of adaptive immune cells, such as T and B lymphocytes, in CRC cell activities will be overviewed.

3.4.1 Tumor Immune and Inflammatory Cell Infiltration

3.4.1.1 Macrophages

In the majority of the tumors, macrophages are a major component of the host leukocytic infiltrate (Condeelis and Pollard 2006). In the 1970s, studies aiming to understand the role of monocytes/macrophages on tumor progression suggested that, at least during the early phases of tumor growth, cancer cells could evade the immune response by releasing factors that diminish monocyte recruitment and activation (Snyderman and Pike 1977; Rhodes et al. 1979).

Nowadays, depending on the ecosystem, macrophages are thought to prevent the establishment and spreading of cancer cells, or alternately to support tumor growth and dissemination (Fig. 3.3). Upon mobilization to the periphery, bone marrow progenitor cells differentiate, depending on the type of stimulators they are exposed to, into dendritic cells or into monocytes, which in turn invade the tissues and differentiate into macrophages. In particular, TAMs are derived from circulating monocytes and recruited to the tumor site by various cytokines and chemokines. For example, VEGF, TGF-β, colony stimulating factors (M-CSF and GM-CSF), CCL2 (MCP-1), and CCL5 produced by stromal or tumor cells can induce the differentiation of incoming monocytes into M1 or M2 macrophages. Such designation mirrors the Th1 and Th2 lymphocyte nomenclature and has been proposed by Mantovani's group to differentiate macrophage populations with distinct functional and molecular phenotypes (Mantovani et al. 2004).

The classical activation into M1 macrophages, induced by INF-γ, bacterial LPS, or TNF-α, leads to the production of pro-inflammatory cytokines, which generally promote cytotoxic T cell activity and reduce tumor growth and metastasis. Alternatively activated, or M2 macrophages, are classified according to the type of stimulators they are exposed to: M2a if induced by IL-4 or IL-13; M2b if induced by immune-complexes and agonists of Toll-like receptors; and M2c if induced by IL-10 or Vitamin D3 (Mantovani et al. 2004). In contrast to M1, M2 macrophages are described to stimulate tissue repair and remodeling, angiogenesis and tumor progression (Fig. 3.3) (Mantovani et al. 2004, 2006; Talmadge et al. 2007). Thus far, this dual activity has only been consistently demonstrated in breast, prostate tumors, and melanomas. In breast tumors, in vitro and in vivo studies revealed macrophages as obligate partners for cancer cell migration, invasion, and metastasis, pointing to them as potential targets for therapy (Condeelis and Pollard 2006). In gastrointestinal tumors, data regarding macrophage density and disease prognosis are scarce and contradictory, and the role of M1 and M2 macrophage populations has never been analyzed in detail. Instead, TAM distribution, based on profiling of lineage marker, has been investigated. Some studies associated high macrophage peritumoral density with improved patient survival, reduced hepatic metastasis, and good prognosis, suggesting their differentiation into an M1 phenotype (Forssell et al. 2007; Zhou et al. 2010; Ohno et al. 2003). One of these studies suggested that anti-tumorigenic activity of TAMs required the direct macrophage-tumor cell contact (Forssell et al. 2007). A more recent study has demonstrated that the expression of CD80, CD86, and HLA-DR macrophage surface receptors, known to be co-stimulatory for T cell activation, was higher in CRC than in normal tissues. To the contrary, the expression of the CD40 surface receptor was less abundant in tumors and considered as a good prognostic marker (Kinouchi et al. 2012). In other cases, the high intratumoral macrophage density, along the invasive margins, in areas of neovascularisation and ECM degradation, was correlated with reduced patient survival and poor prognosis, suggesting their differentiation into an M2 phenotype (Fig. 3.3)

(Oosterling et al. 2005; Ishigami et al. 2003; Bailey et al. 2007). Such contradiction may reflect differences in number, grade, stage, and tumor size, but also in the methods used to assess macrophage distribution and infiltration. Another hypothesis, not confirmed by any of these studies, is that the identified macrophage populations are not necessarily the same. In fact, peritumoral macrophages might be less exposed to tumor-derived cytokines and tumor modulation and, thereby, differentiate into antitumor macrophages, producing cytotoxic molecules (such as ROS, NO and TNF-α) and exerting anti-inflammatory activities. It is possible, however, that when tumors progress by escaping immune surveillance, the tumor microenvironment gets hypoxic, enriched in tumor-derived metabolic products or cytokines, leading to the differentiation of incoming monocytes into pro-tumor macrophages (Mantovani et al. 2004; Erreni et al. 2011).

In CRC, TAMs are described to exert anti-tumoral or pro-tumoral activities (Fig. 3.3). Anti-tumoral activities may occur directly, by inducing tumor cytotoxicity, or indirectly, by modulating the host immune response, and have been suggested to be related with the presence of M1-polarized macrophages (Mantovani et al. 2004; Erreni et al. 2011). TAMs are described to induce tumor cell apoptosis via FAS-ligand mediated pathways, reducing tumor size and metastasis (Sugita et al. 2002). Macrophage production of ROS and NOS, upon T lymphocytes or natural killer (NK) cells stimulation, may also lead to tumor cell death. Tumor cell release of macrophage migration inhibitory factor (MIF) or granulocyte/macrophage-colony stimulating factors (GM-CSF or M-CSF) affect macrophage differentiation, survival, proliferation, migration, and metabolism, promoting phagocytosis, tumor cell lysis and the release of pro-inflammatory cytokines (Pozzi and Weiser 1992; Shinohara et al. 2000).

Upon modulation by the tumor microenvironment, TAMs may share many functional characteristics with M2 macrophages, suppressing the inflammatory response and inducing angiogenesis, tissue remodeling, tumor invasion and metastasis (Fig. 3.3) (Condeelis and Pollard 2006; Mantovani et al. 2006; Ruffell et al. 2012; Sica et al. 2006). Despite these evidences, their role in cancer progression is still controversial.

3.4.1.2 Dendritic Cells

Dendritic cells play (DCs) an important role in the inflammatory process. They are unique antigen-presenting cells (APCs), able to induce primary immune responses, but also capable of promoting immunological tolerance and regulation of T cell-mediated immune responses. Several studies have been performed to evaluate the degree and subsets of DC infiltration in CRC. DCs are more common in normal colon mucosa than in the tumor microenvironment, and nearly absent at metastatic tumors (Schwaab et al. 2001). In another study, lower levels of DCs infiltration in the tumor stroma and of $CD83^+$ DCs at tumor invasive margins were associated with high frequency of distant metastasis and with reduced patient survival (Gulubova et al. 2012). In CRC, mature DC infiltration seems to be enhanced in MSI-high tumors in comparison to their microsatellite stable (MSS) counterparts (Bauer et al. 2011). Interestingly, MSI-high tumors are known to express, as a result of their elevated DNA repair deficiency, high levels of immunogenic molecules and to elicit an exacerbated anti-inflammatory response. Perhaps for this reason, MSI-high tumors rarely metastasize and have a favorable prognosis. Thereby, it has been suggested that the high immunogenicity of MSI-high tumors is related to the pronounced infiltration of dendritic cells, macrophages, and the reduced Foxp3-positive T cells (Tregs). Through their immunesupressive function, Tregs were suggested to impair DC maturation, favoring CRC local immune evasion (Bauer et al. 2011). Concomitantly, in gastrointestinal cancer, impairment of DC maturation was associated with a significant reduction of certain T cell subpopulations, in particular of Thelper lymphocytes and of natural killer (NK) cells (Lissoni et al. 2000). Additional studies revealed that the expression of pro-inflammatory chemokines and the release of antigens resulting from local tumor destruction promoted DC activation and enhanced tumor regression and long-term immunity (Fig. 3.3) (Crittenden et al. 2003). However, all the data aforementioned derive from retrospective and correlative studies and additional research is required in order to formally determine the role of DCs in CRC. A better understanding of the mechanisms responsible for DCs attraction into the tumor microenvironment, counteracting the infiltration and differentiation/activation of other immune cell populations, will elucidate how these cells may modulate the tumor microenvironment in favor of tumor progression and opening new therapeutic strategies.

3.4.1.3 T Lymphocytes

Within the tumor microenvironment, there are numerous tumor-associated antigens (TAA) with the ability to elicit spontaneous T cell responses. This might constitute one of the mechanisms through which cancer cells modulate surrounding immune cells and escape immune surveillance (Nagorsen et al. 2003). At the core and at the invasive margin of CRCs, distinct subsets of T cells may be found and their association with tumor progression and disease outcome has been well-established (Deschoolmeester et al. 2011). Recent experiments performed in mouse models and human patients revealed a significant association between high densities of tumor infiltrating T lymphocytes (TIL) and improved overall survival and prognosis. Several studies have been performed aiming to characterize the distinct subsets of TIL present in CRC. Approximately 80 % of the tumor infiltrating lymphocytes were described to be CD2$^+$, 42 % CD4$^+$, and 27 % CD8$^+$, similar to lamina propria lymphocytes isolated from adjacent colon mucosa. In this study, the major difference between these two subsets of lymphocytes was the reduction of CD8$^+$ T cells, and the diminished expression of IL2 and of transferrin receptors in the lamina propria lymphocytes.

A recent study employing tissue microarrays and automated image analysis evaluated and quantified the densities of CD3$^+$, CD8$^+$, CD45RO$^+$, and FoxP3$^+$ cells within neoplastic areas. From this analysis, the CD45RO$^+$ T cell surface receptor was considered a favorable prognostic biomarker associated with longer patient survival, independent of other molecular, clinical, or pathological factors (Nosho et al.

2010). Additional studies revealed that the combined analysis of $CD8^+$ and $CD45RO^+$ cells in specific tumor areas could be a marker to predict tumor recurrence and survival in patients with early stage CRC (Pages et al. 2009).

Interestingly, the absence of early signs of metastatic invasion, such as lymphovascular emboli, was correlated with a significant increase of the density of memory T cells in situ. Pioneer studies performed by Galon and collaborators demonstrated that an increase in intra-tumoral expression of markers for cytotoxic effector T cells was associated with absence of early metastasis and a decrease in tumor recurrence (Galon et al. 2006). In general, the proportion of tumors with high density of $CD4^+$ and $CD8^+$ memory T cells diminishes with local tumor invasion and metastasis (Halama et al. 2011). Consistently, the proportion of primary tumors with high infiltrates of $CD4^+$ and $CD8^+$ memory T cells, particularly in the center of the tumor, was found to be lower in patients with recurrent tumors. The distribution of these T cell subsets has been mainly described at the invasive margins of colorectal tumor liver metastases (Halama et al. 2011). One of the factors that may explain this antitumor effect is the expression by $CD4^+$ T cells of IFN-γ, a pro-inflammatory cytokine. Indeed, decreased distribution of $CD4^+$ T lymphocytes in the center of CRCs was associated with reduced IFN-γ expression and with the presence of distant metastasis and of advanced clinical stage (Fig. 3.3) (Numata et al. 1991).

T cell function within tumors can also be regulated by TAMs. As high producers of TGF-β, cancer cells and TAMs may induce the differentiation of naïve $CD4^+$ T cells into regulator T cells (Treg), which in turn may suppress the anti-tumor activities of the cytotoxic $CD8^+$ T cells (Izcue et al. 2009). Tregs play a crucial role in homeostasis, preventing autoimmune disorders, by regulating the activity of autoreactive T cells and inducing immune tolerance towards self-antigens, as the ones produced by cancer cells. While Tregs are associated with poor prognosis in ovarian, breast and gastric carcinomas, their role in CRC outcome is still contradictory (Deschoolmeester et al. 2011), although multiple groups have demonstrated that the high density of $FoxP3^+$ Tregs in CRC was associated with improved survival (Ladoire et al. 2011; Salama et al. 2009). These findings suggest that, in CRC, tumor-infiltrating Tregs should be considered as potential allies in the anti-tumor response and therefore not targets for therapy (Fig. 3.3).

3.4.1.4 Natural Killer Cells

In CRC, the high incidence of cytotoxic T cells and of natural killer (NK) cells has been associated with enhanced cancer cell death and improved prognosis (Fig. 3.3) (Deschoolmeester et al. 2011). In a syngeneic rat model of CRC with liver and lung metastasis, NK cells were selectively recruited to the tumors and predominantly towards the stroma surrounding tumor cell nodules. Elimination of cancer cells was then initiated directly by NK cells or by their activation of other immune effectors (Kuppen et al. 2001). Consistent with these cytotoxic functions, it has been reported that decreased numbers of NK cells in pre-operative CRC cell patients was associated with an increased frequency of tumor recurrence (Atreya and Neurath 2008).

3.4.1.5 B Lymphocytes

Infiltrating B cells are the dominant component of inflammation in some cancers, such as ductal carcinoma in situ and invasive breast carcinomas. They express somatic hypermutated antibodies and recognize tumor-associated antigens, such as ganglioside D3. In CRC, B cells were mostly found at the invasive margin of growing tumors and in tertiary lymphoid structures (Dieu-Nosjean et al. 2008), sites of intense immune activity adjacent to tumor nests where proliferating B and mature DCs are in close contact with T cells. The role and the biological impact of intratumoral B cells in CRC are not yet clarified, however. It is possible that B lymphocytes act as antigen-presenting cells and, therefore, may be important for inducing $CD4^+$ and $CD8^+$ memory T cells, counteracting tumor invasion and metastasis (Fig. 3.3) (Deschoolmeester et al. 2011). It is also possible, that B cells promote tumor metastasis by converting resting $CD4^+$ T cells into immune suppressive Treg cells or by activating monocytes into M2 pro-inflammatory macrophages (Fig. 3.3).

Recent mouse models of spontaneous colorectal cancers suggest a destructive role of B lymphocytes, possibly through the production of IL-10, an immune suppressive cytokine, or through the production of IgGs, forming antigen–IgG antibody complexes (Hanahan and Coussens 2012).

3.4.1.6 Myeloid-Derived Suppressor Cells

Myeloid-derived suppressor cells (MDSCs) are classically described as suppressors of the inflammatory response, counteracting tumor cytotoxicity and promoting tumor progression. MDSCs were initially considered to induce immune suppression by inhibiting T cell, NK cell, and DC differentiation and activation (Shurin et al. 2012). In animal models, MDSCs are abundant in tumors and spleens of tumor-bearing mice and described to suppress the proliferation and T cell cytotoxicity and to reduce the responsiveness of $CD4^+$ and $CD8^+$ T cells to IFN-γ stimulation by the production of reactive oxygen (ROS) and nitrogen (NO) species, favoring tumor growth and progression (Mundy-Bosse et al. 2011). In CRC patients, tumor cells promote MDSC recruitment and infiltration and the percentage of MDSC cells was significantly correlated to neutrophil and inversely correlated with lymphocyte counts. Additionally, the presence of MDSC was correlated with worse prognosis and tumor progression (Fig. 3.3) (Ohki et al. 2012; Solito et al. 2011). The induction of MDSC differentiation and inhibition of the interaction of MDSCs with cancer cells are potential strategies for emerging cancer prevention and therapy.

3.4.2 Immune Cell Modulation by Colorectal Cancer Cells

The genetic and epigenetic alterations that occur in tumor cells do not explain the diversity of tumors and responses to therapy. Part of this discrepancy derives from cellular and molecular elements present within the surrounding tumor microenvironment, and from the interactions they establish with resident cancer cells (Galon et al. 2006). To escape immune surveillance and to establish efficient tumors and metastases, cancer cells become resistant to apoptosis, down-regulate antigen-presenting histocompatibility (MHC) complexes, produce immune suppressive cytokines such as TGF-β, express FAS ligands promoting the destruction of immune effector cells (the counterattack hypothesis), and disturb Th1/Th2 responses or modulate immune cells to polarize into pro-tumorigenic immunoregulators (Favre-Felix et al. 2000; Pages et al. 1999; Strand et al. 1996).

The appearance of altered cytokine and chemokine expression is frequently indicative of a reactive tumor stroma and a sign of the establishment of an inflammatory microenvironment. This repertoire of cytokines and chemokines present at the tumor microenvironment influences the recruitment, activation, and function of immune cells. Recently it has been demonstrated that inflammatory mediators present at the tumor microenvironment promote tumor growth, angiogenesis, and escape from immune surveillance by impairing dendritic cell infiltration and maturation (Michielsen et al. 2011). Alternatively, tumor-derived conditioned media may convert immature DCs into regulatory DC (regDC), which in turn suppress the activity of pro-inflammatory T cells, supporting tumor formation (Shurin et al. 2011). Recent studies revealed that cancer cell supernatants were able to modulate the differentiation of human blood-derived monocytes into an M1/M2 mixed phenotype (Caras et al. 2011). The tumor modulation of the monocytic/macrophagic population into an M2 pro-inflammatory subset, with reduced expression of NO and ROS species and of the pro-inflammatory cytokines (IL6, TNF-α) and higher expression of VEGF and matrix metalloproteases (MMPs), has been also previously reported as favoring tumor progression (Mantovani et al. 2004).

3.5 Inflammation in Angiogenesis, Invasion and Metastasis in Colorectal Cancer

The tumor microenvironment is comprised of tumor cells, extracellular matrix components, and stromal cells, including fibroblasts, endothelial cells, and immune cells. Growing evidence demonstrates that the molecular crosstalk established between cancer cells and the surrounding environment has a crucial impact on tumor progression by triggering and modulating invasion-associated activities such as cell-cell adhesion, cell-matrix interactions, growth, survival, proteolysis, motility angiogenesis, invasion, and even metastasis (Mareel and Leroy 2003; Mareel et al. 2009).

During cancer progression, the participation of each of these cell populations may differ. In human tumors, macrophages have been suggested to play an important role in cancer cell migration, invasion, and metastasis (Condeelis and Pollard 2006; Pollard 2004). The analysis of immune cell distribution in colon and breast carcinoma biopsies revealed that TAMs have a crucial role in the endocytosis of local immune complexes, and are active producers of pro-angiogenic factors, such as VEGF-A and VEGF-B, sustaining angiogenesis (Barbera-Guillem et al. 2002).

Interestingly, intrasplenic injection of colon carcinoma cells into syngeneic C57BL/6 mice revealed that the first liver micrometastases established appeared at the region of sinusoids, and were mainly populated by macrophage-derived Kupffer cells. Transitional metastases were characterize by intensive infiltration of macrophages and fibroblasts and were connected by protrusions enriched in fibroblasts, collagen, and endothelial cells. Finally, established metastases were mainly characterized by intense tissue fibrosis and accumulation of tumor cells (Higashi et al. 2002). In stage II and III CRC patients that underwent complete tumor surgical resections, the expression of osteopontin and of CD68, the lineage monocytic/macrophagic marker, co-localized with tumor central areas of high microvascular density. Such co-localization was significantly higher in patients with metachronous liver metastasis, suggesting that osteopontin produced by macrophages might be associated with increased risk of developing liver metastasis (Imano et al. 2011).

3.6 Blocking Inflammation for Colorectal Cancer Prevention and Therapy

Emerging data associating the presence, density, and distribution of certain immune cell populations with tumor progression and clinical outcome support the hypothesis that the adaptive immune response influences the behavior of human tumors and suggest that, by dissecting patient's immune response, novel disease prognostic markers and predictors of therapy may be identified. In the early 1980s, Cameron and Churchill demonstrated that the activation of peripheral blood monocytes with bacterial-derived LPS was cytotoxic towards the malignant, but not against the non-malignant, cell lines tested (Cameron and Churchill 1980). These achievements suggested that the immune system could be exogenously educated to offer anti-tumor protection, opening new perspectives for therapeutic strategies, herein discussed.

3.6.1 Aspirin and Other Non-Steroidal Anti-Inflammatory Drugs as Chemopreventive and Adjuvant Therapies in Colorectal Cancer

Non-steroidal anti-inflammatory drugs (NSAIDs) inhibiting COX protein activity, such as aspirin or other selective COX-2 inhibitors, are commonly used for the treatment of pain and inflammation. Strengthening the role of inflammation as a carcinogenic factor, clinical and experimental data have shown that these anti-inflammatory compounds have potent anti-neoplasic activity, in particular for CRC, protecting against tumor formation and progression (Murff et al. 2011; Burn et al. 2011; Steinbach et al. 2000; Rothwell et al. 2012a).

COX-2 mediates the biosynthesis of prostaglandin E2 (PGE2), a pro-inflammatory molecule that promotes proliferation and angiogenesis, inhibits apoptosis, enhances invasion and modulates immunosuppression (Kraus and Arber 2009; Grosch et al. 2006). COX-2 expression is highly induced under inflammatory conditions and its levels are frequently up-regulated in transformed epithelial and host stroma cells of cancer patients (Asting et al. 2011; Eberhart et al. 1994; Soumaoro et al. 2004). Elevated levels of COX-2 were found in about 50 % of colorectal adenomas and in about 85 % of CRCs (Eberhart et al. 1994; Gupta and Dubois 2001; Marnett and DuBois 2002; Wang and Dubois 2010), and it was associated with worse survival among CRC patients (Ogino et al. 2008). Clinical data have shown that the risk of adenoma formation is decreased in patients that take regularly NSAIDs (Murff et al. 2011). These observations are also valid for patients with hereditary forms of CRC such as FAP and HNPCC (Burn et al. 2011; Steinbach et al. 2000), which have an increased predisposition for neoplasic transformation. In addition, the use of aspirin after CRC diagnosis was reported to be associated with an improvement of CRC survival among individuals with COX-2–positive tumors but not COX-2–negative tumors (Chan et al. 2009). It was also shown that NSAIDs improve disease-free and overall survival of patients receiving chemotherapy (Trifan et al. 2002; Yao et al. 2005; Lin et al. 2005). In addition, high levels of COX-2 expression have been detected in resection specimens after radiotherapy and it has been associated with resistance to radiotherapy and poor prognosis, suggesting that COX-2 inhibition might improve the anti-tumor effect of radiation therapy (de Heer et al. 2007; Bouzourene et al. 2008; Min et al. 2008). Several clinical trials are undergoing in order to better determine the advantages of using NSAIDs as adjuvant therapy (http://clinicaltrials.gov).

The anti-tumor activity of COX inhibitors was recently linked to the capacity of these compounds to inhibit NFκB and JAK3/STAT3 signaling activation and consequently to down-regulate the expression of pro-inflammatory cytokines to a level that inhibits inflammation and carcinogenesis (Vaish and Sanyal 2011; Maihofner et al. 2003). In addition, the NSAID Sulindac was shown to inhibit proliferation and to induce apoptosis of CRC cells by downregulating the WNT signaling pathway through inhibition of β-catenin expression, nuclear translocation and subsequent activation of its downstream targets (Boon et al. 2004; Koornstra et al. 2005; Gardner et al. 2004). The capacity of Sulindac to modulate the WNT pathway was also linked to inhibition of metastatic spread of CRC cell lines (Stein et al. 2011). This anti-metastatic effect is not exclusive of Sulindac, since other NSAIDs were previously shown to abrogate CRC invasion and metastases in in vitro and in vivo models (Yao et al. 2003, 2004, 2005), as well as in a recent study using patient information from randomized controlled trials (Rothwell et al. 2012b). Furthermore, NSAIDs also have the capacity to inhibit cancer progression by suppressing angiogenesis and to decrease vascular permeability by inhibiting both VEGF expression and function (Ruegg et al. 2003) and integrin αVβ3-mediated Rac activation signaling (Ruegg et al. 2003; Dormond et al. 2001). Additionally, Sulindac and Celecoxib were observed to inhibit angiogenesis by interfering with PI3K/PTEN/AKT signaling, the canonical WNT/β-catenin signaling, regulation of MMPs activation, and by inhibition of inflammatory response via suppressing nitric oxide production (Vaish and Sanyal 2012).

Regardless of all the beneficial anti-cancer effects of NSAIDs, prolonged use of these anti-inflammatory compounds has been associated with an increased risk of breast and hematological cancers, cardiac disease, gastrointestinal bleeding, and kidney failure (Vinogradova et al. 2011). Regular use of NSAIDs might be recommended for individuals with a family history of CRC or of other conditions associated with CRC development such as obesity, sedentary lifestyle, red meat consumption, cigarette smokers and type 2 diabetes patients; however cautions should be taken when recommending it to those who are at average risk due to abovementioned side effects (Fuchs 2011).

Interestingly, a lot of controversy exists about the beneficial use of NSAIDs as chemopreventive or adjuvant therapy for patients with CAC. COX-2 expression was increased in IBD patients, as well as in inflamed tissues of IL-10-knockout and AOM/DSS mice (Greten et al. 2004; Wang and Dubois 2010; Ishikawa and Herschman 2010; Shattuck-Brandt et al. 2000; Singer et al. 1998). In some experimental models of colitis, the use of NSAIDs was associated with an exacerbation of the inflammatory reaction, however (Greten et al. 2004; Morteau et al. 2000). In addition, despite a significant elevation of COX-2 expression in AOM/DSS-induced colon tumors of wild-type mice, similar tumors developed in AOM/DSS-treated Cox-2- and Cox-1-knockout mice (Ishikawa and Herschman 2010). On the other hand, others have shown that NSAIDs inhibited both dysplasia and cancer in DSS-treated mice or rats, supporting a chemopreventive activity of NSAIDs against colitis-associated tumorigenesis (Inoue et al. 2008; Mukawa et al. 2008; Takeda et al. 2004). Further studies are needed in order to understand the role of COX-2 in CAC, and the beneficial effect of NSAIDs in the treatment of these patients.

3.6.2 Immunotherapy

Currently, novel strategies to actively modulate the immune system in order to prime cells or to boost the activity of tumor-associated immune cells have been proposed to improve the outcome of disease in cancer patients (Nespoli et al. 2012). In patients with CRC, preoperative subcutaneous injection of IL-2 was effective in counteracting post-operative immunosuppression related to surgical stress by affecting both phenotype and function of resident dendritic cells (DC) and T-cells, skewing local immunity towards a more immunogenic one (Nespoli et al. 2012). In addition, systemic delivery of chitosan (CS)-tripolyphosphate (TPP)/IL-12 nanoparticles was shown to be an efficient therapeutic strategy against CRC liver metastasis achieving a significant reduction of the number and volume of CRC liver metastasis foci. Mechanistically, CS-TPP/IL-12 nanoparticles were shown to recruit and induce infiltration of NK and T cells, which were most likely the effector cells that mediated tumor metastasis inhibition during CS-TPP/IL-12 immunotherapy (Xu et al. 2012).

In addition, the molecular identification of TAA has created new possibilities for antigen-specific immunotherapy for patients with advanced cancer, including CRC

(Okuno et al. 2012; Palucka and Banchereau 2012). Recombinant vaccines containing the carcinoembryonic antigen (CEA) gene and DCs loaded with CEA peptide were administered to patients with CEA-elevated CRC but the clinical responses to this therapeutic strategy were limited, even when administered in combination with chemotherapeutic agents (Okuno et al. 2012; Liu et al. 2004; Weihrauch et al. 2005).

Other, more specific, approaches are currently being tested in clinical trials. This includes a personalized protocol that consists in determining the repertoires of peptide-specific cytotoxic T lymphocytes (CTL) precursors in each patient, followed by administration of multiple peptides with higher CTL precursor frequency (Itoh and Yamada 2006). Several clinical trials using this approach were conducted, and the results show that personalized peptide vaccinations achieved prompt and strong activation of CTL with definite clinical benefits for certain cancer patients (lung and gynecologic cancer, and malignant glioma). This methodology has been recently tested in CRC patients and the preliminary results of such clinical trials were very promising regarding specific T cell responses to peptides and survival benefits, especially when in combination with chemotherapy (Okuno et al. 2012; Takahashi et al. 2012). Another strategy envisages the use of factors that diminish the recruitment and differentiation of certain tumor-related immune cells. This is the case of the macrophage colony-stimulating factor 1 (CSF-1), the cytokine produced by breast cancer cells, stimulating monocyte to macrophage differentiation, and described to delay breast cancer invasion and metastasis in the Polyomavirus-middle T (PyMT) transgenic mouse model of mammary cancer (Lin et al. 2001; Wyckoff et al. 2004). Several inhibitors have been developed to target CSF-1 and currently phase 1 clinical trials are recruiting patients with advanced metastatic cancers, including CRC (Baay et al. 2011).

3.7 Concluding Remarks

The evidence accumulated until now pinpoint inflammation as a potent intestinal tumor inductor, and therapeutic strategies aiming at inhibiting inflammation or potentiate the immunologic reaction against tumor cells represent a promising, but still underdeveloped, strategy to prevent CRC progression. Further studies are therefore required in order to better elucidate the intricate relationship between CRC and the distinct immune cell populations, and the molecular mechanisms by which they may modulate each other responses. These studies will represent an important step towards the designing of more effective therapeutic approaches by combining tailored therapies targeting important CRC-related pathways with immunotherapies promoting an anti-tumorigenic microenvironment. Since cancer is a product of both genetic instability and the selective pressure exerted by the micro-environment, attacking cancer in these two fronts will possibly be a more efficient way to improve patient survival.

References

Alkim C, Sakiz D, Alkim H, Livaoglu A, Kendir T, Demirsoy H, Erdem L, Akbayir N, Sokmen M (2012) Thrombospondin-1 and VEGF in inflammatory bowel disease. Libyan J Med 7

Andres PG, Beck PL, Mizoguchi E, Mizoguchi A, Bhan AK, Dawson T, Kuziel WA, Maeda N, MacDermott RP, Podolsky DK et al (2000) Mice with a selective deletion of the CC chemokine receptors 5 or 2 are protected from dextran sodium sulfate-mediated colitis: lack of CC chemokine receptor 5 expression results in a NK1.1+ lymphocyte-associated Th2-type immune response in the intestine. J Immunol 164(12):6303–6312

Askling J, Dickman PW, Karlen P, Brostrom O, Lapidus A, Lofberg R, Ekbom A (2001) Family history as a risk factor for colorectal cancer in inflammatory bowel disease. Gastroenterology 120(6):1356–1362

Asting AG, Caren H, Andersson M, Lonnroth C, Lagerstedt K, Lundholm K (2011) COX-2 gene expression in colon cancer tissue related to regulating factors and promoter methylation status. BMC Cancer 11:238

Atreya I, Neurath MF (2008) Immune cells in colorectal cancer: prognostic relevance and therapeutic strategies. Expert Rev Anticancer Ther 8(4):561–572

Atreya R, Mudter J, Finotto S, Mullberg J, Jostock T, Wirtz S, Schutz M, Bartsch B, Holtmann M, Becker C et al (2000) Blockade of interleukin 6 trans signaling suppresses T-cell resistance against apoptosis in chronic intestinal inflammation: evidence in crohn disease and experimental colitis in vivo. Nat Med 6(5):583–588

Aust DE, Haase M, Dobryden L, Markwarth A, Lohrs U, Wittekind C, Baretton GB, Tannapfel A (2005) Mutations of the BRAF gene in ulcerative colitis-related colorectal carcinoma. Int J Cancer 115(5):673–677

Baay M, Brouwer A, Pauwels P, Peeters M, Lardon F (2011) Tumor cells and tumor-associated macrophages: secreted proteins as potential targets for therapy. Clin Dev Immunol 2011:565187

Bailey C, Negus R, Morris A, Ziprin P, Goldin R, Allavena P, Peck D, Darzi A (2007) Chemokine expression is associated with the accumulation of tumour associated macrophages (TAMs) and progression in human colorectal cancer. Clin Exp Metastasis 24(2):121–130

Balkwill F, Mantovani A (2001) Inflammation and cancer: back to Virchow? Lancet 357(9255):539–545

Barbera-Guillem E, Nyhus JK, Wolford CC, Friece CR, Sampsel JW (2002) Vascular endothelial growth factor secretion by tumor-infiltrating macrophages essentially supports tumor angiogenesis, and IgG immune complexes potentiate the process. Cancer Res 62(23):7042–7049

Bauer K, Michel S, Reuschenbach M, Nelius N, von Knebel DM, Kloor M (2011) Dendritic cell and macrophage infiltration in microsatellite-unstable and microsatellite-stable colorectal cancer. Fam Cancer 10(3):557–565

Becker C, Fantini MC, Schramm C, Lehr HA, Wirtz S, Nikolaev A, Burg J, Strand S, Kiesslich R, Huber S et al (2004) TGF-beta suppresses tumor progression in colon cancer by inhibition of IL-6 trans-signaling. Immunity 21(4):491–501

Becker C, Fantini MC, Wirtz S, Nikolaev A, Lehr HA, Galle PR, Rose-John S, Neurath MF (2005) IL-6 signaling promotes tumor growth in colorectal cancer. Cell Cycle 4(2):217–220

Berg DJ, Davidson N, Kuhn R, Muller W, Menon S, Holland G, Thompson-Snipes L, Leach MW, Rennick D (1996) Enterocolitis and colon cancer in interleukin-10-deficient mice are associated with aberrant cytokine production and CD4(+) TH1-like responses. J Clin Invest 98(4):1010–1020

Bernstein CN, Blanchard JF, Kliewer E, Wajda A (2001) Cancer risk in patients with inflammatory bowel disease: a population-based study. Cancer 91(4):854–862

Bollrath J, Greten FR (2009) IKK/NF-kappaB and STAT3 pathways: central signalling hubs in inflammation-mediated tumour promotion and metastasis. EMBO Rep 10(12):1314–1319

Bollrath J, Phesse TJ, von Burstin VA, Putoczki T, Bennecke M, Bateman T, Nebelsiek T, Lundgren-May T, Canli O, Schwitalla S et al (2009) gp130-mediated Stat3 activation in enterocytes regulates cell survival and cell-cycle progression during colitis-associated tumorigenesis. Cancer Cell 15(2):91–102

Bonecchi R, Locati M, Mantovani A (2011) Chemokines and cancer: a fatal attraction. Cancer Cell 19(4):434–435

Boon EM, Keller JJ, Wormhoudt TA, Giardiello FM, Offerhaus GJ, van der Neut R, Pals ST (2004) Sulindac targets nuclear beta-catenin accumulation and Wnt signalling in adenomas of patients with familial adenomatous polyposis and in human colorectal cancer cell lines. Br J Cancer 90(1):224–229

Borrello MG, Degl'Innocenti D, Pierotti MA (2008) Inflammation and cancer: the oncogene-driven connection. Cancer Lett 267(2):262–270

Botelho MC, Machado JC, Brindley PJ, Correia da Costa JM (2011) Targeting molecular signaling pathways of Schistosoma haemotobium infection in bladder cancer. Virulence 2(4):267–279

Bouzourene H, Yan P, Sandmeier D, Zouhair A, Matter M, Vuilleumier H, Coucke P (2008) The role of COX-2 in rectal cancer treated with preoperative radiotherapy. Virchows Arch 452(5):499–505

Brackmann S, Andersen SN, Aamodt G, Roald B, Langmark F, Clausen OP, Aadland E, Fausa O, Rydning A, Vatn MH (2009) Two distinct groups of colorectal cancer in inflammatory bowel disease. Inflamm Bowel Dis 15(1):9–16

Burn J, Gerdes AM, Macrae F, Mecklin JP, Moeslein G, Olschwang S, Eccles D, Evans DG, Maher ER, Bertario L et al (2011) Long-term effect of aspirin on cancer risk in carriers of hereditary colorectal cancer: an analysis from the CAPP2 randomised controlled trial. Lancet 378(9809):2081–2087

Cameron DJ, Churchill WH (1980) Cytotoxicity of human macrophages for tumor cells: enhancement by bacterial lipopolysaccharides (LPS). J Immunol 124(2):708–712

Canavan C, Abrams KR, Mayberry J (2006) Meta-analysis: colorectal and small bowel cancer risk in patients with Crohn's disease. Aliment Pharmacol Ther 23(8):1097–1104

Capel E, Flejou JF, Hamelin R (2007) Assessment of MLH1 promoter methylation in relation to gene expression requires specific analysis. Oncogene 26(54):7596–7600

Caras I, Tucureanu C, Lerescu L, Pitica R, Melinceanu L, Neagu S, Salageanu A (2011) Influence of tumor cell culture supernatants on macrophage functional polarization: in vitro models of macrophage-tumor environment interaction. Tumori 97(5):647–654

Cavallo F, De Giovanni C, Nanni P, Forni G, Lollini PL (2011) 2011: the immune hallmarks of cancer. Cancer Immunol Immunother 60(3):319–326

Chan AT, Ogino S, Fuchs CS (2009) Aspirin use and survival after diagnosis of colorectal cancer. JAMA 302(6):649–658

Choi PM, Zelig MP (1994) Similarity of colorectal cancer in Crohn's disease and ulcerative colitis: implications for carcinogenesis and prevention. Gut 35(7):950–954

Condeelis J, Pollard JW (2006) Macrophages: obligate partners for tumor cell migration, invasion, and metastasis. Cell 124(2):263–266

Coussens LM, Werb Z (2002) Inflammation and cancer. Nature 420(6917):860–867

Crittenden M, Gough M, Harrington K, Olivier K, Thompson J, Vile RG (2003) Expression of inflammatory chemokines combined with local tumor destruction enhances tumor regression and long-term immunity. Cancer Res 63(17):5505–5512

Crohn BB, Rosenberg H (1925) The sigmoidoscopy picture of chronic ulcerative colitis (non-specific). Am J Med Sci 170:220–228

Cummings JR, Ahmad T, Geremia A, Beckly J, Cooney R, Hancock L, Pathan S, Guo C, Cardon LR, Jewell DP (2007a) Contribution of the novel inflammatory bowel disease gene IL23R to disease susceptibility and phenotype. Inflamm Bowel Dis 13(9):1063–1068

Cummings JR, Cooney R, Pathan S, Anderson CA, Barrett JC, Beckly J, Geremia A, Hancock L, Guo C, Ahmad T et al (2007b) Confirmation of the role of ATG16L1 as a Crohn's disease susceptibility gene. Inflamm Bowel Dis 13(8):941–946

Danese S, Sans M, de la Motte C, Graziani C, West G, Phillips MH, Pola R, Rutella S, Willis J, Gasbarrini A et al (2006) Angiogenesis as a novel component of inflammatory bowel disease pathogenesis. Gastroenterology 130(7):2060–2073

Danese S, Malesci A, Vetrano S (2011) Colitis-associated cancer: the dark side of inflammatory bowel disease. Gut 60(12):1609–1610

de Heer P, Gosens MJ, de Bruin EC, Dekker-Ensink NG, Putter H, Marijnen CA, van den Brule AJ, van Krieken JH, Rutten HJ, Kuppen PJ et al (2007) Cyclooxygenase 2 expression in rectal cancer is of prognostic significance in patients receiving preoperative radiotherapy. Clin Cancer Res 13(10):2955–2960

Delaunoit T, Limburg PJ, Goldberg RM, Lymp JF, Loftus EV Jr (2006) Colorectal cancer prognosis among patients with inflammatory bowel disease. Clin Gastroenterol Hepatol 4(3): 335–342

Deng G, Chen A, Hong J, Chae HS, Kim YS (1999) Methylation of CpG in a small region of the hMLH1 promoter invariably correlates with the absence of gene expression. Cancer Res 59(9):2029–2033

Deschoolmeester V, Baay M, Lardon F, Pauwels P, Peeters M (2011) Immune cells in colorectal cancer: prognostic relevance and role of MSI. Cancer Microenviron 4(3):377–392

Dieu-Nosjean MC, Antoine M, Danel C, Heudes D, Wislez M, Poulot V, Rabbe N, Laurans L, Tartour E, de Chaisemartin L et al (2008) Long-term survival for patients with non-small-cell lung cancer with intratumoral lymphoid structures. J Clin Oncol 26(27):4410–4417

Dominitzki S, Fantini MC, Neufert C, Nikolaev A, Galle PR, Scheller J, Monteleone G, Rose-John S, Neurath MF, Becker C (2007) Cutting edge: trans-signaling via the soluble IL-6R abrogates the induction of FoxP3 in naive CD4+CD25 T cells. J Immunol 179(4):2041–2045

Dormond O, Foletti A, Paroz C, Ruegg C (2001) NSAIDs inhibit alpha V beta 3 integrin-mediated and Cdc42/Rac-dependent endothelial-cell spreading, migration and angiogenesis. Nat Med 7(9):1041–1047

Eaden J (2004) Review article: colorectal carcinoma and inflammatory bowel disease. Aliment Pharmacol Ther 20(Suppl 4):24–30

Eberhart CE, Coffey RJ, Radhika A, Giardiello FM, Ferrenbach S, DuBois RN (1994) Up-regulation of cyclooxygenase 2 gene expression in human colorectal adenomas and adenocarcinomas. Gastroenterology 107(4):1183–1188

Eckmann L, Nebelsiek T, Fingerle AA, Dann SM, Mages J, Lang R, Robine S, Kagnoff MF, Schmid RM, Karin M et al (2008) Opposing functions of IKKbeta during acute and chronic intestinal inflammation. Proc Natl Acad Sci U S A 105(39):15058–15063

Edelblum KL, Goettel JA, Koyama T, McElroy SJ, Yan F, Polk DB (2008) TNFR1 promotes tumor necrosis factor-mediated mouse colon epithelial cell survival through RAF activation of NF-kappaB. J Biol Chem 283(43):29485–29494

Ekbom A, Helmick C, Zack M, Adami HO (1990) Ulcerative colitis and colorectal cancer. A population-based study. N Engl J Med 323(18):1228–1233

Erreni M, Mantovani A, Allavena P (2011) Tumor-associated macrophages (TAM) and inflammation in colorectal cancer. Cancer Microenviron 4(2):141–154

Favre-Felix N, Fromentin A, Hammann A, Solary E, Martin F, Bonnotte B (2000) Cutting edge: the tumor counterattack hypothesis revisited—colon cancer cells do not induce T cell apoptosis via the Fas (CD95, APO-1) pathway. J Immunol 164(10):5023–5027

Fearon ER, Vogelstein B (1990) A genetic model for colorectal tumorigenesis. Cell 61(5): 759–767

Fina D, Sarra M, Fantini MC, Rizzo A, Caruso R, Caprioli F, Stolfi C, Cardolini I, Dottori M, Boirivant M et al (2008) Regulation of gut inflammation and th17 cell response by interleukin-21. Gastroenterology 134(4):1038–1048

Fingerle-Rowson G, Petrenko O, Metz CN, Forsthuber TG, Mitchell R, Huss R, Moll U, Muller W, Bucala R (2003) The p53-dependent effects of macrophage migration inhibitory factor revealed by gene targeting. Proc Natl Acad Sci U S A 100(16):9354–9359

Fleisher AS, Esteller M, Harpaz N, Leytin A, Rashid A, Xu Y, Liang J, Stine OC, Yin J, Zou TT et al (2000) Microsatellite instability in inflammatory bowel disease-associated neoplastic lesions is associated with hypermethylation and diminished expression of the DNA mismatch repair gene, hMLH1. Cancer Res 60(17):4864–4868

Foran E, Garrity-Park MM, Mureau C, Newell J, Smyrk TC, Limburg PJ, Egan LJ (2010) Upregulation of DNA methyltransferase-mediated gene silencing, anchorage-independent growth, and migration of colon cancer cells by interleukin-6. Mol Cancer Res 8(4):471–481

Forssell J, Oberg A, Henriksson ML, Stenling R, Jung A, Palmqvist R (2007) High macrophage infiltration along the tumor front correlates with improved survival in colon cancer. Clin Cancer Res 13(5):1472–1479
Frysz-Naglak D, Fryc B, Klimacka-Nawrot E, Mazurek U, Suchecka W, Kajor M, Kurek J, Stadnicki A (2011) Expression, localization and systemic concentration of vascular endothelial growth factor (VEGF) and its receptors in patients with ulcerative colitis. Int Immunopharmacol 11(2):220–225
Fuchs CS (2011) The role of aspirin in prevention and management of colorectal cancer. Clin Adv Hematol Oncol 9(4):330–332
Galon J, Costes A, Sanchez-Cabo F, Kirilovsky A, Mlecnik B, Lagorce-Pages C, Tosolini M, Camus M, Berger A, Wind P et al (2006) Type, density, and location of immune cells within human colorectal tumors predict clinical outcome. Science 313(5795):1960–1964
Gardner SH, Hawcroft G, Hull MA (2004) Effect of nonsteroidal anti-inflammatory drugs on beta-catenin protein levels and catenin-related transcription in human colorectal cancer cells. Br J Cancer 91(1):153–163
Gelfand M, Weinberg RW, Castle WM (1967) Relation between carcinoma of the bladder and infestation with Schistosoma haematobium. Lancet 1(7502):1249–1251
Goel GA, Kandiel A, Achkar JP, Lashner B (2011) Molecular pathways underlying IBD-associated colorectal neoplasia: therapeutic implications. Am J Gastroenterol 106(4):719–730
Grady WM, Carethers JM (2008) Genomic and epigenetic instability in colorectal cancer pathogenesis. Gastroenterology 135(4):1079–1099
Greten FR, Eckmann L, Greten TF, Park JM, Li ZW, Egan LJ, Kagnoff MF, Karin M (2004) IKKbeta links inflammation and tumorigenesis in a mouse model of colitis-associated cancer. Cell 118(3):285–296
Grivennikov SI, Karin M (2010) Inflammation and oncogenesis: a vicious connection. Curr Opin Genet Dev 20(1):65–71
Grivennikov SI, Karin M (2011) Inflammatory cytokines in cancer: tumour necrosis factor and interleukin 6 take the stage. Ann Rheum Dis 70(Suppl 1):i104–i108
Grivennikov S, Karin E, Terzic J, Mucida D, Yu GY, Vallabhapurapu S, Scheller J, Rose-John S, Cheroutre H, Eckmann L et al (2009) IL-6 and Stat3 are required for survival of intestinal epithelial cells and development of colitis-associated cancer. Cancer Cell 15(2):103–113
Grivennikov SI, Greten FR, Karin M (2010) Immunity, inflammation, and cancer. Cell 140(6):883–899
Grosch S, Maier TJ, Schiffmann S, Geisslinger G (2006) Cyclooxygenase-2 (COX-2)-independent anticarcinogenic effects of selective COX-2 inhibitors. J Natl Cancer Inst 98(11):736–747
Guerra C, Schuhmacher AJ, Canamero M, Grippo PJ, Verdaguer L, Perez-Gallego L, Dubus P, Sandgren EP, Barbacid M (2007) Chronic pancreatitis is essential for induction of pancreatic ductal adenocarcinoma by K-Ras oncogenes in adult mice. Cancer Cell 11(3):291–302
Gulubova MV, Ananiev JR, Vlaykova TI, Yovchev Y, Tsoneva V, Manolova IM (2012) Role of dendritic cells in progression and clinical outcome of colon cancer. Int J Colorectal Dis 27(2):159–169
Gupta RA, Dubois RN (2001) Colorectal cancer prevention and treatment by inhibition of cyclooxygenase-2. Nat Rev Cancer 1(1):11–21
Gupta RB, Harpaz N, Itzkowitz S, Hossain S, Matula S, Kornbluth A, Bodian C, Ullman T (2007) Histologic inflammation is a risk factor for progression to colorectal neoplasia in ulcerative colitis: a cohort study. Gastroenterology 133(4):1099–1105; quiz 1340–1091
Hahn MA, Hahn T, Lee DH, Esworthy RS, Kim BW, Riggs AD, Chu FF, Pfeifer GP (2008) Methylation of polycomb target genes in intestinal cancer is mediated by inflammation. Cancer Res 68(24):10280–10289
Halama N, Michel S, Kloor M, Zoernig I, Benner A, Spille A, Pommerencke T, von Knebel DM, Folprecht G, Luber B et al (2011) Localization and density of immune cells in the invasive margin of human colorectal cancer liver metastases are prognostic for response to chemotherapy. Cancer Res 71(17):5670–5677

Hamilton KE, Simmons JG, Ding S, Van Landeghem L, Lund PK (2011) Cytokine induction of tumor necrosis factor receptor 2 is mediated by STAT3 in colon cancer cells. Mol Cancer Res 9(12):1718–1731
Hanahan D, Coussens LM (2012) Accessories to the crime: functions of cells recruited to the tumor microenvironment. Cancer Cell 21(3):309–322
Harpaz N, Polydorides AD (2010) Colorectal dysplasia in chronic inflammatory bowel disease: pathology, clinical implications, and pathogenesis. Arch Pathol Lab Med 134(6):876–895
Hartnett L, Egan LJ (2012) Inflammation, DNA methylation and colitis-associated cancer. Carcinogenesis 33(4):723–731
Henle W, Henle G (1973) Evidence for an oncogenic potential of the Epstein-Barr virus. Cancer Res 33(6):1419–1423
Herbeuval JP, Lelievre E, Lambert C, Dy M, Genin C (2004) Recruitment of STAT3 for production of IL-10 by colon carcinoma cells induced by macrophage-derived IL-6. J Immunol 172(7):4630–4636
Higashi N, Ishii H, Fujiwara T, Morimoto-Tomita M, Irimura T (2002) Redistribution of fibroblasts and macrophages as micrometastases develop into established liver metastases. Clin Exp Metastasis 19(7):631–638
Holzmann K, Klump B, Borchard F, Hsieh CJ, Kuhn A, Gaco V, Gregor M, Porschen R (1998) Comparative analysis of histology, DNA content, p53 and Ki-ras mutations in colectomy specimens with long-standing ulcerative colitis. Int J Cancer 76(1):1–6
Hosking R (2012) Cancer and the immune system. Cell 149(1):5–7
Hussain SP, Hofseth LJ, Harris CC (2003) Radical causes of cancer. Nat Rev Cancer 3(4):276–285
Imano M, Okuno K, Itoh T, Satou T, Ishimaru E, Yasuda T, Hida J, Imamoto H, Takeyama Y, Shiozaki H (2011) Osteopontin induced by macrophages contribute to metachronous liver metastases in colorectal cancer. Am Surg 77(11):1515–1520
Inoue T, Hirata I, Murano M (2008) Effects of nimesulide, a cyclooxygenase-2 selective inhibitor, on colitis induced tumors. Inflammopharmacology 16(1):36–39
Ionov Y, Peinado MA, Malkhosyan S, Shibata D, Perucho M (1993) Ubiquitous somatic mutations in simple repeated sequences reveal a new mechanism for colonic carcinogenesis. Nature 363(6429):558–561
Ishigami S, Natsugoe S, Tokuda K, Nakajo A, Okumura H, Matsumoto M, Miyazono F, Hokita S, Aikou T (2003) Tumor-associated macrophage (TAM) infiltration in gastric cancer. Anticancer Res 23(5A):4079–4083
Ishikawa TO, Herschman HR (2010) Tumor formation in a mouse model of colitis-associated colon cancer does not require COX-1 or COX-2 expression. Carcinogenesis 31(4):729–736
Ishitsuka T, Kashiwagi H, Konishi F (2001) Microsatellite instability in inflamed and neoplastic epithelium in ulcerative colitis. J Clin Pathol 54(7):526–532
Issa JP, Ahuja N, Toyota M, Bronner MP, Brentnall TA (2001) Accelerated age-related CpG island methylation in ulcerative colitis. Cancer Res 61(9):3573–3577
Itoh K, Yamada A (2006) Personalized peptide vaccines: a new therapeutic modality for cancer. Cancer Sci 97(10):970–976
Itzkowitz SH, Yio X (2004) Inflammation and cancer IV. Colorectal cancer in inflammatory bowel disease: the role of inflammation. Am J Physiol Gastrointest Liver Physiol 287(1):G7–G17
Izcue A, Coombes JL, Powrie F (2009) Regulatory lymphocytes and intestinal inflammation. Annu Rev Immunol 27:313–338
Jass JR (2005) Serrated adenoma of the colorectum and the DNA-methylator phenotype. Nat Clin Pract Oncol 2(8):398–405
Jauch D, Martin M, Schiechl G, Kesselring R, Schlitt HJ, Geissler EK, Fichtner-Feigl S (2011) Interleukin 21 controls tumour growth and tumour immunosurveillance in colitis-associated tumorigenesis in mice. Gut 60(12):1678–1686
Karin M, Greten FR (2005) NF-kappaB: linking inflammation and immunity to cancer development and progression. Nat Rev Immunol 5(10):749–759

Katsurano M, Niwa T, Yasui Y, Shigematsu Y, Yamashita S, Takeshima H, Lee MS, Kim YJ, Tanaka T, Ushijima T (2012) Early-stage formation of an epigenetic field defect in a mouse colitis model, and non-essential roles of T- and B-cells in DNA methylation induction. Oncogene 31(3):342–351
Keith B, Simon MC (2008) Tumor angiogenesis. In: Mendelsohn J, Howley P, Israel MA, Gray JW, Thompson CB (eds) The molecular basis of cancer, 3rd edn. Elsevier, Philadelphia
Kinouchi M, Miura K, Mizoi T, Ishida K, Fujibuchi W, Sasaki H, Ohnuma S, Saito K, Katayose Y, Naitoh T et al (2012) Infiltration of CD40-positive tumor-associated macrophages indicates a favorable prognosis in colorectal cancer patients. Hepatogastroenterology 60(121):83–88
Kiran RP, Khoury W, Church JM, Lavery IC, Fazio VW, Remzi FH (2010) Colorectal cancer complicating inflammatory bowel disease: similarities and differences between Crohn's and ulcerative colitis based on three decades of experience. Ann Surg 252(2):330–335
Kollias G (2004) Modeling the function of tumor necrosis factor in immune pathophysiology. Autoimmun Rev 3(Suppl 1):S24–S25
Kominsky DJ, Keely S, MacManus CF, Glover LE, Scully M, Collins CB, Bowers BE, Campbell EL, Colgan SP (2011) An endogenously anti-inflammatory role for methylation in mucosal inflammation identified through metabolite profiling. J Immunol 186(11):6505–6514
Konishi K, Shen L, Wang S, Meltzer SJ, Harpaz N, Issa JP (2007) Rare CpG island methylator phenotype in ulcerative colitis-associated neoplasias. Gastroenterology 132(4):1254–1260
Koornstra JJ, Rijcken FE, Oldenhuis CN, Zwart N, van der Sluis T, Hollema H, deVries EG, Keller JJ, Offerhaus JA, Giardiello FM et al (2005) Sulindac inhibits beta-catenin expression in normal-appearing colon of hereditary nonpolyposis colorectal cancer and familial adenomatous polyposis patients. Cancer Epidemiol Biomarkers Prev 14(7):1608–1612
Kornfeld D, Ekbom A, Ihre T (1997) Is there an excess risk for colorectal cancer in patients with ulcerative colitis and concomitant primary sclerosing cholangitis? A population based study. Gut 41(4):522–525
Kraus S, Arber N (2009) Inflammation and colorectal cancer. Curr Opin Pharmacol 9(4):405–410
Kuhn R, Lohler J, Rennick D, Rajewsky K, Muller W (1993) Interleukin-10-deficient mice develop chronic enterocolitis. Cell 75(2):263–274
Kuppen PJ, Gorter A, Hagenaars M, Jonges LE, Giezeman-Smits KM, Nagelkerke JF, Fleuren G, van de Velde CJ (2001) Role of NK cells in adoptive immunotherapy of metastatic colorectal cancer in a syngeneic rat model. Immunol Rev 184:236–243
Ladoire S, Martin F, Ghiringhelli F (2011) Prognostic role of FOXP3+ regulatory T cells infiltrating human carcinomas: the paradox of colorectal cancer. Cancer Immunol Immunother 60(7):909–918
Lau KS, Haigis KM (2009) Non-redundancy within the RAS oncogene family: insights into mutational disparities in cancer. Mol Cells 28(4):315–320
Li Y, de Haar C, Chen M, Deuring J, Gerrits MM, Smits R, Xia B, Kuipers EJ, van der Woude CJ (2010) Disease-related expression of the IL6/STAT3/SOCS3 signalling pathway in ulcerative colitis and ulcerative colitis-related carcinogenesis. Gut 59(2):227–235
Lin EY, Nguyen AV, Russell RG, Pollard JW (2001) Colony-stimulating factor 1 promotes progression of mammary tumors to malignancy. J Exp Med 193(6):727–740
Lin J, Hsiao PW, Chiu TH, Chao JI (2005) Combination of cyclooxygenase-2 inhibitors and oxaliplatin increases the growth inhibition and death in human colon cancer cells. Biochem Pharmacol 70(5):658–667
Lissoni P, Brivio F, Ferrante R, Vigore L, Vaghi M, Fumagalli E, Bucovec R, Malugani F, Fumagalli L (2000) Circulating immature and mature dendritic cells in relation to lymphocyte subsets in patients with gastrointestinal tract cancer. Int J Biol Markers 15(1):22–25
Liu KJ, Wang CC, Chen LT, Cheng AL, Lin DT, Wu YC, Yu WL, Hung YM, Yang HY, Juang SH et al (2004) Generation of carcinoembryonic antigen (CEA)-specific T-cell responses in HLA-A*0201 and HLA-A*2402 late-stage colorectal cancer patients after vaccination with dendritic cells loaded with CEA peptides. Clin Cancer Res 10(8):2645–2651

Lovato P, Brender C, Agnholt J, Kelsen J, Kaltoft K, Svejgaard A, Eriksen KW, Woetmann A, Odum N (2003) Constitutive STAT3 activation in intestinal T cells from patients with Crohn's disease. J Biol Chem 278(19):16777–16781

Lyda MH, Noffsinger A, Belli J, Fenoglio-Preiser CM (2000) Microsatellite instability and K-ras mutations in patients with ulcerative colitis. Hum Pathol 31(6):665–671

Maihofner C, Charalambous MP, Bhambra U, Lightfoot T, Geisslinger G, Gooderham NJ (2003) Expression of cyclooxygenase-2 parallels expression of interleukin-1beta, interleukin-6 and NF-kappaB in human colorectal cancer. Carcinogenesis 24(4):665–671

Mantovani A, Sica A, Sozzani S, Allavena P, Vecchi A, Locati M (2004) The chemokine system in diverse forms of macrophage activation and polarization. Trends Immunol 25(12):677–686

Mantovani A, Schioppa T, Porta C, Allavena P, Sica A (2006) Role of tumor-associated macrophages in tumor progression and invasion. Cancer Metastasis Rev 25(3):315–322

Mantovani A, Allavena P, Sica A, Balkwill F (2008) Cancer-related inflammation. Nature 454(7203):436–444

Mareel M, Leroy A (2003) Clinical, cellular, and molecular aspects of cancer invasion. Physiol Rev 83(2):337–376

Mareel M, Oliveira MJ, Madani I (2009) Cancer invasion and metastasis: interacting ecosystems. Virchows Arch 454(6):599–622

Marnett LJ, DuBois RN (2002) COX-2: a target for colon cancer prevention. Annu Rev Pharmacol Toxicol 42:55–80

Matsumoto S, Hara T, Mitsuyama K, Yamamoto M, Tsuruta O, Sata M, Scheller J, Rose-John S, Kado S, Takada T (2010) Essential roles of IL-6 trans-signaling in colonic epithelial cells, induced by the IL-6/soluble-IL-6 receptor derived from lamina propria macrophages, on the development of colitis-associated premalignant cancer in a murine model. J Immunol 184(3):1543–1551

McCabe MT, Brandes JC, Vertino PM (2009) Cancer DNA methylation: molecular mechanisms and clinical implications. Clin Cancer Res 15(12):3927–3937

Meira LB, Bugni JM, Green SL, Lee CW, Pang B, Borenshtein D, Rickman BH, Rogers AB, Moroski-Erkul CA, McFaline JL et al (2008) DNA damage induced by chronic inflammation contributes to colon carcinogenesis in mice. J Clin Invest 118(7):2516–2525

Michielsen AJ, Hogan AE, Marry J, Tosetto M, Cox F, Hyland JM, Sheahan KD, O'Donoghue DP, Mulcahy HE, Ryan EJ et al (2011) Tumour tissue microenvironment can inhibit dendritic cell maturation in colorectal cancer. PLoS One 6(11):e27944

Min BS, Choi YJ, Pyo HR, Kim H, Seong J, Chung HC, Rha SY, Kim NK (2008) Cyclooxygenase-2 expression in pretreatment biopsy as a predictor of tumor responses after preoperative chemoradiation in rectal cancer. Arch Surg 143(11):1091–1097; discussion 1097

Mizoguchi E, Mizoguchi A, Takedatsu H, Cario E, de Jong YP, Ooi CJ, Xavier RJ, Terhorst C, Podolsky DK, Bhan AK (2002) Role of tumor necrosis factor receptor 2 (TNFR2) in colonic epithelial hyperplasia and chronic intestinal inflammation in mice. Gastroenterology 122(1):134–144

Moody CA, Laimins LA (2010) Human papillomavirus oncoproteins: pathways to transformation. Nat Rev Cancer 10(8):550–560

Morteau O, Morham SG, Sellon R, Dieleman LA, Langenbach R, Smithies O, Sartor RB (2000) Impaired mucosal defense to acute colonic injury in mice lacking cyclooxygenase-1 or cyclooxygenase-2. J Clin Invest 105(4):469–478

Mukawa K, Fujii S, Tominaga K, Yoshitake N, Abe A, Kono T, Sekikawa A, Fukui H, Ichikawa K, Tomita S et al (2008) Inhibitory effects of the cyclooxygenase-2 inhibitor, etodolac, on colitis-associated tumorigenesis in p53-deficient mice treated with dextran sulfate sodium. Oncol Rep 19(2):393–399

Mundy-Bosse BL, Lesinski GB, Jaime-Ramirez AC, Benninger K, Khan M, Kuppusamy P, Guenterberg K, Kondadasula SV, Chaudhury AR, La Perle KM et al (2011) Myeloid-derived suppressor cell inhibition of the IFN response in tumor-bearing mice. Cancer Res 71(15):5101–5110

Munkholm P (2003) Review article: the incidence and prevalence of colorectal cancer in inflammatory bowel disease. Aliment Pharmacol Ther 18(Suppl 2):1–5

Murff HJ, Shrubsole MJ, Chen Z, Smalley WE, Chen H, Shyr Y, Ness RM, Zheng W (2011) Nonsteroidal anti-inflammatory drug use and risk of adenomatous and hyperplastic polyps. Cancer Prev Res (Phila) 4(11):1799–1807

Musso A, Dentelli P, Carlino A, Chiusa L, Repici A, Sturm A, Fiocchi C, Rizzetto M, Pegoraro L, Sategna-Guidetti C et al (2005) Signal transducers and activators of transcription 3 signaling pathway: an essential mediator of inflammatory bowel disease and other forms of intestinal inflammation. Inflamm Bowel Dis 11(2):91–98

Nagorsen D, Scheibenbogen C, Marincola FM, Letsch A, Keilholz U (2003) Natural T cell immunity against cancer. Clin Cancer Res 9(12):4296–4303

Nespoli L, Uggeri F, Romano F, Nespoli A, Brivo F, Fumagalli L, Sargenti M, Gianotti L (2012) Modulation of systemic and intestinal immune response by interleukin-2 therapy in gastrointestinal surgical oncology. Personal experience in the context of current knowledge and future perspectives. Anticancer Res 32(3):989–996

Neufert C, Becker C, Neurath MF (2007) An inducible mouse model of colon carcinogenesis for the analysis of sporadic and inflammation-driven tumor progression. Nat Protoc 2(8): 1998–2004

Nosho K, Baba Y, Tanaka N, Shima K, Hayashi M, Meyerhardt JA, Giovannucci E, Dranoff G, Fuchs CS, Ogino S (2010) Tumour-infiltrating T-cell subsets, molecular changes in colorectal cancer, and prognosis: cohort study and literature review. J Pathol 222(4):350–366

Nuako KW, Ahlquist DA, Mahoney DW, Schaid DJ, Siems DM, Lindor NM (1998) Familial predisposition for colorectal cancer in chronic ulcerative colitis: a case–control study. Gastroenterology 115(5):1079–1083

Numata A, Minagawa T, Asano M, Nakane A, Katoh H, Tanabe T (1991) Functional evaluation of tumor-infiltrating mononuclear cells. Detection of endogenous interferon-gamma and tumor necrosis factor-alpha in human colorectal adenocarcinomas. Cancer 68(9):1937–1943

O'Brien MJ, Yang S, Mack C, Xu H, Huang CS, Mulcahy E, Amorosino M, Farraye FA (2006) Comparison of microsatellite instability, CpG island methylation phenotype, BRAF and KRAS status in serrated polyps and traditional adenomas indicates separate pathways to distinct colorectal carcinoma end points. Am J Surg Pathol 30(12):1491–1501

O'Hagan HM, Wang W, Sen S, Destefano Shields C, Lee SS, Zhang YW, Clements EG, Cai Y, Van Neste L, Easwaran H et al (2011) Oxidative damage targets complexes containing DNA methyltransferases, SIRT1, and polycomb members to promoter CpG Islands. Cancer Cell 20(5):606–619

Ogino S, Kirkner GJ, Nosho K, Irahara N, Kure S, Shima K, Hazra A, Chan AT, Dehari R, Giovannucci EL et al (2008) Cyclooxygenase-2 expression is an independent predictor of poor prognosis in colon cancer. Clin Cancer Res 14(24):8221–8227

Ohki S, Shibata M, Gonda K, Machida T, Shimura T, Nakamura I, Ohtake T, Koyama Y, Suzuki S, Ohto H et al (2012) Circulating myeloid-derived suppressor cells are increased and correlate to immune suppression, inflammation and hypoproteinemia in patients with cancer. Oncol Rep 28(2):453–458

Ohno S, Inagawa H, Dhar DK, Fujii T, Ueda S, Tachibana M, Suzuki N, Inoue M, Soma G, Nagasue N (2003) The degree of macrophage infiltration into the cancer cell nest is a significant predictor of survival in gastric cancer patients. Anticancer Res 23(6D):5015–5022

Okayasu I, Hatakeyama S, Yamada M, Ohkusa T, Inagaki Y, Nakaya R (1990) A novel method in the induction of reliable experimental acute and chronic ulcerative colitis in mice. Gastroenterology 98(3):694–702

Okuno K, Sugiura F, Itoh K, Yoshida K, Tsunoda T, Nakamura Y (2012) Recent advances in active specific cancer vaccine treatment for colorectal cancer. Curr Pharm Biotechnol 13(8): 1439–1445

Olaru AV, Cheng Y, Agarwal R, Yang J, David S, Abraham JM, Yu W, Kwon JH, Lazarev M, Brant SR et al (2012) Unique patterns of CpG island methylation in inflammatory bowel disease-associated colorectal cancers. Inflamm Bowel Dis 18(4):641–648

Oliveira C, Westra JL, Arango D, Ollikainen M, Domingo E, Ferreira A, Velho S, Niessen R, Lagerstedt K, Alhopuro P et al (2004) Distinct patterns of KRAS mutations in colorectal carcinomas according to germline mismatch repair defects and hMLH1 methylation status. Hum Mol Genet 13(19):2303–2311

Onizawa M, Nagaishi T, Kanai T, Nagano K, Oshima S, Nemoto Y, Yoshioka A, Totsuka T, Okamoto R, Nakamura T et al (2009) Signaling pathway via TNF-alpha/NF-kappaB in intestinal epithelial cells may be directly involved in colitis-associated carcinogenesis. Am J Physiol Gastrointest Liver Physiol 296(4):G850–G859

Oosterling SJ, van der Bij GJ, Meijer GA, Tuk CW, van Garderen E, van Rooijen N, Meijer S, van der Sijp JR, Beelen RH, van Egmond M (2005) Macrophages direct tumour histology and clinical outcome in a colon cancer model. J Pathol 207(2):147–155

Ouaissi M, Maggiori L, Alves A, Giger U, Sielezneff I, Valleur P, Sastre B, Panis Y (2011) Colorectal cancer complicating inflammatory bowel disease: a comparative study of Crohn's disease vs ulcerative colitis in 34 patients. Colorectal Dis 13(6):684–688

Pages F, Vives V, Sautes-Fridman C, Fossiez F, Berger A, Cugnenc PH, Tartour E, Fridman WH (1999) Control of tumor development by intratumoral cytokines. Immunol Lett 68(1): 135–139

Pages F, Kirilovsky A, Mlecnik B, Asslaber M, Tosolini M, Bindea G, Lagorce C, Wind P, Marliot F, Bruneval P et al (2009) In situ cytotoxic and memory T cells predict outcome in patients with early-stage colorectal cancer. J Clin Oncol 27(35):5944–5951

Palucka K, Banchereau J (2012) Cancer immunotherapy via dendritic cells. Nat Rev Cancer 12(4):265–277

Peek RM Jr, Blaser MJ (2002) Helicobacter pylori and gastrointestinal tract adenocarcinomas. Nat Rev Cancer 2(1):28–37

Pickert G, Neufert C, Leppkes M, Zheng Y, Wittkopf N, Warntjen M, Lehr HA, Hirth S, Weigmann B, Wirtz S et al (2009) STAT3 links IL-22 signaling in intestinal epithelial cells to mucosal wound healing. J Exp Med 206(7):1465–1472

Pollard JW (2004) Tumour-educated macrophages promote tumour progression and metastasis. Nat Rev Cancer 4(1):71–78

Popivanova BK, Kitamura K, Wu Y, Kondo T, Kagaya T, Kaneko S, Oshima M, Fujii C, Mukaida N (2008) Blocking TNF-alpha in mice reduces colorectal carcinogenesis associated with chronic colitis. J Clin Invest 118(2):560–570

Popivanova BK, Kostadinova FI, Furuichi K, Shamekh MM, Kondo T, Wada T, Egashira K, Mukaida N (2009) Blockade of a chemokine, CCL2, reduces chronic colitis-associated carcinogenesis in mice. Cancer Res 69(19):7884–7892

Pozzi LA, Weiser WY (1992) Human recombinant migration inhibitory factor activates human macrophages to kill tumor cells. Cell Immunol 145(2):372–379

Rhodes J, Bishop M, Benfield J (1979) Tumor surveillance: how tumors may resist macrophage-mediated host defense. Science 203(4376):179–182

Rizzo A, Pallone F, Monteleone G, Fantini MC (2011) Intestinal inflammation and colorectal cancer: a double-edged sword? World J Gastroenterol 17(26):3092–3100

Roberts-Thomson IC, Fon J, Uylaki W, Cummins AG, Barry S (2011) Cells, cytokines and inflammatory bowel disease: a clinical perspective. Expert Rev Gastroenterol Hepatol 5(6):703–716

Rothwell PM, Price JF, Fowkes FG, Zanchetti A, Roncaglioni MC, Tognoni G, Lee R, Belch JF, Wilson M, Mehta Z et al (2012a) Short-term effects of daily aspirin on cancer incidence, mortality, and non-vascular death: analysis of the time course of risks and benefits in 51 randomised controlled trials. Lancet 379(9826):1602–1612

Rothwell PM, Wilson M, Price JF, Belch JF, Meade TW, Mehta Z (2012b) Effect of daily aspirin on risk of cancer metastasis: a study of incident cancers during randomised controlled trials. Lancet 379(9826):1591–1601

Ruegg C, Zaric J, Stupp R (2003) Non steroidal anti-inflammatory drugs and COX-2 inhibitors as anti-cancer therapeutics: hypes, hopes and reality. Ann Med 35(7):476–487

Ruffell B, Affara NI, Coussens LM (2012) Differential macrophage programming in the tumor microenvironment. Trends Immunol 33(3):119–126

Rutter M, Saunders B, Wilkinson K, Rumbles S, Schofield G, Kamm M, Williams C, Price A, Talbot I, Forbes A (2004) Severity of inflammation is a risk factor for colorectal neoplasia in ulcerative colitis. Gastroenterology 126(2):451–459

Rutter MD, Saunders BP, Wilkinson KH, Rumbles S, Schofield G, Kamm MA, Williams CB, Price AB, Talbot IC, Forbes A (2006) Thirty-year analysis of a colonoscopic surveillance program for neoplasia in ulcerative colitis. Gastroenterology 130(4):1030–1038

Salama P, Phillips M, Grieu F, Morris M, Zeps N, Joseph D, Platell C, Iacopetta B (2009) Tumor-infiltrating FOXP3+ T regulatory cells show strong prognostic significance in colorectal cancer. J Clin Oncol 27(2):186–192

Sanchez JA, Dejulius KL, Bronner M, Church JM, Kalady MF (2011) Relative role of methylator and tumor suppressor pathways in ulcerative colitis-associated colon cancer. Inflamm Bowel Dis 17(9):1966–1970

Scaldaferri F, Vetrano S, Sans M, Arena V, Straface G, Stigliano E, Repici A, Sturm A, Malesci A, Panes J et al (2009) VEGF-A links angiogenesis and inflammation in inflammatory bowel disease pathogenesis. Gastroenterology 136(2):585–595 e585

Schirbel A, Fiocchi C (2010) Inflammatory bowel disease: established and evolving considerations on its etiopathogenesis and therapy. J Dig Dis 11(5):266–276

Schreiber RD, Old LJ, Smyth MJ (2011) Cancer immunoediting: integrating immunity's roles in cancer suppression and promotion. Science 331(6024):1565–1570

Schulmann K, Mori Y, Croog V, Yin J, Olaru A, Sterian A, Sato F, Wang S, Xu Y, Deacu E et al (2005) Molecular phenotype of inflammatory bowel disease-associated neoplasms with microsatellite instability. Gastroenterology 129(1):74–85

Schwaab T, Weiss JE, Schned AR, Barth RJ Jr (2001) Dendritic cell infiltration in colon cancer. J Immunother 24(2):130–137

Sepulveda AR, Aisner DL (2010) Molecular basis of diseases of the gastrointestinal tract—essential concepts in molecular pathology. In: Coleman WB, Tsongalis GJ (eds) Essential concepts in molecular pathology, 1st edn. Academic, Burlington

Shattuck-Brandt RL, Varilek GW, Radhika A, Yang F, Washington MK, DuBois RN (2000) Cyclooxygenase 2 expression is increased in the stroma of colon carcinomas from IL-10(−/−) mice. Gastroenterology 118(2):337–345

Shetty K, Rybicki L, Brzezinski A, Carey WD, Lashner BA (1999) The risk for cancer or dysplasia in ulcerative colitis patients with primary sclerosing cholangitis. Am J Gastroenterol 94(6):1643–1649

Shinohara H, Yano S, Bucana CD, Fidler IJ (2000) Induction of chemokine secretion and enhancement of contact-dependent macrophage cytotoxicity by engineered expression of granulocyte-macrophage colony-stimulating factor in human colon cancer cells. J Immunol 164(5):2728–2737

Shurin MR, Naiditch H, Zhong H, Shurin GV (2011) Regulatory dendritic cells: new targets for cancer immunotherapy. Cancer Biol Ther 11(11):988–992

Shurin GV, Ouellette CE, Shurin MR (2012) Regulatory dendritic cells in the tumor immunoenvironment. Cancer Immunol Immunother 61(2):223–230

Sica A, Schioppa T, Mantovani A, Allavena P (2006) Tumour-associated macrophages are a distinct M2 polarised population promoting tumour progression: potential targets of anti-cancer therapy. Eur J Cancer 42(6):717–727

Singer II, Kawka DW, Schloemann S, Tessner T, Riehl T, Stenson WF (1998) Cyclooxygenase 2 is induced in colonic epithelial cells in inflammatory bowel disease. Gastroenterology 115(2):297–306

Snyderman R, Pike MC (1977) Macrophage migratory dysfunction in cancer. A mechanism for subversion of surveillance. Am J Pathol 88(3):727–739

Soetikno RM, Lin OS, Heidenreich PA, Young HS, Blackstone MO (2002) Increased risk of colorectal neoplasia in patients with primary sclerosing cholangitis and ulcerative colitis: a meta-analysis. Gastrointest Endosc 56(1):48–54

Solito S, Falisi E, Diaz-Montero CM, Doni A, Pinton L, Rosato A, Francescato S, Basso G, Zanovello P, Onicescu G et al (2011) A human promyelocytic-like population is responsible for the immune suppression mediated by myeloid-derived suppressor cells. Blood 118(8): 2254–2265

Soumaoro LT, Uetake H, Higuchi T, Takagi Y, Enomoto M, Sugihara K (2004) Cyclooxygenase-2 expression: a significant prognostic indicator for patients with colorectal cancer. Clin Cancer Res 10(24):8465–8471

Sparmann A, Bar-Sagi D (2004) Ras-induced interleukin-8 expression plays a critical role in tumor growth and angiogenesis. Cancer Cell 6(5):447–458

Stein U, Arlt F, Smith J, Sack U, Herrmann P, Walther W, Lemm M, Fichtner I, Shoemaker RH, Schlag PM (2011) Intervening in beta-catenin signaling by sulindac inhibits S100A4-dependent colon cancer metastasis. Neoplasia 13(2):131–144

Steinbach G, Lynch PM, Phillips RK, Wallace MH, Hawk E, Gordon GB, Wakabayashi N, Saunders B, Shen Y, Fujimura T et al (2000) The effect of celecoxib, a cyclooxygenase-2 inhibitor, in familial adenomatous polyposis. N Engl J Med 342(26):1946–1952

Stolfi C, Rizzo A, Franze E, Rotondi A, Fantini MC, Sarra M, Caruso R, Monteleone I, Sileri P, Franceschilli L et al (2011) Involvement of interleukin-21 in the regulation of colitis-associated colon cancer. J Exp Med 208(11):2279–2290

Strand S, Hofmann WJ, Hug H, Muller M, Otto G, Strand D, Mariani SM, Stremmel W, Krammer PH, Galle PR (1996) Lymphocyte apoptosis induced by CD95 (APO-1/Fas) ligand-expressing tumor cells—a mechanism of immune evasion? Nat Med 2(12):1361–1366

Sugimoto K (2008) Role of STAT3 in inflammatory bowel disease. World J Gastroenterol 14(33):5110–5114

Sugimoto K, Ogawa A, Mizoguchi E, Shimomura Y, Andoh A, Bhan AK, Blumberg RS, Xavier RJ, Mizoguchi A (2008) IL-22 ameliorates intestinal inflammation in a mouse model of ulcerative colitis. J Clin Invest 118(2):534–544

Sugita J, Ohtani H, Mizoi T, Saito K, Shiiba K, Sasaki I, Matsuno S, Yagita H, Miyazawa M, Nagura H (2002) Close association between Fas ligand (FasL; CD95L)-positive tumor-associated macrophages and apoptotic cancer cells along invasive margin of colorectal carcinoma: a proposal on tumor-host interactions. Jpn J Cancer Res 93(3):320–328

Sumimoto H, Imabayashi F, Iwata T, Kawakami Y (2006) The BRAF-MAPK signaling pathway is essential for cancer-immune evasion in human melanoma cells. J Exp Med 203(7):1651–1656

Svennevig JL, Lunde OC, Holter J (1982) In situ analysis of the inflammatory cell infiltrates in colon carcinomas and in the normal colon wall. Acta Pathol Microbiol Immunol Scand A 90(2):131–137

Svrcek M, Cosnes J, Beaugerie L, Parc R, Bennis M, Tiret E, Flejou JF (2007) Colorectal neoplasia in Crohn's colitis: a retrospective comparative study with ulcerative colitis. Histopathology 50(5):574–583

Tahara T, Inoue N, Hisamatsu T, Kashiwagi K, Takaishi H, Kanai T, Watanabe M, Ishii H, Hibi T (2005) Clinical significance of microsatellite instability in the inflamed mucosa for the prediction of colonic neoplasms in patients with ulcerative colitis. J Gastroenterol Hepatol 20(5):710–715

Takahashi N, Ohkuri T, Homma S, Ohtake J, Wakita D, Togashi Y, Kitamura H, Todo S, Nishimura T (2012) First clinical trial of cancer vaccine therapy with artificially synthesized helper/killer-hybrid epitope long peptide of MAGE-A4 cancer antigen. Cancer Sci 103(1):150–153

Takeda K, Kaisho T, Yoshida N, Takeda J, Kishimoto T, Akira S (1998) Stat3 activation is responsible for IL-6-dependent T cell proliferation through preventing apoptosis: generation and characterization of T cell-specific Stat3-deficient mice. J Immunol 161(9):4652–4660

Takeda K, Clausen BE, Kaisho T, Tsujimura T, Terada N, Forster I, Akira S (1999) Enhanced Th1 activity and development of chronic enterocolitis in mice devoid of Stat3 in macrophages and neutrophils. Immunity 10(1):39–49

Takeda J, Kitajima K, Fujii S, Horiuchi H, Hori H, Chibana Y, Okuyama T, Tominaga K, Ichikawa K, Ono Y et al (2004) Inhibitory effects of etodolac, a selective COX-2 inhibitor, on the occurrence of tumors in colitis-induced tumorigenesis model in rats. Oncol Rep 11(5):981–985

Talmadge JE, Donkor M, Scholar E (2007) Inflammatory cell infiltration of tumors: Jekyll or Hyde. Cancer Metastasis Rev 26(3–4):373–400

Tanaka T, Kohno H, Suzuki R, Yamada Y, Sugie S, Mori H (2003) A novel inflammation-related mouse colon carcinogenesis model induced by azoxymethane and dextran sodium sulfate. Cancer Sci 94(11):965–973

Terzic J, Grivennikov S, Karin E, Karin M (2010) Inflammation and colon cancer. Gastroenterology 138(6):2101–2114 e2105
Torres J, de Chambrun GP, Itzkowitz S, Sachar DB, Colombel JF (2011) Review article: colorectal neoplasia in patients with primary sclerosing cholangitis and inflammatory bowel disease. Aliment Pharmacol Ther 34(5):497–508
Triantafillidis JK, Merikas E, Georgopoulos F (2011) Current and emerging drugs for the treatment of inflammatory bowel disease. Drug Des Devel Ther 5:185–210
Trifan OC, Durham WF, Salazar VS, Horton J, Levine BD, Zweifel BS, Davis TW, Masferrer JL (2002) Cyclooxygenase-2 inhibition with celecoxib enhances antitumor efficacy and reduces diarrhea side effect of CPT-11. Cancer Res 62(20):5778–5784
Tsiolakidou G, Koutroubakis IE, Tzardi M, Kouroumalis EA (2008) Increased expression of VEGF and CD146 in patients with inflammatory bowel disease. Dig Liver Dis 40(8):673–679
Umetani N, Sasaki S, Watanabe T, Shinozaki M, Matsuda K, Ishigami H, Ueda E, Muto T (1999) Genetic alterations in ulcerative colitis-associated neoplasia focusing on APC, K-ras gene and microsatellite instability. Jpn J Cancer Res 90(10):1081–1087
Vaish V, Sanyal SN (2011) Chemopreventive effects of NSAIDs on cytokines and transcription factors during the early stages of colorectal cancer. Pharmacol Rep 63(5):1210–1221
Vaish V, Sanyal SN (2012) Role of Sulindac and Celecoxib in the regulation of angiogenesis during the early neoplasm of colon: Exploring PI3-K/PTEN/Akt pathway to the canonical Wnt/beta-catenin signaling. Biomed Pharmacother 66(5):354–367
Velho S, Moutinho C, Cirnes L, Albuquerque C, Hamelin R, Schmitt F, Carneiro F, Oliveira C, Seruca R (2008) BRAF, KRAS and PIK3CA mutations in colorectal serrated polyps and cancer: primary or secondary genetic events in colorectal carcinogenesis? BMC Cancer 8:255
Vinogradova Y, Coupland C, Hippisley-Cox J (2011) Exposure to cyclooxygenase-2 inhibitors and risk of cancer: nested case–control studies. Br J Cancer 105(3):452–459
Waldner MJ, Wirtz S, Jefremow A, Warntjen M, Neufert C, Atreya R, Becker C, Weigmann B, Vieth M, Rose-John S et al (2010) VEGF receptor signaling links inflammation and tumorigenesis in colitis-associated cancer. J Exp Med 207(13):2855–2868
Wang D, Dubois RN (2010) The role of COX-2 in intestinal inflammation and colorectal cancer. Oncogene 29(6):781–788
Weihrauch MR, Ansen S, Jurkiewicz E, Geisen C, Xia Z, Anderson KS, Gracien E, Schmidt M, Wittig B, Diehl V et al (2005) Phase I/II combined chemoimmunotherapy with carcinoembryonic antigen-derived HLA-A2-restricted CAP-1 peptide and irinotecan, 5-fluorouracil, and leucovorin in patients with primary metastatic colorectal cancer. Clin Cancer Res 11(16):5993–6001
Wick EC, Leblanc RE, Ortega G, Robinson C, Platz E, Pardoll DM, Iacobuzio-Donahue C, Sears CL (2012) Shift from pStat6 to pStat3 predominance is associated with inflammatory bowel disease-associated dysplasia. Inflamm Bowel Dis 18(7):1267–1274
Wirtz S, Neurath MF (2007) Mouse models of inflammatory bowel disease. Adv Drug Deliv Rev 59(11):1073–1083
Wirtz S, Neufert C, Weigmann B, Neurath MF (2007) Chemically induced mouse models of intestinal inflammation. Nat Protoc 2(3):541–546
Wiseman H, Halliwell B (1996) Damage to DNA by reactive oxygen and nitrogen species: role in inflammatory disease and progression to cancer. Biochem J 313(Pt 1):17–29
Woodman CB, Collins SI, Young LS (2007) The natural history of cervical HPV infection: unresolved issues. Nat Rev Cancer 7(1):11–22
Wyckoff J, Wang W, Lin EY, Wang Y, Pixley F, Stanley ER, Graf T, Pollard JW, Segall J, Condeelis J (2004) A paracrine loop between tumor cells and macrophages is required for tumor cell migration in mammary tumors. Cancer Res 64(19):7022–7029
Xie J, Itzkowitz SH (2008) Cancer in inflammatory bowel disease. World J Gastroenterol 14(3):378–389
Xu Q, Guo L, Gu X, Zhang B, Hu X, Zhang J, Chen J, Wang Y, Chen C, Gao B et al (2012) Prevention of colorectal cancer liver metastasis by exploiting liver immunity via chitosan-TPP/nanoparticles formulated with IL-12. Biomaterials 33(15):3909–3918

Yao M, Kargman S, Lam EC, Kelly CR, Zheng Y, Luk P, Kwong E, Evans JF, Wolfe MM (2003) Inhibition of cyclooxygenase-2 by rofecoxib attenuates the growth and metastatic potential of colorectal carcinoma in mice. Cancer Res 63(3):586–592

Yao M, Lam EC, Kelly CR, Zhou W, Wolfe MM (2004) Cyclooxygenase-2 selective inhibition with NS-398 suppresses proliferation and invasiveness and delays liver metastasis in colorectal cancer. Br J Cancer 90(3):712–719

Yao M, Zhou W, Sangha S, Albert A, Chang AJ, Liu TC, Wolfe MM (2005) Effects of nonselective cyclooxygenase inhibition with low-dose ibuprofen on tumor growth, angiogenesis, metastasis, and survival in a mouse model of colorectal cancer. Clin Cancer Res 11(4):1618–1628

Young LS, Murray PG (2003) Epstein-Barr virus and oncogenesis: from latent genes to tumours. Oncogene 22(33):5108–5121

Zhou Q, Peng RQ, Wu XJ, Xia Q, Hou JH, Ding Y, Zhou QM, Zhang X, Pang ZZ, Wan DS et al (2010) The density of macrophages in the invasive front is inversely correlated to liver metastasis in colon cancer. J Transl Med 8:13

Chapter 4
Importance of the Niche: Wnt Signaling and Stem Cell Plasticity in Intestinal Homeostasis and Disease

Owen J. Sansom and Inke Näthke

4.1 Introduction/Summary

Colorectal cancer (CRC) is the second most common cause of cancer deaths in the Western world. Its molecular pathology is well characterized and has taught us much about the normal physiology of the tissue that lines the gut lumen and gives rise to tumors. Central to our understanding of what drives cancer in this tissue was the discovery that mutations in the adenomatous polyposis coli gene (*APC*) are common to most tumors and that both familial and sporadic cancers in gut tissue are driven by mutations in *APC* (Su et al. 1992; Moser et al. 1992). A key function that explained the high penetrance of *APC* mutations in this tissue is its role in mediating signaling by Wnt (Polakis 1999). Specifically important is the ability of APC to act as a scaffold in assembling a protein complex that regulates the availability of β-catenin, which in turn regulates the activity of transcription factors that drive the expression of genes important for proliferation and differentiation (Rubinfeld et al. 1996; Polakis 2007). Signaling by Wnt has emerged not only as a key factor in driving the initiation of CRC but is also as a crucial regulator of normal tissue maintenance in the gut epithelium (Polakis 2007; Bejsovec 1999; Muncan et al. 2006). In particular, it plays an important role in specifying cell fate in this highly dynamic tissue. The APC protein is involved in many other functions that include regulation of cytoskeletal proteins to affect migration (Näthke et al. 1996; Kroboth et al. 2007), contribution to apoptosis (Steigerwald et al. 2005; Brocardo et al. 2008), nuclear import

O.J. Sansom (✉)
Cancer Research UK Beatson Laboratories, Garscube Estate, Switchback Road, Glasgow, G61 1BD, UK
e-mail: o.sansom@beatson.gla.ac.uk

I. Näthke (✉)
Division of Cell and Developmental Biology, University of Dundee, Dow Street, Dundee, DD1 5EH, UK
e-mail: inke@lifesci.dundee.ac.uk

K.M. Haigis (ed.), *Molecular Pathogenesis of Colorectal Cancer*,
DOI 10.1007/978-1-4614-8412-7_4,

(Murawala et al. 2009), and mitotic spindles and related checkpoints (Dikovskaya et al. 2004, 2007; Draviam et al. 2006). In the background of the increased proliferation and decreased differentiation that accompanies mutations of APC, loss of the normal execution of these additional functions renders cells particularly prone to accumulate additional mutations that confer a significant survival advantage to APC mutant cells in gut epithelial tissue (McCartney and Näthke 2008).

In this chapter, we summarize the mechanisms that underpin normal Wnt signaling, the contribution it makes to the maintenance of the niche that supports stem cells in normal gut and also in tumors. What emerges from this discussion is that Wnt is crucial for normal maintenance of epithelial tissue by supporting the proliferative niche that drives normal homeostasis in gut epithelium. Although we focus exclusively on Wnt signaling in this chapter, it is important to acknowledge that it does not operate in isolation. In fact, all the other signaling pathways that commonly contribute to tissue morphogenesis and homeostasis, most notably BMP, Notch, HH, EGF, TGF, and PDGF, also contribute significantly to the processes discussed below (Crosnier et al. 2006). The integration of the complex interactions and feedback loops between these pathways ultimately not only produce spatially and temporally regulated outputs that underpin the tissue organization in the intestinal tract but also permit the immense plasticity that characterizes this tissue and the cells populated by it.

4.2 Intestinal Stem Cell Wnt Signaling

Like other regenerative tissues, gut epithelium is highly dynamic and constantly replaced from progenitor cells that reside in the bottom of tissue invaginations that act as highly specialized biochemical and mechanical niches (Fig. 4.1). The gut epithelium is formed by a simple, single layer of epithelial cells that is constantly regenerated from progenitor cells that are compartmentalized within highly ordered arrays of tissue invaginations called crypts of Lieberkühn (Fig. 4.1b) (Wright and Alison 1984). These invaginations provide discrete units that are tightly packed into a regular pattern that line the gut. In the small intestine, they connect to villi, long finger-like structures that vary in size and density along the stomach–rectum axis (200–500 μm height in mouse and 500–1,500 μm in human) (Wright and Alison 1984). In mice, individual crypts have been estimated to contain about 14–18 stem cells (Snippert et al. 2010) that reside near its bottom and continually replenish cells that are exfoliated from the surface epithelium or villus tips. However, the lack of a universally accepted marker for stem cells in this tissue has made it difficult to measure this number accurately (Kaur and Potten 2011). Nonetheless, it is clear that this arrangement and the dynamics of this tissue create an epithelial conveyor belt that is extremely rapidly turned over. In the adult mouse, each intestinal crypt produces about 25 cells per hour, whereas crypts in the colon, where there are no villi, produce about 16 cells per hour (Cheng and Bjerknes 1983; Totafurno et al. 1987). This cell birth is balanced by the loss of about 240 ± 66 cells each hour from a villus,

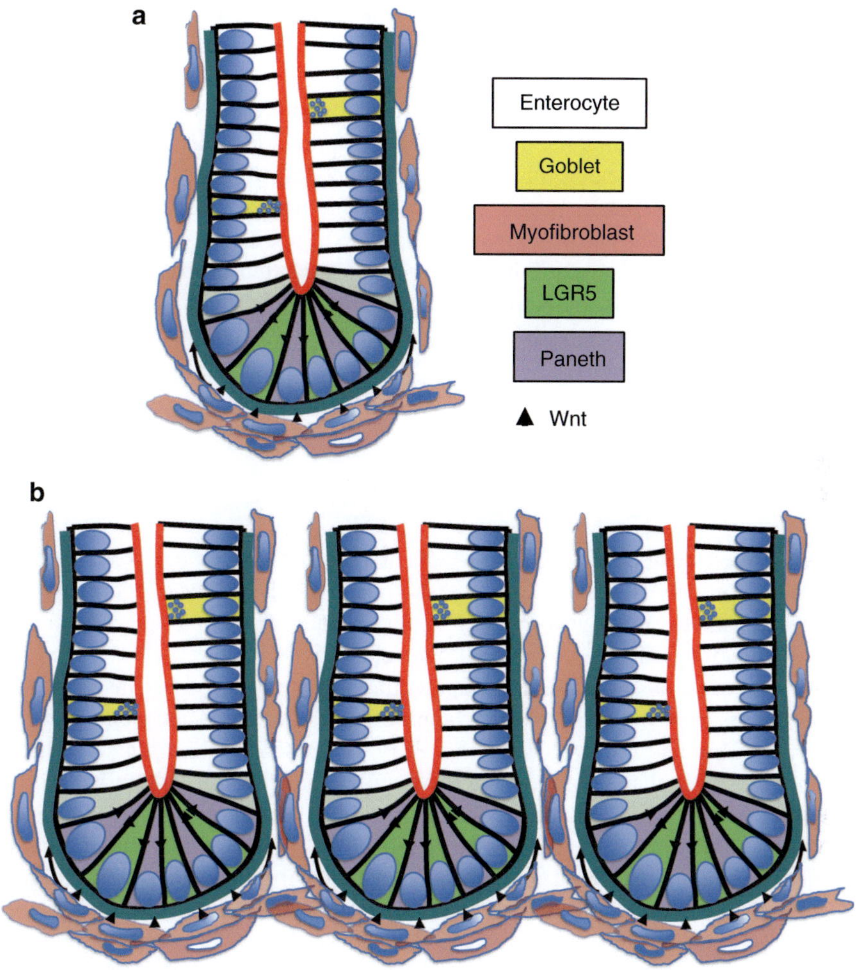

Fig. 4.1 (**a**) Schematic of an intestinal crypt showing the major cell types. In this tissue, a basement membrane (*green solid line*) surrounds and supports the crypt epithelium. Wnt and other growth factors (indicated by *arrows* and *arrowheads*) are provided by myofibroblasts that surround each crypt. Paneth cells also provide growth factors including Wnt (*arrowheads*) to the neighboring LGR5 positive cells. (**b**) Individual crypts are tightly packed into regular arrays

which has about 2,000–8,000 cells, depending on its position in the long axis of the intestine (Wright and Irwin 1982). This means that 3–10 % of the cells on a villus are shed every hour and that the estimated 1.1 million crypts in the mouse intestine (Potten and Loeffler 1990) produce about 25 million cells per hour. These staggering numbers illustrate the incredibly high turnover of this tissue. Given the fact that the gut epithelium continues to grow throughout our entire lifetime, it is not surprising that, even with exceedingly low endogenous rates of mutations, tumorigenesis increases with age.

The Wnt signaling pathway impacts on gut tissue, and particularly stem cell maintenance, most prominently. Wnt proteins are small, heavily disulfide bonded, secreted proteins that interact with surface receptors called Frizzled to stimulate the stabilization of β-catenin (Polakis 2012). At the centre of this process is a large protein complex that contains the APC tumor suppressor protein and Axin, both acting as scaffolding proteins, plus a variety of regulating proteins including GSK3β and CK1a (Polakis 2012). In the absence of Wnt signals, this complex recruits β-catenin and facilitates its phosphorylation by GSK3β. This acts as a signal for the ubiquitylation of β-catenin and its subsequent degradation by the proteasome. In response to Wnt signals, GSK3β is inactivated so that β-catenin is not degraded, accumulates and becomes available to regulate transcription of genes important for differentiation (Polakis 2012).

4.3 Stem Cells

The exact identity of the ultimate stem cells in intestinal and colonic tissue has been debated vigorously (Kaur and Potten 2011). And which marker of the many that have been used to definitively identify stem cells in this tissue is ‘the right one’ is not our intention to debate or decide (Potten et al. 2009).

We propose that the diversity in potential stem cell markers illustrates that there are distinct, but partially overlapping populations of stem cells that can exchange (Tian and Finley 2012). It is likely that position within the stem niche dictates the exact characteristics of a given cell at any given time. In other words, the exact properties of a given stem cell, and thus the complement of markers it carries, at any one time are likely to be dictated by its position in the niche because of the unique combination of chemical and mechanical signals present in different positions. For instance, while crypt structures can be established in vitro from single individual LGR-5 positive cells, crypts lacking LGR-5 columnar basal crypt cells can be replenished from an alternative population of stem cells that express Bmi-1 (Tian and Finley 2012; Sato et al. 2009). These observations indicate that there may be not only different pools of progenitor cells with some distinct but also overlapping properties that can exchange places with each other, as recently suggested (Takeda et al. 2011). These pools normally exist in equilibrium with each other, but the exact stoichiometry between them may fluctuate and depend on the developmental state of a crypt as suggested by a recent model of stem cell pool maintenance. This model suggests that, early in a crypt’s lifetime, the number of stem cells may expand beyond the steady-state value via symmetric divisions and then be maintained and adjusted by asymmetric divisions and differentiation (Itzkovitz et al. 2012). Together, these data support the idea that there can be active mutual exchange between populations of cells that reside at different positions within the crypt. This may explain why it is difficult to resolve the controversy about the ultimate stem cell marker for this tissue (Kaur and Potten 2011; Potten et al. 2009).

In this context, it is important to mention that the ability of LGR5 positive cells to seed crypts ex vivo is greatly enhanced by their association with Paneth cells (Sato et al. 2010). These secretory cells produce antimicrobials and are found directly adjacent to LGR5 positive cells at the base of intestinal crypts (Fig. 4.1). In fact, the packing of cells in this highly curved part of crypts favors direct contact between LGR5 and Paneth cells. Although Paneth cells are absent in the colon, LGR5 positive cells still are interspersed with Paneth-like cells, suggesting that this alternate packing is important for this structure. Paneth cells do not only secrete antimicrobials but they also secrete Wnt pathway ligands, and this is likely to be a key factor in their ability to support the 'stemness' of their neighbors (Sato et al. 2010). However, Wnt ligands are also provided by the myofibroblasts that surround each crypt (Fig. 4.1) (Neal and Potten 1981), which may be able to explain the recent finding that crypts without Paneth cells are viable (Kim et al. 2012). Nonetheless, the normal presence of Paneth cells, the highly regular, intermingling arrangement they adopt, and the fact that similar features are adopted by colonic crypts, although full-fledged Paneth cells are absent in colon, suggests that they contribute to optimal tissue function. It is likely that the balance of growth and other regulatory factors, together with the mechanical environment created by the combination of cell types in the epithelium and surrounding tissue, helps to create a biochemical and mechanical niche that allows for the enormous plasticity required.

The importance of Wnt signaling for this compartment is further illustrated by the fact that ablating TCF4, the transcription factor that is a key target for Wnt regulation, eliminates stem cells in the gut epithelium in mice (Korinek et al. 1998). In Drosophila, Wnt also is an important factor in regulating stem cell activity in the gut. However, in this case, nutritional cues can also contribute directly to the size of the stem cell compartment (O'Brien et al. 2011). Niche production of insulin in response to higher food intake drives increased stem cell number and overall growth that is reversed upon food withdrawal. This, again, illustrates the plasticity of gut tissue and suggests that stem cell populations can be modified in response to environmental or physiological cues and do not exist as invariable static populations (O'Brien et al. 2011).

4.4 Cell of Origin for CRC

Despite the ongoing debate about the precise nature of the intestinal stem cell, the availability of knock-in mice where an inducible Cre recombinase is inserted into an intestinal stem cell marker locus has allowed us to establish that intestinal stem cells can act as cells of origin for CRC or (at the very least) intestinal adenomas in the mouse. For instance, Barker et al. showed that Lgr5 can act as a highly efficient cell of origin. Genetic deletion of APC using *Lgr5CRE*^{ER} in the adult murine intestine caused rapid formation multiple adenomas in as little as 20 days (Barker et al. 2009). Thus, loss or mutation of both copies of APC within a single stem cell could be sufficient to initiate tumorigenesis. Similar results have been obtained using the CD133 and BMI promoters in the mouse intestine (Sangiorgi and Capecchi 2008; Zhu et al. 2009). One caveat in these studies is that it normally takes 2–3 days for Cre-mediated excision of the *APC* gene to produce increased β-catenin signaling. Because intestinal stem cells produce daughter cells relatively rapidly, it is still possible that some of the adenomas in these models derive from daughters of Lgr5 cells that were produced while deletion of the *APC* gene was still ongoing.

Another question that arises from these studies is whether the situation is the same in human CRC. For many years, our model of human cancer was the so-called Vogelgram, with tumors arising over a number of years accompanied by the accumulation of further mutations in KRAS, P53, SMAD4, etc. (Fearon and Vogelstein 1990). Moreover, discussions have ranged over 'bottom-up' and 'top-down' models for CRC (Shih et al. 2001; Preston et al. 2003). These models are based on histological observations from pathologists who found tumors on the luminal surface of the colon with all crypts beneath appearing normal. The 'top-down' model is difficult to reconcile with the position of stem cells at the base of the crypt: How could a tumor that initiated at the top of crypt be initiated from a stem cell? Most of the work describing top-down tumors was performed on formalin fixed histological cross sections, which cannot distinguish the possibility that one single aberrant crypt was the cell of origin for the tumor. The advent of intestinal wholemount imaging should provide much improved resolution of this process (Appleton et al. 2009). In mice, an alternative approach can be taken to determine if non-stem cells can initiate tumors. Specifically, a marker gene that is not expressed in the intestinal stem cells can be utilized to assess if deleting *APC* specifically in non-stem cells can lead to tumors. However, this is not without difficulty in this system. Intestinal turnover is rapid (3–5 days) and it takes 3 days for APC to be deleted. That means that by the time the APC protein has turned over and β-catenin signaling is activated, most differentiated cells will be sloughed off into the intestine. Therefore, marker genes are required that are expressed in transit amplifying or early lineage secretory cells, but not stem cells to answer this question. Given the difficulty in finding markers not expressed in stem cells, we used an alternative approach, which relied on the fact that stem cells are located at the base of intestinal crypts. Specifically, we used a cytochrome P450cre (AHCRE) transgenic mouse that permits Cre induction in both ISC and transit amplifying cells. In this case, instead of injecting the inducer intraperitoneally, which induces recombination in both the stem cells and the transit amplifying cells, mice were orally gavaged with low doses of the inducer, which preferentially caused gene deletion at the top of crypts and not in stem cell zones (Barker et al. 2009). In contrast to stem cell deletion, mice developed numerous small lesions (or microadenomas), which only progressed to adenomas at long latencies (Barker et al. 2009). This proof of principle data suggested that intestinal tumors can be initiated in non-stem cell compartments although with kinetics that are distinct from stem cell-mediated tumorigenesis. Therefore, one can produce a model where the initial APC mutation is within a stem cell population and this repopulates the entire crypt. Following a second stem cell mutation, rapid tumorigenesis occurs while a mutation outside the stem cell zone will produce a lesion that more rarely accumulates the additional mutations that allow tumor progression. An important prediction from this model is that changes that modify intestinal homeostasis to allow cells to persist longer before they are sloughed off create increased opportunities for non-stem cell mutations to persist, and hence be transformative. This may explain the large number factors that can impinge on CRC: CRC is affected by diet and inflammation, ulcerative colitis carries a 1 in 3 lifetime risk of CRC, while aspirin administration can reduce CRC risk in high-risk groups by half (Rothwell et al. 2010). These ideas are supported by recent sleeping beauty studies that revealed many events that can cooperate with APC loss in intestinal tumorigenesis (March et al. 2011). One potential reason for this effect may be the ability of such extra mutations to expand the numbers of cells of origin.

4.5 Mouse Models of APC and Relationship to Wnt Signaling Strength

Within human CRC, a specific spectrum of mutations within the APC tumor suppressor gene occurs. Mouse models are starting to reveal that the precise nature of these mutations may confer important properties, especially in terms of Wnt signaling and ISCs.

The first APC mouse mutant was produced through an ENU mutagenesis screen and designated the APC^{Min} allele (Su et al. 1992). Here, APC has a point mutation at codon 850 that leads to a premature truncation and expression of an N-terminal fragment that cannot bind to β-catenin and Axin (Fig. 4.2). In the heterozygous state, this 'Min' (Multiple Intestinal neoplasia) mutation produced mice that rapidly developed intestinal adenomas that show loss of the remaining wild-type allele (Luongo et al. 1994). The importance of the deregulation of Wnt signaling resulting from APC loss for intestinal adenoma formation was demonstrated by the APC^{1638T} allele (Fig. 4.2) (Smits et al. 1999). This allele produced a truncated APC protein that still retained the ability to degrade β-catenin and, importantly, mice were not prone to intestinal tumorigenesis. More recently, Lewis and colleagues generated an APC allele that produced a protein that could not turnover β-catenin but retained interactions sites for other binding partners of APC such as microtubules (Fig. 4.2) (Lewis et al. 2012). These mice rapidly developed intestinal cancer, highlighting the importance of deregulated Wnt signaling for tumor initiation. However, in none of these studies has the effect of the truncated APC protein fragment on the wild-type protein been examined in detail. Interactions between truncated and full-length APC proteins are likely to alter at least a subset of protein interactions. Indeed, N-terminal APC fragment on their own bind full-length APC protein (Fig. 4.2) and have dominant effects on a number of cellular processes (Li et al. 2008; Nelson et al. 2012) suggesting that the exact nature and extent of the predisposition to cancer induced by their presence will be affected by the details of the mutation.

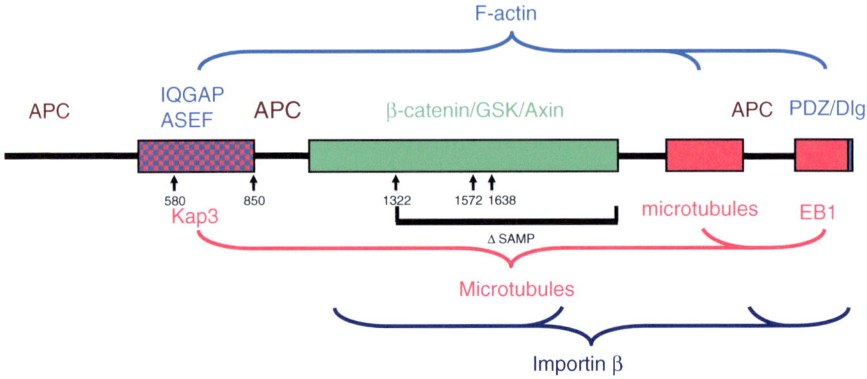

Fig. 4.2 Schematic of APC sequence. Binding partners are shown in various colors with brackets and correspondingly *colored boxes* indicating the regions where they bind to APC. The position of the nonsense mutations that produce truncated APC into the mouse models discussed in the text are indicated in *black* with relevant numbers indicating the position where the resulting APC protein is truncated. The region deleted in the **Δ**SAMP model (Lewis et al. 2012) is shown by a *black horizontal bar* below the relevant region

The previously described alleles have provided important functional insights into tumor initiation following APC mutation but have not addressed the precise APC mutations that occur in CRC. In human tumors, most APC mutations (ca. 60 %) occur within similar locations within the gene: the so-called mutation cluster region (Miyoshi et al. 1992). Moreover, a truncated protein product is normally produced that is stable and can still bind but not degrade β-catenin, as it lacks the ability to bind Axin (Fig. 4.2). Therefore, it is possible that these mutations introduce other effects in addition to simple loss of function and Wnt signaling activation, which predicts that such mutations differ in their effect from complete loss-of-function alleles. The 'just right' hypothesis of β-catenin signaling following APC loss is also part of this prediction and proposes that the precise levels of Wnt deregulation is important for tumorigenesis (Albuquerque et al. 2002). Again, mouse models have provided excellent evidence that the relative level of Wnt signaling activation is key for tumorigenic outcome. The APC^{580S} allele is a conditional allele that is hypomorphic due to the retention of neomycin cassette in an intron (Shibata et al. 1997) (Fig. 4.2). Mice carrying two copies of this allele are not prone to develop intestinal tumors but exhibit changes in liver zonation, which ultimately leads to hepatocellular carcinoma (Buchert et al. 2010). Another hypomorphic APC allele, APC^{1572T}, leads to a modest increase in Wnt signaling and predisposes to breast but not intestinal cancer in the mouse (Gaspar et al. 2009). Interestingly, complete deletion of one APC allele led to enhanced tumor formation relative to APC^{Min}, despite lower Wnt signaling activity (Cheung et al. 2010). Most relevant for human CRC is the recently generated APC^{1322T} allele (Pollard et al. 2009) (Fig. 4.2). This allele mimics one of the most common mutations of CRC, where the retained APC fragment can bind β-catenin, but not Axin, so cannot target β-catenin for degradation. Importantly, mice heterozygous for this allele develop tumors more readily than the $APC^{Min/+}$ mice. Nonetheless, the amplitude of Wnt target gene activation was decreased and the majority of Wnt target genes were activated to a lesser extent than in the $APC^{Min/+}$ mouse (Lewis et al. 2010). However, higher levels of the ISC marker genes, such as Lgr5 and Musashi-1, were detected in these tumors, suggestive of an altered repertoire of activated Wnt target genes. Interestingly, these adenomas also contained more Paneth cells, which may have contributed to the increased stem cell number (see above).

Taken together, these studies highlight the complex relationship between APC mutation and the resultant phenotype. Although deregulation of Wnt signaling is directly associated with CRC, the precise level of activation appears to be important in the tumor that ensues. Further study is required to understand precisely the underlying reasons.

4.6 Cancer Stem Cells and Wnt Signaling

Our increased knowledge about normal intestinal stem cells has raised the question of how similar these are to 'cancer stem cell' populations. The concept of the cancer stem cell hypothesis has been particularly attractive in CRC, as tumors retain many differentiated features suggestive of a hierarchal structure, with stem cells producing daughter cells that have some capacity to differentiate. Prior to the identification of the intestinal stem cell signature, tumor transplantation experiments in nude mice identified a number of potential cancer stem cell markers such as CD133, CD44, and CD24, which characterize tumor-initiating capacity. Moreover, other studies linked cancer stem cells to increased levels of Wnt signaling, epithelial–mesenchymal transition (EMT), and invasive fronts (Brabletz et al. 2001, 2005; Vermeulen et al. 2010). More than 10 years ago, Thomas Brabletz noted that invasive fronts of CRC had higher levels of β-catenin than the centre of the tumor (Brabletz et al. 1998). This phenomenon became known as the 'β-catenin' paradox, referring to the fact that most CRC have lost normal APC and hence should have high β-catenin levels throughout the tumor. However, β-catenin levels are not uniform throughout the tumor. Not only were invasive fronts associated with the highest levels of β-catenin, they also had low levels of E-Cadherin and high levels of the EMT markers Zeb1, Snail and Twist suggesting that EMT is important for metastatic spread of tumor cells. Discussions still remain over how much this reflects a 'bona fide' EMT or just the expression of markers of EMT. Importantly, in CRC, markers of EMT at invasive fronts correlate with poor prognosis (Spaderna et al. 2006). Further interest in EMT has developed, as markers for EMT are also associated with the cancer stem cell phenotype (Mani et al. 2008). For instance, ZEB1 expression can confer stem cell properties in part through the regulation of the mir200 family (Wellner et al. 2009). The pertinence of these findings to colon cancer stem cells has recently come to the forefront. Vermuelen and colleagues suggested that specific levels of Wnt signaling confer the cancer stem cell phenotype (Vermeulen et al. 2010). Therefore, CRC cells that have the highest levels of Wnt signaling become cancer stem cells. The authors suggested that mesenchymal cells express factors such as HGF, which can activate Wnt signaling in the nearby tumor cells. Thus, those tumor cells in closest proximity to the mesenchymal cells would adopt a cancer stem cell fate and hence the tumor stroma is acting as the cancer stem cell niche (also see above).

Importantly, in normal intestine, the ISCs are associated with the highest levels of Wnt signaling. Consistently, ISC markers, such as Lgr5, EphB2/3, and ASCL2, are targets of the Wnt signaling pathway, but only Paneth and crypt columnar cells show nuclear β-catenin when stained using IHC (van Es et al. 2005; Barker et al. 2007). A study by the Batlle group appeared to reconcile these findings by showing that colon cancer stem cells express the ISC signature, that they can form tumors in transplantation experiments, and that resultant tumors contained a complete repertoire of differentiation. Furthermore, the presence of these cancer stem cell markers within CRC tumors conferred predicted disease relapse (Wellner et al. 2009; Merlos-Suarez et al. 2011). Unfortunately, the consensus offered by these observations has become more complicated again. One caveat is that the CSC were identified using the EphB receptor, which, although a component of the ISC, is down-regulated in later CRC and metastasis (de Sousa et al. 2011). Although a recent study confirmed that the ISC is associated with CSC, it also showed that in later, more aggressive disease these ISC signature genes are methylated and the consequent down-regulation was associated with a poor prognosis (and poorly differentiated tumors) (de Sousa et al. 2011).

Given the heterogeneity of human cancers, it is perhaps not surprising that different studies produce disparate results. Moreover, much of this work was based on flow sorting of cells isolated from tissues. It is entirely possible that removal of cells from their niche caused changes in the expression of these markers. Important new information may be gained by examining expression on a cell-by-cell basis at the protein level in human tissues rather than by simply looking at relative message levels in heterogeneous samples.

Novel approaches are required to target cells expressing ISC markers at different stages of the tumorigenic process to define their purpose. If they are depleted, but the niche remains the same, will other cells now switch their fate to replace the missing components, or are they redundant in later stages tumors? The normal ISC compartment appears to be very plastic so the same may be true for CSCs (see above).

In CRC, the stage of disease (1–4) is still the best indicator of prognosis. It is important to identify the subset of early stage patients with a poor prognosis (e.g. stage 2) as these may benefit most from therapy. Apart from this group of patients, it is much more important to find ‘predictive’ markers, i.e. those markers that delineate response to therapy rather than patient outcome. Therefore, rather on fixating on whether tumors with high ISC signatures confer a good or poor prognosis, the functional relevance of the expression of the set of genes that defines this signature for the tumor that carries them is a far more important question to be tackled.

4.7 Implications for Therapy and Diagnosis

The idea that entire tumor genomes can provide information about the ability of a tumor to respond to certain types of therapy and can predict prognostic value is attractive. However, solid tumors like those that arise in the colonic epithelium contain many different types of cells that may maintain the ability to adopt a number of different fates depending on their environment. So any sample from a biopsy is only a single window in time and space that may not accurately reflect the types of cells that therapy has to target to be successful. Cell-by-cell information, and the changes that occur over time and in response to therapy, will be much more informative in this regard. Furthermore, the flexibility and plasticity of the normal stem cell niche may also operate for CSCs making it difficult to target the stem cells because they are defined in a non-cell-autonomous manner by a niche. Specifically, if normal stem cells can partially differentiate and then re-differentiate why should the CRC stem cell not be able to do the same in response to niche factors?

These ideas illustrate the importance of continuing to apply a wide spectrum of approaches and techniques to understand the factors that contribute to normal tissue maintenance in gut epithelium. Identifying the molecular mechanisms that underpin the tissue changes that accompany the different stage of tumor initiation and progression will continue to help us to understand how best to detect, prognose, prevent, and treat this extremely common disease.

References

Albuquerque C, Breukel C, van der Luijt R, Fidalgo P, Lage P, Slors FJ, Leitao CN, Fodde R, Smits R (2002) The 'just-right' signaling model: APC somatic mutations are selected based on a specific level of activation of the beta-catenin signaling cascade. Hum Mol Genet 11:1549–1560

Appleton PL, Quyn AJ, Swift S, Nathke I (2009) Preparation of wholemount mouse intestine for high-resolution three-dimensional imaging using two-photon microscopy. J Microsc 234:196–204

Barker N, van Es JH, Kuipers J, Kujala P, van den Born M, Cozijnsen M, Haegebarth A, Korving J, Begthel H, Peters PJ, Clevers H (2007) Identification of stem cells in small intestine and colon by marker gene Lgr5. Nature 449:1003–1007

Barker N, Ridgway RA, van Es JH, van de Wetering M, Begthel H, van den Born M, Danenberg E, Clarke AR, Sansom OJ, Clevers H (2009) Crypt stem cells as the cells-of-origin of intestinal cancer. Nature 457:608–611

Bejsovec A (1999) Signal transduction: Wnt signalling shows its versatility. Curr Biol 9:R684–R687

Brabletz T, Jung A, Hermann K, Gunther K, Hohenberger W, Kirchner T (1998) Nuclear overexpression of the oncoprotein beta-catenin in colorectal cancer is localized predominantly at the invasion front. Pathol Res Pract 194:701–704

Brabletz T, Jung A, Reu S, Porzner M, Hlubek F, Kunz-Schughart LA, Knuechel R, Kirchner T (2001) Variable beta-catenin expression in colorectal cancers indicates tumor progression driven by the tumor environment. Proc Natl Acad Sci U S A 98:10356–10361

Brabletz T, Jung A, Spaderna S, Hlubek F, Kirchner T (2005) Opinion: migrating cancer stem cells—an integrated concept of malignant tumour progression. Nat Rev Cancer 5:744–749

Brocardo M, Lei Y, Tighe A, Taylor SS, Mok MT, Henderson BR (2008) Mitochondrial targeting of adenomatous polyposis coli protein is stimulated by truncating cancer mutations: regulation of Bcl-2 and implications for cell survival. J Biol Chem 283:5950–5959
Buchert M, Athineos D, Abud HE, Burke ZD, Faux MC, Samuel MS, Jarnicki AG, Winbanks CE, Newton IP, Meniel VS, Suzuki H, Stacker SA, Nathke IS, Tosh D, Huelsken J, Clarke AR, Heath JK, Sansom OJ, Ernst M (2010) Genetic dissection of differential signaling threshold requirements for the Wnt/beta-catenin pathway in vivo. PLoS Genet 6:e1000816
Cheng H, Bjerknes M (1983) Cell production in mouse intestinal epithelium measured by stathmokinetic flow cytometry and Coulter particle counting. Anat Rec 207:427–434
Cheung AF, Carter AM, Kostova KK, Woodruff JF, Crowley D, Bronson RT, Haigis KM, Jacks T (2010) Complete deletion of Apc results in severe polyposis in mice. Oncogene 29:1857–1864
Crosnier C, Stamataki D, Lewis J (2006) Organizing cell renewal in the intestine: stem cells, signals and combinatorial control. Nat Rev Genet 7:349–359
de Sousa EMF, Colak S, Buikhuisen J, Koster J, Cameron K, de Jong JH, Tuynman JB, Prasetyanti PR, Fessler E, van den Bergh SP, Rodermond H, Dekker E, van der Loos CM, Pals ST, van de Vijver MJ, Versteeg R, Richel DJ, Vermeulen L, Medema JP (2011) Methylation of cancer-stem-cell-associated Wnt target genes predicts poor prognosis in colorectal cancer patients. Cell Stem Cell 9:476–485
Dikovskaya D, Newton IP, Näthke IS (2004) The adenomatous polyposis coli protein is required for the formation of robust spindles formed in CSF Xenopus extracts. Mol Biol Cell 15:2978–2991
Dikovskaya D, Schiffmann D, Newton IP, Oakley A, Kroboth K, Sansom O, Jamieson TJ, Meniel V, Clarke A, Näthke IS (2007) Loss of APC induces polyploidy as a result of a combination of defects in mitosis and apoptosis. J Cell Biol 176:183–195
Draviam VM, Shapiro I, Aldridge B, Sorger PK (2006) Misorientation and reduced stretching of aligned sister kinetochores promote chromosome missegregation in EB1- or APC-depleted cells. EMBO J 25:2814–2827
Fearon ER, Vogelstein B (1990) A genetic model for colorectal tumorigenesis. Cell 61:759–767
Gaspar C, Franken P, Molenaar L, Breukel C, van der Valk M, Smits R, Fodde R (2009) A targeted constitutive mutation in the APC tumor suppressor gene underlies mammary but not intestinal tumorigenesis. PLoS Genet 5:e1000547
Itzkovitz S, Blat IC, Jacks T, Clevers H, van Oudenaarden A (2012) Optimality in the development of intestinal crypts. Cell 148:608–619
Kaur P, Potten CS (2011) The interfollicular epidermal stem cell saga: sensationalism versus reality check. Exp Dermatol 20:697–702
Kim TH, Escudero S, Shivdasani RA (2012) Intact function of Lgr5 receptor-expressing intestinal stem cells in the absence of Paneth cells. Proc Natl Acad Sci U S A 109(10):3932–3937
Korinek V, Barker N, Moerer P, van Donselaar E, Huls G, Peters PJ, Clevers H (1998) Depletion of epithelial stem-cell compartments in the small intestine of mice lacking Tcf-4. Nat Genet 1998(19):379–383
Kroboth K, Newton IP, Kita K, Dikovskaya D, Zumbrunn J, Waterman-Storer CM, Näthke IS (2007) Lack of adenomatous polyposis coli protein correlates with a decrease in cell migration and overall changes in microtubule stability. Mol Biol Cell 18:910–918
Lewis A, Segditsas S, Deheragoda M, Pollard P, Jeffery R, Nye E, Lockstone H, Davis H, Clark S, Stamp G, Poulsom R, Wright N, Tomlinson I (2010) Severe polyposis in Apc(1322T) mice is associated with submaximal Wnt signalling and increased expression of the stem cell marker Lgr5. Gut 59:1680–1686
Lewis A, Davis H, Deheragoda M, Pollard P, Nye E, Jeffery R, Segditsas S, East P, Poulsom R, Stamp G, Wright N, Tomlinson I (2012) The C-terminus of Apc does not influence intestinal adenoma development or progression. J Pathol 226:73–83
Li Z, Kroboth K, Newton IP, Nathke IS (2008) Novel self-association of the APC molecule affects APC clusters and cell migration. J Cell Sci 121:1916–1925
Luongo C, Moser AR, Gledhill S, Dove WF (1994) Loss of *APC* in intestinal adenomas from Min mice. Cancer Res 54:5947–5952

Mani SA, Guo W, Liao MJ, Eaton EN, Ayyanan A, Zhou AY, Brooks M, Reinhard F, Zhang CC, Shipitsin M, Campbell LL, Polyak K, Brisken C, Yang J, Weinberg RA (2008) The epithelial-mesenchymal transition generates cells with properties of stem cells. Cell 133:704–715

March HN, Rust AG, Wright NA, ten Hoeve J, de Ridder J, Eldridge M, van der Weyden L, Berns A, Gadiot J, Uren A, Kemp R, Arends MJ, Wessels LF, Winton DJ, Adams DJ (2011) Insertional mutagenesis identifies multiple networks of cooperating genes driving intestinal tumorigenesis. Nat Genet 43:1202–1209

McCartney BM, Näthke IS (2008) Cell regulation by the Apc protein Apc as master regulator of epithelia. Curr Opin Cell Biol 20:186–193

Merlos-Suarez A, Barriga FM, Jung P, Iglesias M, Cespedes MV, Rossell D, Sevillano M, Hernando-Momblona X, da Silva-Diz V, Munoz P, Clevers H, Sancho E, Mangues R, Batlle E (2011) The intestinal stem cell signature identifies colorectal cancer stem cells and predicts disease relapse. Cell Stem Cell 8:511–524

Miyoshi Y, Nagase H, Ando H, Horii A, Ichii S, Nakatsuru S, Aoki T, Miki Y, Mori T, Nakamura Y (1992) Somatic mutations of the APC gene in colorectal tumors: mutation cluster region in the APC gene. Hum Mol Genet 1:229–233

Moser AR, Dove WF, Roth KA, Gordon JI (1992) The *Min* (multiple intestinal neoplasia) mutation: its effect on gut epithelial cell differentiation and interaction with a modifier system. J Cell Biol 116:1517–1526

Muncan V, Sansom OJ, Tertoolen L, Phesse TJ, Begthel H, Sancho E, Cole AM, Gregorieff A, de Alboran IM, Clevers H, Clarke AR (2006) Rapid loss of intestinal crypts upon conditional deletion of the Wnt/Tcf-4 target gene c-Myc. Mol Cell Biol 26:8418–8426

Murawala P, Tripathi MM, Vyas P, Salunke A, Joseph J (2009) Nup358 interacts with APC and plays a role in cell polarization. J Cell Sci 122:3113–3122

Näthke IS, Adams CL, Polakis P, Sellin JH, Nelson WJ (1996) The adenomatous polyposis coli tumor suppressor protein localizes to plasma membrane sites involved in active cell migration. J Cell Biol 134:165–179

Neal JV, Potten CS (1981) Description and basic cell kinetics of the murine pericryptal fibroblast sheath. Gut 22:19–24

Nelson SA, Li Z, Newton IP, Fraser D, Milne RE, Martin DM, Schiffmann D, Yang X, Dormann D, Weijer CJ, Appleton PL, Nathke IS (2012) Tumorigenic fragments of APC cause dominant defects in directional cell migration in multiple model systems. Dis Model Mech 5(6):940–947

O'Brien LE, Soliman SS, Li X, Bilder D (2011) Altered modes of stem cell division drive adaptive intestinal growth. Cell 147:603–614

Polakis P (1999) The oncogenic activation of b-catenin. Curr Opin Genet Dev 9:15–21

Polakis P (2007) The many ways of Wnt in cancer. Curr Opin Genet Dev 17:45–51

Polakis P (2012) Wnt signaling in cancer. Cold Spring Harb Perspect Biol 4(5). pii: a008052

Pollard P, Deheragoda M, Segditsas S, Lewis A, Rowan A, Howarth K, Willis L, Nye E, McCart A, Mandir N, Silver A, Goodlad R, Stamp G, Cockman M, East P, Spencer-Dene B, Poulsom R, Wright N, Tomlinson I (2009) The Apc 1322T mouse develops severe polyposis associated with submaximal nuclear beta-catenin expression. Gastroenterology 136:2204–2213. e1–13

Potten CS, Loeffler M (1990) Stem cells: attributes, cycles, spirals, pitfalls and uncertainties. Lessons for and from the crypt. Development 110:1001–1020

Potten CS, Gandara R, Mahida YR, Loeffler M, Wright NA (2009) The stem cells of small intestinal crypts: where are they? Cell Prolif 42:731–750

Preston SL, Wong WM, Chan AO, Poulsom R, Jeffery R, Goodlad RA, Mandir N, Elia G, Novelli M, Bodmer WF, Tomlinson IP, Wright NA (2003) Bottom-up histogenesis of colorectal adenomas: origin in the monocryptal adenoma and initial expansion by crypt fission. Cancer Res 63:3819–3825

Rothwell PM, Wilson M, Elwin CE, Norrving B, Algra A, Warlow CP, Meade TW (2010) Long-term effect of aspirin on colorectal cancer incidence and mortality: 20-year follow-up of five randomised trials. Lancet 376:1741–1750

Rubinfeld B, Albert I, Porfiri E, Fiol C, Munemitsu S, Polakis P (1996) Binding of GSK3beta to the APC-beta-catenin complex and regulation of complex assembly. Science 272:1023–1026

Sangiorgi E, Capecchi MR (2008) Bmi1 is expressed in vivo in intestinal stem cells. Nat Genet 40:915–920
Sato T, Vries RG, Snippert HJ, van de Wetering M, Barker N, Stange DE, van Es JH, Abo A, Kujala P, Peters PJ, Clevers H (2009) Single Lgr5 stem cells build crypt-villus structures in vitro without a mesenchymal niche. Nature 459:262–265
Sato T, van Es JH, Snippert HJ, Stange DE, Vries RG, van den Born M, Barker N, Shroyer NF, van de Wetering M, Clevers H (2010) Paneth cells constitute the niche for Lgr5 stem cells in intestinal crypts. Nature 469:415–418
Shibata H, Toyama K, Shioya H, Ito M, Hirota M, Hasegawa S, Matsumoto H, Takano H, Akiyama T, Toyoshima K, Kanamaru R, Kanegae Y, Saito I, Nakamura Y, Shiba K, Noda T (1997) Rapid colorectal adenoma formation initiated by conditional targeting of the Apc gene. Science 278:120–123
Shih IM, Wang TL, Traverso G, Romans K, Hamilton SR, Ben-Sasson S, Kinzler KW, Vogelstein B (2001) Top-down morphogenesis of colorectal tumors. Proc Natl Acad Sci U S A 98:2640–2645
Smits R, Kielman MF, Breukel C, Zurcher C, Neufeld K, Jagmohan-Changur S, Hofland N, van Dijk J, White R, Edelmann W, Kucherlapati R, Khan PM, Fodde R (1999) Apc1638T: a mouse model delineating critical domains of the adenomatous polyposis coli protein involved in tumorigenesis and development. Genes Dev 13:1309–1321
Snippert HJ, van der Flier LG, Sato T, van Es JH, van den Born M, Kroon-Veenboer C, Barker N, Klein AM, van Rheenen J, Simons BD, Clevers H (2010) Intestinal crypt homeostasis results from neutral competition between symmetrically dividing Lgr5 stem cells. Cell 143:134–144
Spaderna S, Schmalhofer O, Hlubek F, Berx G, Eger A, Merkel S, Jung A, Kirchner T, Brabletz T (2006) A transient, EMT-linked loss of basement membranes indicates metastasis and poor survival in colorectal cancer. Gastroenterology 131:830–840
Steigerwald K, Behbehani GK, Combs KA, Barton MC, Groden J (2005) The APC tumor suppressor promotes transcription-independent apoptosis in vitro. Mol Cancer Res 3:78–89
Su LK, Kinzler KW, Vogelstein B, Preisinger AC, Moser A, Luongo C, Gould KA, Dove WF (1992) Multiple intestinal neoplasia caused by mutations in the murine homolog of the APC gene. Science 256:668–670
Takeda N, Jain R, LeBoeuf MR, Wang Q, Lu MM, Epstein JA (2011) Interconversion between intestinal stem cell populations in distinct niches. Science 334:1420–1424
Tian G, Finley D (2012) Cell biology: destruction deconstructed. Nature 482:170–171
Totafurno J, Bjerknes M, Cheng H (1987) The crypt cycle. Crypt and villus production in the adult intestinal epithelium. Biophys J 52:279–294
van Es JH, Jay P, Gregorieff A, van Gijn ME, Jonkheer S, Hatzis P, Thiele A, van den Born M, Begthel H, Brabletz T, Taketo MM, Clevers H (2005) Wnt signalling induces maturation of Paneth cells in intestinal crypts. Nat Cell Biol 7:381–386
Vermeulen L, De Sousa EMF, van der Heijden M, Cameron K, de Jong JH, Borovski T, Tuynman JB, Todaro M, Merz C, Rodermond H, Sprick MR, Kemper K, Richel DJ, Stassi G, Medema JP (2010) Wnt activity defines colon cancer stem cells and is regulated by the microenvironment. Nat Cell Biol 12:468–476
Wellner U, Schubert J, Burk UC, Schmalhofer O, Zhu F, Sonntag A, Waldvogel B, Vannier C, Darling D, zur Hausen A, Brunton VG, Morton J, Sansom O, Schuler J, Stemmler MP, Herzberger C, Hopt U, Keck T, Brabletz S, Brabletz T (2009) The EMT-activator ZEB1 promotes tumorigenicity by repressing stemness-inhibiting microRNAs. Nat Cell Biol 11:1487–1495
Wright NA, Alison M (1984) The biology of epithelial cell populations. Clarendon, Oxford
Wright NA, Irwin M (1982) The kinetics of villus cell populations in the mouse small intestine. I. Normal villi: the steady state requirement. Cell Tissue Kinet 15:595–609
Zhu L, Gibson P, Currle DS, Tong Y, Richardson RJ, Bayazitov IT, Poppleton H, Zakharenko S, Ellison DW, Gilbertson RJ (2009) Prominin 1 marks intestinal stem cells that are susceptible to neoplastic transformation. Nature 457:603–607

Chapter 5
Mutational Activation of *KRAS* and *BRAF* in Colorectal Cancer

Katherine H. Pedone, Jennifer L. Sells, and Channing J. Der

Abstract The failure of farnesyltransferase inhibitors to show antitumor activity against *KRAS*-mutant malignancies diminished enthusiasm for efforts to develop anti-Ras inhibitors for cancer treatment. However, two recent developments have rekindled interest in these endeavors. First, genome-wide exome sequencing verified that mutational activation of the *KRAS* gene is the most prevalent oncogene mutation in colorectal cancer (CRC). Second, a major step toward the application of personalized medicine for CRC was taken when mutant *KRAS* was established as a prognostic marker for resistance to epidermal growth factor receptor monoclonal antibody therapy. Thus, there is renewed and considerable interest in understanding the role of *KRAS* mutation in CRC progression and growth and in developing pharmacologic approaches for blocking aberrant K-Ras protein function for CRC treatment. Since the K-Ras protein itself is considered "undruggable," current strategies to develop anti-K-Ras inhibitors have focused on antagonists of K-Ras downstream effector signaling. The frequent mutational activation of *BRAF*, which is mutually with *KRAS* activation, suggests that the encoded B-Raf serine/threonine kinase and activation of the ERK mitogen-activated protein kinase cascade is a key driver of mutant K-Ras-dependent CRC growth. In this review, we summarize the importance of mutant K-Ras and B-Raf in CRC growth and current efforts in targeting the Raf-MEK-ERK cascade for CRC treatment.

Keywords Akt • GTPase • Epidermal growth factor receptor • ERK • MEK • Mitogen-activated protein kinase • Phosphatidylinositol 3-kinase • Raf

---

K.H. Pedone (✉) • J.L. Sells • C.J. Der
University of North Carolina at Chapel Hill, Lineberger Comprehensive Cancer Center, Chapel Hill, NC 27599, USA
e-mail: katherine_pedone@med.unc.edu; jsells@med.unc.edu; cjder@med.unc.edu

K.M. Haigis (ed.), *Molecular Pathogenesis of Colorectal Cancer*,
DOI 10.1007/978-1-4614-8412-7_5,

5.1 Introduction

While surgery remains the most effective treatment for colorectal cancer (CRC), treatment for metastatic CRC (mCRC; 40–50 % of newly diagnosed patients) remains highly ineffective. The prognosis for patients with mCRC remains poor, with a median overall survival of 18–21 months and a 5-year survival rate of only 8 % (Markowitz and Bertagnolli 2009; Wilson et al. 2010; Fearon 2011a). Conventional chemotherapy with cytotoxic drugs remains the foundation for first-line treatment of mCRC, with limited advances in the development of signal transduction-targeted therapies for this cancer. Recent genome-wide sequencing studies verified that *KRAS* is the most frequently mutated oncogene in CRC (Sjoblom et al. 2006; Wood et al. 2007). Additionally, *KRAS* mutation status is now utilized to identify patients who will not be responsive to anti-epidermal growth factor receptor (EGFR) therapy (Allegra et al. 2009). Thus, understanding the role of aberrant K-Ras signaling in CRC growth and the development of anti-K-Ras inhibitors provide promise for more effective therapies for CRC. In this chapter, we summarize the biochemistry and cell biology of K-Ras and the status of efforts to develop anti-K-Ras inhibitors, in particular, pharmacologic antagonists of the Raf-MEK-ERK mitogen-activated protein kinase (MAPK) cascade.

5.2 The Genetic Basis of CRC

CRC is the second leading cause of cancer deaths in the USA, with an estimated 141,210 new cases and 49,380 deaths for both sexes in 2011 (Siegel et al. 2011). Worldwide, CRC is the third most commonly diagnosed cancer in males and the second in females, with greater than 1.2 million new cancer cases and 608,700 deaths estimated in 2008 (Jemal et al. 2011). The *RAS* oncogenes (*HRAS*, *KRAS*, and *NRAS*) comprise the first genes found mutated in human cancer in 1982 (Cox and Der 2010). CRC was among the earliest examples of *RAS* mutational activation in patient-derived tumors and was the first tumor model in which high frequency mutagenic Ras activation was detected (Shimizu et al. 1983; Bos et al. 1987; Forrester et al. 1987). Together with the subsequent identification of *APC*, *TP53*, and *SMAD4* tumor suppressor genes loss in CRC, in 1990, a genetic model for CRC tumor progression was defined (Fearon and Vogelstein 1990), with subsequent findings providing further refinement of this model (Fig. 5.1a) (Fearon 2011b).

With advances in DNA sequencing technology and the arrival of cancer genome sequencing, two key components of Ras effector signaling were linked to CRC (Fig. 5.1b). One of the most important discoveries was the identification of *BRAF* activating mutations in human cancers in 2002 (Davies et al. 2002), with CRC having the third highest frequency (11 %) after melanoma (60 %) and papillary thyroid cancer (45 %) (Davies et al. 2002; Rajagopalan et al. 2002). This was followed in 2004 by genomic sequencing focused on phosphatidylinositol 3-kinases (PI3Ks) and the discovery of activating mutations in the gene encoding the p110 alpha (p110α) catalytic subunit of class 1A PI3Ks, with high frequencies seen in CRC (11 %) (Samuels et al. 2004).

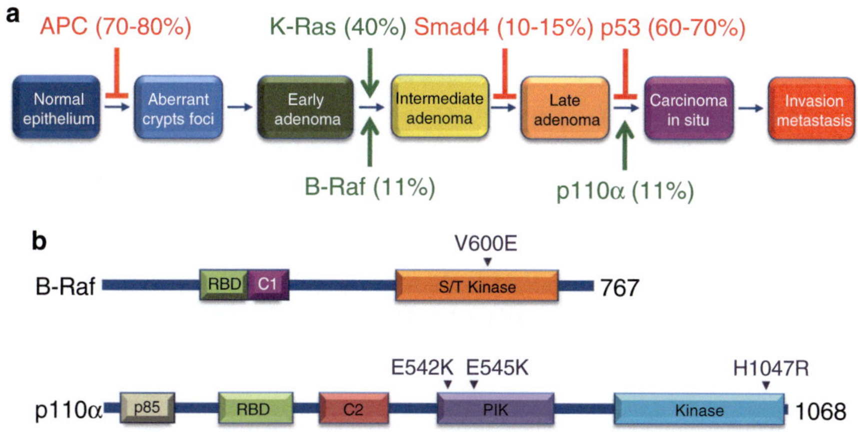

Fig. 5.1 Genetic basis of CRC progression. (**a**) CRCs arise through well-defined clinical stages associated with characteristic mutations. Colon cancer begins as a benign adenomatous polyp, which develops into an advanced adenoma with high-grade dysplasia and then progresses to an invasive cancer that is confined within the wall of the colon (stages I and II) and are curable by surgery, spread to regional lymph nodes (stage III; most cases curable by surgery and adjuvant chemotherapy) and then metastasize to distant sites that include the liver (stage IV; usually incurable). This progression is associated with mutational inactivation of tumor suppressor genes or mutational activation of oncogenes. Loss of APC or activation of **β**-catenin and the Wnt signaling pathway is the key initiating event. Mutational activation of K-Ras or its effector B-Raf leads to activation of the ERK MAPK pathway. Mutational activation of another K-Ras effector, the p110**α** catalytic subunit of PI3K, can occur together K-Ras activation. Smad4 loss leads to inactivation of the TGF**β** tumor suppressor pathway. Inactivation of p53 and loss of cell cycle checkpoint control is associated with progression to invasive carcinomas. The frequency of mutations was taken from a recent review (Fearon 2011b) or from COSMIC (*BRAF* and *PIK3CA*). (**b**) Mutational activation of Ras effectors in CRC. The V600E mutation is the predominant mutation found in CRC (98 %; COSMIC) and is located in the B-Raf kinase domain. The three predominant mutations (73 % of total) in p110**α** in CRC are located either in the helical (E542K and E545K) or kinase (H1047R) domain. Domain structure was determined in SMART (http://smart.embl-heidelberg.de/)

Subsequent exome sequencing of 11 CRCs (18,191 genes) identified >100 novel mutated genes, where most occurred in only a low percentage (<5 %) of CRC (Sjoblom et al. 2006; Wood et al. 2007). Thus, the genetic heterogeneity of CRC is considerably more complex than what the more uniform tumor histology would suggest. These studies also revealed that there were only a handful of genes mutated in a large proportion (>10 %) of CRCs. These "mountains" were genes already known to be mutated in CRC, with *KRAS* verified as the most frequently mutated oncogene. Taken together, these discoveries established the importance of the proteins encoded by these genes, the K-Ras small GTPase and its two key effector targets, the B-Raf serine/threonine kinase and the PI3K lipid kinase, as important targets for the development of molecularly targeted therapies for this prevalent and aggressive disease.

CRC is divided into two distinct molecular categories, those characterized with microsatellite instability (MSI; ~15 %) due to a deficient mismatch-repair (MMR) system and are hypermutated, but have a relatively normal cytogenetic karyotype, and those that are microsatellite stable (MSS), non-hypermutated but are chromosomally unstable. MSI CRCs have a better prognosis than MSS CRC. Interestingly, a recent exome sequencing of 224 CRC tumors and normal pairs found high frequency mutation of *KRAS* (43 %), *NRAS* (9 %), and *PIK3CA* (18 %) in non-hypermutated CRC, while *BRAF* (46 %) mutations were high in hypermutated CRC (Cancer Genome Atlas Network 2012). Thus, *RAS* and *BRAF* activation may not be functionally equivalent genetic events and are associated with the development of distinct subsets of CRCs.

5.3 Ras Proteins: A Family of GDP-GTP-Regulated Binary On-Off Switches

The Ras (Rat sarcoma) small GTPases were identified initially as the gene products of the transforming elements of oncogenic Harvey and Kirsten retroviruses that caused rapid sarcoma formation in rats (Cox and Der 2010). The viral H-*ras* and K-*ras* genes represented cellular *HRAS* and *KRAS* genes transduced from the host rat genome and the linkage to human cancer came with their detection as mutated and transforming genes in gene transfer assays. The third *RAS* gene, *NRAS*, was identified initially as a transforming gene in genomic DNA isolated from a human neuroblastoma cell line, but subsequently found mutated in a spectrum of different human cancers.

Ras proteins are well recognized as master regulators and transducers of diverse intracellular signals for mitogenic growth, differentiation, and apoptosis. Ras proteins are key signaling nodes, activated downstream of receptor tyrosine kinases that are critical drivers of cancer growth. Activated Ras proteins then regulate the activities of cytoplasmic signaling networks that control cell proliferation and survival. This central role translates into a heavy cellular dependence on Ras activity, one that is sufficient to drive oncogenesis when deregulated.

5.3.1 RAS *Gene Transcription*

RAS genes are expressed ubiquitously. *KRAS* is alternatively spliced into two transcripts due to alternative fourth exon utilization (*KRAS4A* and *KRAS4B*) (Fig. 5.2a). *KRAS4A* and *KRAS4B* encode proteins that differ at residues 151–153 and 165–188/189. *KRAS4B* is the predominant or exclusive transcript expressed in most normal tissue, although a 1:1 ratio was seen in normal mouse colonic tissue (Patek et al. 2008; Luo et al. 2010). Whereas a *K-Ras* deficiency causes mouse embryonic lethality (Johnson et al. 1997; Koera et al. 1997), a deficiency in *K-Ras4A* alone is dispensable for normal development (Plowman et al. 2003). Both *N-Ras* and *H-Ras* are also dispensable for normal mouse development and growth (Esteban et al. 2001). That *H-Ras* expressed from the *K-Ras* locus can restore normal development, but exhibits cardiac defects, suggests overlapping and distinct developmental functions of H-Ras and K-Ras (Potenza et al. 2005).

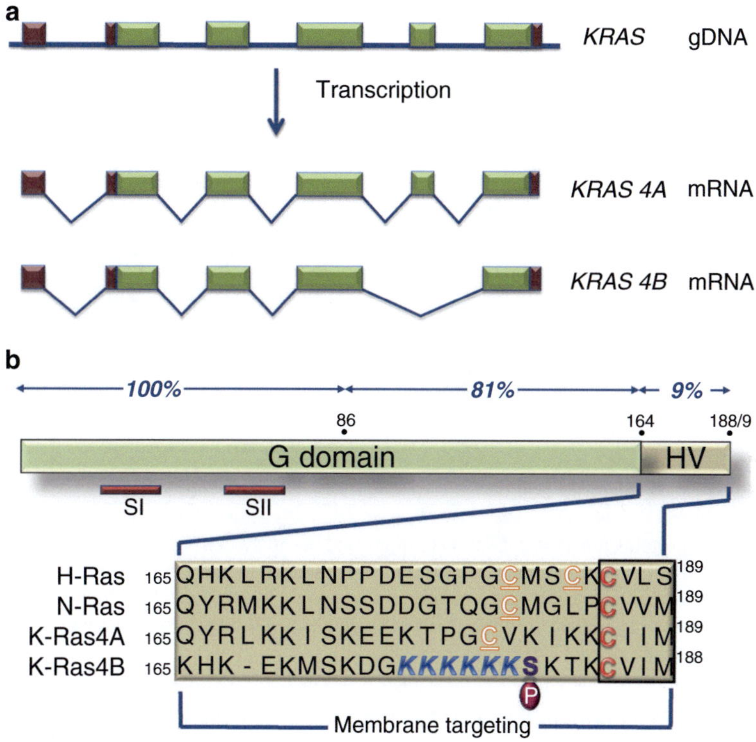

Fig. 5.2 Ras isoforms. (**a**) *KRAS* encodes two related isoforms through alternative RNA splicing. *KRAS* mRNA is alternatively spliced into two transcripts due to alternative fourth codon utilization, 4A and 4B, that encode 39 and 38 amino acids, respectively. The *KRAS4A* and *KRAS4B* mRNA transcripts encode highly identical proteins that differ at C-terminal residues 151–153 and 165–188/189. (**b**) Ras proteins share 100 % (residues 1–86) or 81 % (residues 87–164) sequence identity in the N-terminal G domain that contains the GTP binding and hydrolysis activity. Within the G domain are the switch I and II domains that change in conformation during GDP-GTP cycling. Ras isoforms diverge primarily in C-terminal hypervariable (HV) sequences involved in membrane association and subcellular localization (9 % identity). All terminate with the CAAX prenylation motif that signals for posttranslation modifications, including covalent addition of a farnesyl isoprenoid lipid to the cysteine residue of the CAAX motif. Palmitoylated cysteines (H-Ras, N-Ras and K-Ras4A) or polybasic sequences (K-Ras4B) immediately upstream of the CAAX motif act as second signals that are essential for full plasma membrane association. K-Ras4B is phosphorylated at S181 by protein kinase C

5.3.2 *Ras Protein Structure and GDP-GTP Regulation*

The three human *RAS* genes encode four highly related 188–189 amino acid Ras proteins (82–90 % sequence identity): H-Ras, N-Ras, K-Ras4A, and K-Ras4B (Fig. 5.2b). Among the Ras isoforms, the primary sequence of the G domain (residues 1–164) is remarkably similar, with 100 % identity within the first 89 amino

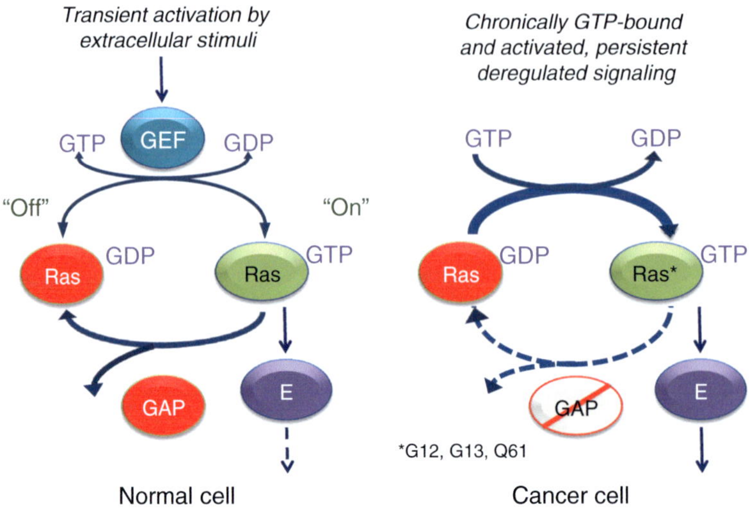

Fig. 5.3 Ras proteins function as GDP-GTP-regulated on-off switches. The Ras-GTP bound state is active and the Ras-GDP-bound form is inactive, due to differential affinity for effector binding. Normal Ras proteins are activated transiently and reversibly by extracellular stimuli. Mutated Ras proteins are impaired in their intrinsic and GAP-stimulated GTP hydrolysis activities, leading to accumulation of the active GTP-bound form

acids and 80 % identity in the next 78 amino acids. The three-dimensional structure of the G domain contains the GDP/GTP-binding pocket and GTP catalytic activity. Across the protein, two loop-like structures in the G domain, switch I (residues 32–38) and switch 2 (residues 59–67), are affected most dramatically according to the presence of either the di- or triphosphate nucleotide in the binding pocket (Vetter and Wittinghofer 2001). The switch loop structures therefore reflect the nucleotide-bound state of the enzyme and are involved in the interaction of small GTPases with effectors and regulatory molecules (Cox and Der 2010).

The elemental mechanism of Ras activation is a highly regulated GDP-GTP cycle controlled by guanine nucleotide exchange and GTP hydrolysis (Fig. 5.3) (Vetter and Wittinghofer 2001). The monomeric GTPase is an active signaling enzyme in the GTP-bound state. Intrinsic hydrolysis activity specified by the G domain works to return Ras to its basal, GDP-bound conformation where it rests until specific stimuli induce GTP binding, and the cycle begins again. The GTP-bound state is the active state due to its preferential binding affinity for effectors. In this way, the timeframe when active Ras can interact with effector molecules is strictly limited to the duration of the transient GTP-bound conformation. The intrinsic rates of GDP dissociation and GTP hydrolysis of Ras are very slow processes that do not allow it to operate effectively in temporally regulated signaling activities. Instead, there are two different classes of Ras-selective regulatory proteins that increase the rates of these steps: guanine nucleotide exchange factors (RasGEFs; e.g., Sos1) promote nucleotide exchange and GTPase-activating proteins (RasGAPs; e.g., neurofibromin) stimulate GTP hydrolysis (Figs. 5.3 and 5.4) (Vigil et al. 2010).

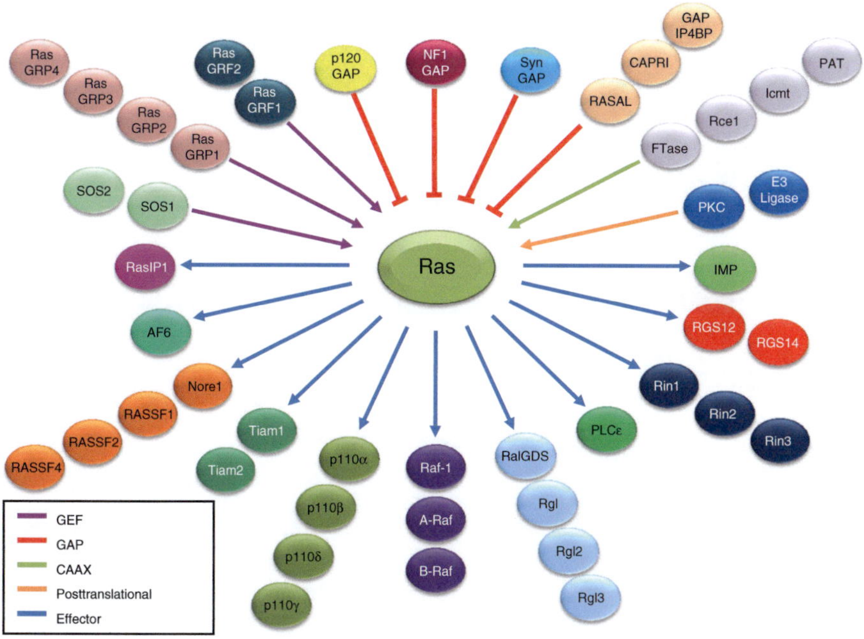

Fig. 5.4 Ras interacting proteins. Regulators (GAPs and GEFs), posttranslational modifiers and downstream effectors of Ras. Most have been characterized for interaction with H-Ras and only speculated to interact with K-Ras

In normal quiescent cells, wild-type Ras exists predominantly in its inactive GDP-bound state. Upon growth factor stimulation, Ras is activated transiently by GEF-mediated nucleotide exchange and then quickly cycles back to the inactive GDP-bound state through GAP activity. Each class of proteins interacts directly with either GDP-bound Ras (GEFs) or GTP-bound Ras (GAPs) to stabilize the structure for GTP loading or hydrolysis, respectively. As a result, the cycling kinetics of Ras are enhanced by several orders of magnitude. In this way, regulators of Ras manage the efficiency and dynamics of Ras-mediated signal transduction. Issues with mitogenic imbalance arise then when Ras proteins develop internal mechanisms to evade this regulatory control paradigm.

5.3.3 Ras Controls Intracellular Signaling from Cellular Membranes

In addition to GTP-binding, a second essential requirement for Ras protein function is association with the cytosolic side of the plasma and intracellular membrane structures. Ras proteins are synthesized initially as cytosolic and inactive proteins. All Ras isoforms then undergo a sequence of three posttranslational processing

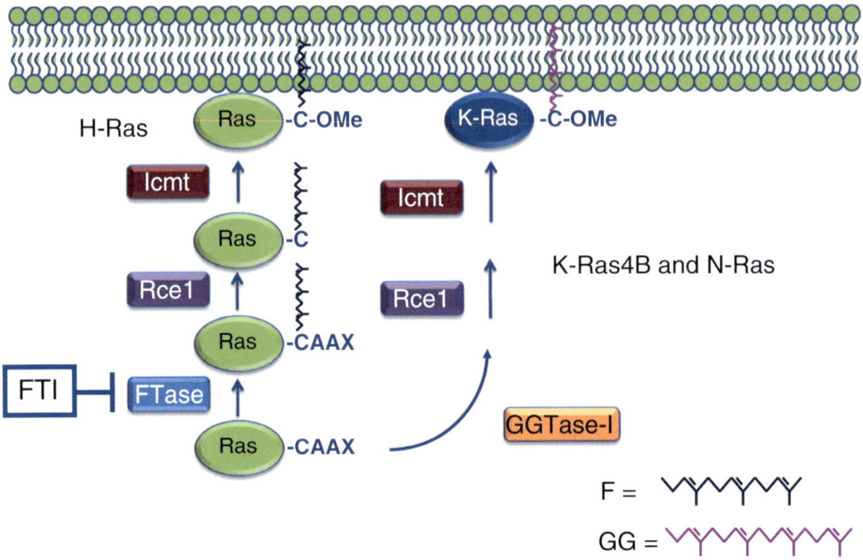

Fig. 5.5 Ras posttranslational processing and membrane association. All Ras proteins terminate with C-terminal CAAX motifs that signal for a series of three posttranslational modifications catalyzed by the cytosolic FTase and the Golgi-associated Rce1 and Icmt enzymes. Although all Ras proteins are normally modified with a farnesyl isoprenoid, K-Ras4B and N-Ras undergo alternative prenylation and modification with addition of a geranylgeranyl isoprenoid that is catalyzed by geranylgeranyltransferase-I when FTase activity is inhibited by FTase inhibitors (FTI)

events at the extreme C-terminus, a region called the CAAX (C=cysteine, A=aliphatic amino acid, X=terminal residue) box (Figs. 5.2b and 5.5) (Berndt et al. 2011). First, unmodified Ras interacts with the cytosolic enzyme farnesyltransferase (FTase) to incorporate the 15-carbon farnesyl isoprenoid lipid at the cysteine residue of the invariant CAAX box. The farnesyl group is attached via a stable thioether linkage and helps mediate Ras interactions with cellular membranes. Prenylated Ras proteins then are subject to proteolysis of the three terminal residues of the CAAX box by Ras converting enzyme 1 (Rce1). Finally, the protein becomes carboxylmethylated by isoprenylcysteine-carboxyl-methyltransferase (Icmt) at the now terminal farnesylated cysteine. Both Rce1 and Icmt reside at the endoplasmic reticulum (ER) outer membrane. Upon maturation at the ER, Ras proteins then transit to the plasma membrane.

Although the CAAX-signaled modifications are necessary, they not sufficient to facilitate full Ras plasma membrane association. Additional membrane targeting elements reside in the 24/25 amino acid hypervariable (HV) sequences immediately upstream of the CAAX box. The Ras proteins exhibit only 8 % sequence identity in the HV region. This strong HV sequence divergence dictates isoform-distinct localization to other membrane compartments as well as posttranslational mechanisms that dynamically regulate Ras subcellular localization.

Following prenylation, H-Ras, N-Ras, and K-Ras4A are trafficked to the Golgi apparatus where they undergo posttranslational modification by a 16-carbon palmitate fatty acid at one or two cysteine residues upstream of the farnesylated cysteine (Figs. 5.2b and 5.5). In contrast to the irreversible and permanent farnesylation modification, palmitoylation is a reversible modification that helps the proteins move to and away from membranes depending on their lipidation state. K-Ras4B, on the other hand, has a polybasic sequence upstream and adjacent to the site of prenylation. This region of concentrated positive charge helps mediate interactions with negatively charged membrane lipids.

Although it was originally believed that Ras proteins signal exclusively from the plasma membrane, it is now understood that different Ras isoforms have overlapping but distinct patterns of plasma and endo-membrane attachment (Ahearn et al. 2011). While K-Ras4B operates mainly at the plasma membrane, H-Ras, N-Ras, and K-Ras4A also function at the Golgi, ER, and endosomes. Since all Ras isoforms undergo identical prenylation processing, their separate localization profiles ultimately depend on protein–protein interactions, secondary lipidation events, and other posttranslational modifications including ubiquitination and phosphorylation (de la Vega et al. 2011; Fehrenbacher et al. 2009). For example, K-Ras4B is phosphorylated by protein kinase C at S181 within the HV sequence (Fig. 5.2b) (Bivona et al. 2006). This modification promotes rapid dissociation of K-Ras4B from the plasma and association with intracellular membranes, including the outer membrane of mitochondria where phosphorylation at S181 promotes K-Ras4B interaction with Bcl-XL, leading to apoptosis.

5.3.4 The Ras Superfamily

Ras proteins are the founding members of a superfamily of 20–25 kDa small GTPases (Wennerberg et al. 2005). Based on sequence identity and function, the superfamily is divided into five subfamilies: Ras, Rho, Rab, Arf, and Ran. All Ras superfamily proteins function as GDP-GTP-regulated molecular on-off switches. However, they diverge in the GEFs and GAPs that regulate their GDP-GTP cycles and their effectors. Most undergo posttranslational modification by lipids that are essential for their subcellular localization and membrane interactions. In addition to their biochemical connections with Ras, some are also linked with Ras via their involvement with shared signaling networks. In particular, as described below, some are activated by Ras effector signaling networks (e.g., Ral, Rheb, and Rac).

5.3.5 Ras Effector Signaling

Active Ras-GTP binds preferentially to at least 11 functionally distinct classes of effectors (Fig. 5.4). Of these, cell culture and mouse model studies support the requirement of at least five for mutant *RAS*-dependent oncogenesis. For example, mice deficient in the RalGDS (Gonzalez-Garcia et al. 2005), Tiam1 (Malliri et al. 2002), or phospholipase C epsilon (Ikuta et al. 2008) exhibit normal development, but show impaired tumorigenesis when subjected to the 7,12-dimethylbenzanthracene (DMBA)/12-O-tetradecanoylphorbol-13-acetate (TPA) carcinogenic treatment, which induces *HRAS*-mutant squamous cell carcinoma in the skin. Similarly, mice harboring a variant of p110α with point mutations that affect Ras binding, but not catalytic activity, were impaired in DMBA/TPA treatment mutant *HRAS*-induced skin and *KRAS*-induced lung tumorigenesis (Gupta et al. 2007). As described below, the mutational activation of B-Raf and p110α in cancer provides strong validation of the importance of the Raf and PI3K effectors in Ras-dependent cancer growth.

The best studied Ras effectors are the Raf serine/threonine protein kinases (Raf-1, A-Raf, and B-Raf). Raf proteins are recruited to the plasma membrane by GAPs where, after a series of activating events including multisite phosphorylation, they phosphorylate and activate the MEK1 and MEK2 dual specificity protein kinases (Fig. 5.6). Activated MEK1/2 then phosphorylate and activate the ERK1 and ERK2 MAPKs. Activated ERK1/2 then translocate to the nucleus where they phosphorylate a large spectrum of proteins, including Ets family transcription factors. The transcriptional output of ERK1/2 activity contributes to changes in cell cycle progression, cell morphology, and differentiation that are important for normal cellular homeostasis, and when deregulated, contribute to neoplastic growth.

Since MEK1/2 are the only known substrates of Raf and ERK1/2 are the only known substrates of MEK1/2, the Raf-MEK-ERK cascade is often depicted as a simple linear signaling pathway. However, this three-component module is considerably more complex and is the core backbone of a complex signaling network with considerable cross-talk with other MAPK signaling networks. There is additional regulation at the level of each protein kinase by both positive and negative regulatory components. For example, there are scaffold proteins (e.g., KSR) that bind to one or more components of this pathway to influence spatially distinct activity and influence the precise components of different MAPK complexes (Brown and Sacks 2009; Udell et al. 2011). The Raf isoforms can form homo- and hetero-dimers, their kinase activities are regulated by both positive and negative phosphorylation events through the activities of protein kinases (Src, protein kinase A) and phosphatases (PP2A, PP1) and through protein interactions (e.g., 14-3-3, RKIP) (Matallanas et al. 2011). MEK1/2 are phosphorylated and activated by other protein kinases (Cot/Tpl2 and Mos). ERK-selective dual specificity protein phosphatases (e.g., DUSP6/MKP3) can dephosphorylate and inactivate ERK1/2 (Bermudez et al. 2010). There are numerous feedback mechanisms initiated by ERK activation that lead to inactivation of upstream activators of the pathway. The recent screen for ERK MAPK interacting proteins provides further evidence for the complex interactions of the ERK MAPK cascade (Bandyopadhyay et al. 2010).

Although there has been some evidence for non-MEK substrates for Raf, current evidence indicates that MEK1/2 are the primary effectors in Raf-dependent transformation and tumorigenesis (Vakiani and Solit 2011). There is also limited evidence that Raf may have non-kinase functions. One recent study with isogenic human colon cancer lines, however, showed that *KRAS*-mutant cells were

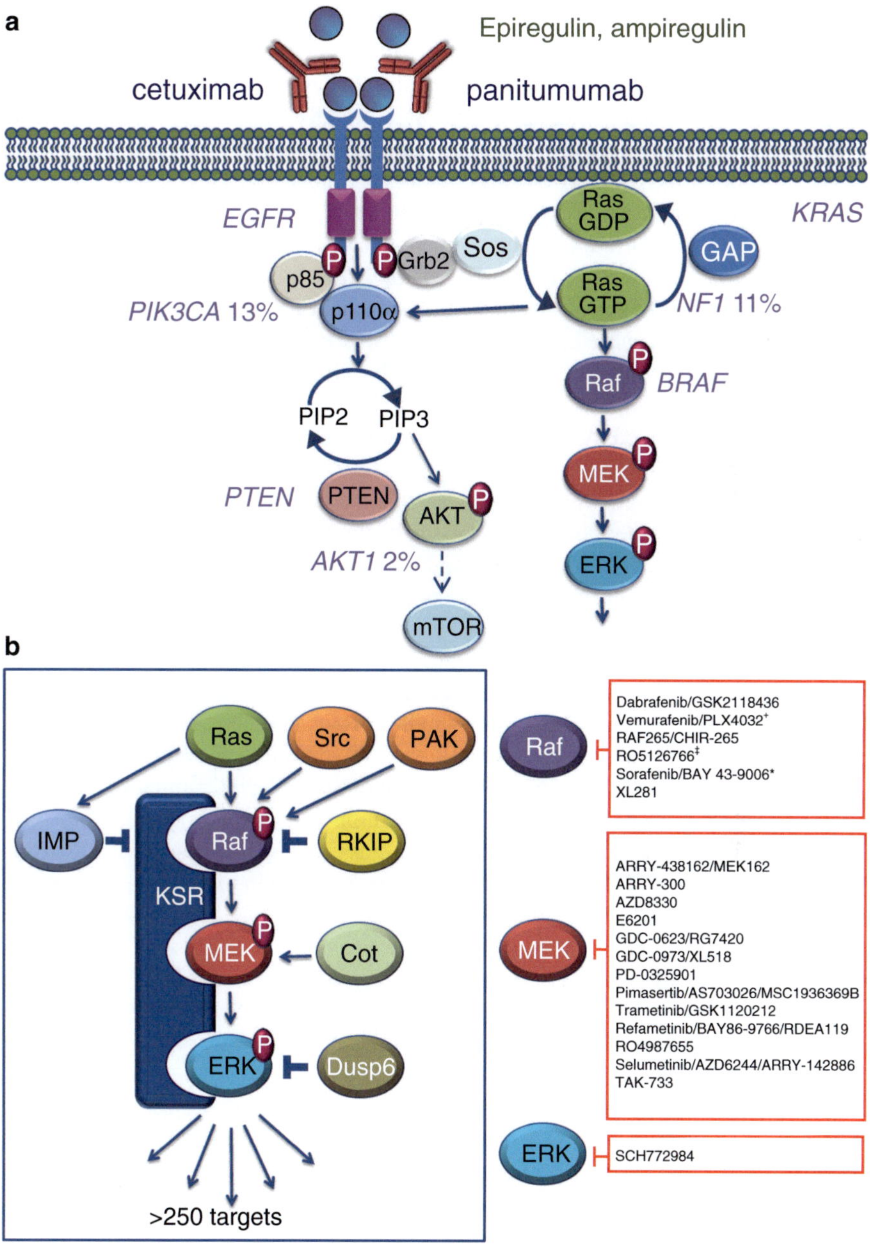

Fig. 5.6 Inhibitors of EGFR and Raf signaling. (**a**) Anti-EGFR therapies for CRC. Two monoclonal antibodies that target the extracellular domain of the EGFR are used alone or together with cytotoxic drugs for CRC treatment. Response may be correlated not with EGFR expression alone, but with increased mRNA expression of genes encoding EGFR ligands. CRCs with *KRAS* mutations strongly correlate with nonresponse and used to eliminate CRC patients from anti-EGFR treatment. Mutations in *NRAS*, *BRAF*, *PIK3CA*, and *PTEN* may also define nonresponse. (**b**) The Raf-MEK-ERK signaling network. (**c**) Inhibitors of the Raf-MEK-ERK signaling cascade currently under clinical evaluation. Compiled from clinicaltrials.gov

hypersensitive to small molecule inhibition of Raf, but were insensitive to MEK inhibition, despite complete abrogation of ERK1/2 activation (Haigis et al. 2008). The notion that K-Ras requires Raf in cancer, but not downstream MEK1/2 and ERK1/2, is novel, and whether or not Raf signals in a noncanonical, MEK-independent manner in *KRAS*-mutant CRC will require further investigation.

The second best characterized Ras effectors are the p110 catalytic subunits of the class IA (α, β, and δ) and IB (γ) PI3K lipid kinases. p110α/β/δ form heterodimeric complexes with one of five regulatory subunits (p85α, p55α, p50α, p85β, or p55γ), whereas p110γ complexes with a p101 or p84 regulatory subunit. The normally cytosolic class I PI3Ks are activated by recruitment to the plasma membrane by activated receptor tyrosine kinases or by activated Ras. Induction of PI3K leads to conversion of the membrane phospholipid phosphatidylinositol 4,5-bisphosphate (PIP_2) to the second messenger phosphatidylinositol 3,4,5-trisphosphate (PIP_3). This reaction is reversed by the PTEN tumor suppressor, which acts as a lipid phosphatase that converts PIP_3 back to PIP_2.

Formation of the membrane-associated PIP_3 can regulate a multitude of proteins, of which the Akt serine/threonine kinase is of the most important for cancer. PIP_3 binds to the pleckstrin homology domains of Akt and PDK1, with PDK1 phosphorylating and activating Akt. Akt phosphorylates and regulates the activities of a spectrum of substrates that regulate cell survival and growth. These include inhibition of the proapoptotic Bcl-2 family members Bad and Bax, and the Tsc2 tumor suppressor. Tsc2 forms a heterodimeric complex with Tsc1 and the Tsc1/2 complex acts as a GAP for the Rheb Ras family small GTPase. Akt phosphorylation of Tsc2 inactivates the GAP activity, leading to Rheb-GTP formation and Rheb activation of mTORC1 complex. mTORC1 phosphorylation of the eukaryotic initiation factor 4E and the ribosomal S6 protein kinase, leading to increased protein synthesis.

The third-best characterized Ras effector class includes GEFs for the Ras-like RalA and RalB small GTPases, members of the Ras branch of the Ras superfamily (Wennerberg et al. 2005). Although RalA and RalB share significant sequence (82 %) and biochemical identity, they commonly exhibit distinct, sometimes antagonistic functions in cancer (Bodemann and White 2008; Neel et al. 2011). For example, whereas RalA promotes, RalB antagonizes CRC anchorage-independent growth. Key effectors of Ral include the Sec5 and Exo84 subunits of the exocyst complex and RalBP1/RLIP76.

Rac1, a Rho family small GTPase, is the fourth-best characterized Ras effector. Rho family GTPases comprise a major branch of the Ras superfamily and are best known for their regulation of actin cytoskeletal organization. In particular, Rac activation promotes actin polymerization and the leading edge of migrating cells, causing membrane ruffling and promoting cell movement. Rac can be activated by Ras activation through several possible mechanisms, including direct binding to the effector Tiam1 or through PI3K activation and PIP_3 activation of Rac-selective GEFs. The importance of Rac1 in RAS-induced oncogenesis is demonstrated by the impairment of *KRAS*-induced lung and pancreatic tumorigenesis in mouse models of cancer (Kissil et al. 2007; Heid et al. 2011). Rac effector functions important in cancer growth may include the PAK1 serine/threonine kinase (Ong et al. 2011).

5.4 Aberrant K-Ras Function in CRC

Deregulated K-Ras function in CRC occurs primarily through direct mutational activation. Indirect mechanisms of activation of K-Ras signaling can occur through persistent activation of the EGFR or by mutational activation of K-Ras downstream effectors.

5.4.1 Mutational Activation of K-Ras

Mutational activation of the three *RAS* genes occurs with high frequency across the spectrum of all human cancers (33 %; COSMIC). However, the majority of mutations are found in *KRAS* (87.2 %), followed by *NRAS* (9.8 %), with *HRAS* rarely mutated (3.0 %) (Fig. 5.7a). In CRC, data compiled in the COSMIC database indicate that *KRAS* is the most commonly mutated (99.8 %), whereas *NRAS* is rarely mutated (0.2 %), and *HRAS* mutations are never seen. However, data from several studies suggest that the frequency of *NRAS* mutations may be higher (2.6–2.9 %) (De Roock et al. 2010b; Vakiani et al. 2012). *KRAS* mutations are found in hyperplastic polyps (Chan et al. 2003) indicating its occurrence early in tumor progression. *KRAS* mutations were more prevalent in adenomas compared with primary carcinomas (68.6 % vs. 42.7 %, respectively), whereas similar mutation frequency has been estimated for CRC primary and metastatic tumors (Bos et al. 1987; Forrester et al. 1987; Cejas et al. 2009; Santini et al. 2008; Vakiani et al. 2012). As with CRC, *KRAS* is the predominant isoform mutated in pancreatic ductal adenocarcinoma (PDAC) and non-small cell lung cancer (NSCLC). However, the frequency of specific activation mutations is quite variable for each cancer. For CRC, 98.7 % of the lesions are point missense mutations in codons encoding G12 (78.4 %) or G13 (20.3 %), with very infrequent mutation of Q61 (0.8 %; Fig. 5.7b, c). In contrast, 98.9 % of *KRAS* mutations in PDAC are seen at G12, with only 0.7 % G13 mutation and 0.4 % Q61 mutations seen. NSCLC is intermediate in mutation frequencies, with 90.4 % G12 and 6.8 % G13 mutations seen. Finally, 0.2 % of CRC tumors harbor mutations at A146 and, to a lesser extent, K117, both highly conserved residues in a consensus guanine nucleotide binding motif (Edkins et al. 2006; Janakiraman et al. 2010). These atypical mutations were present in a mutually exclusive manner with codon 12, 13, and 61 mutations in patient samples, and these mutant proteins exhibited increased Ras-GTP steady-state levels when expressed in cells, although each was less potent than RasG12D in this assay (Janakiraman et al. 2010). A146 mutations are not seen in PDAC or NSCLC and rarely seen in other cancers. Whether these different patterns of mutation frequencies reflect the different mutagenic insults that each tissue is subjected to, or instead, reflect tissue-specific differences in the biological potencies of the different K-Ras-mutant proteins remains to be determined.

In addition to differences in mutation frequency *NRAS* mutations may arise later in tumor progression than mutant *KRAS* (Vogelstein et al. 1988; Irahara et al. 2010).

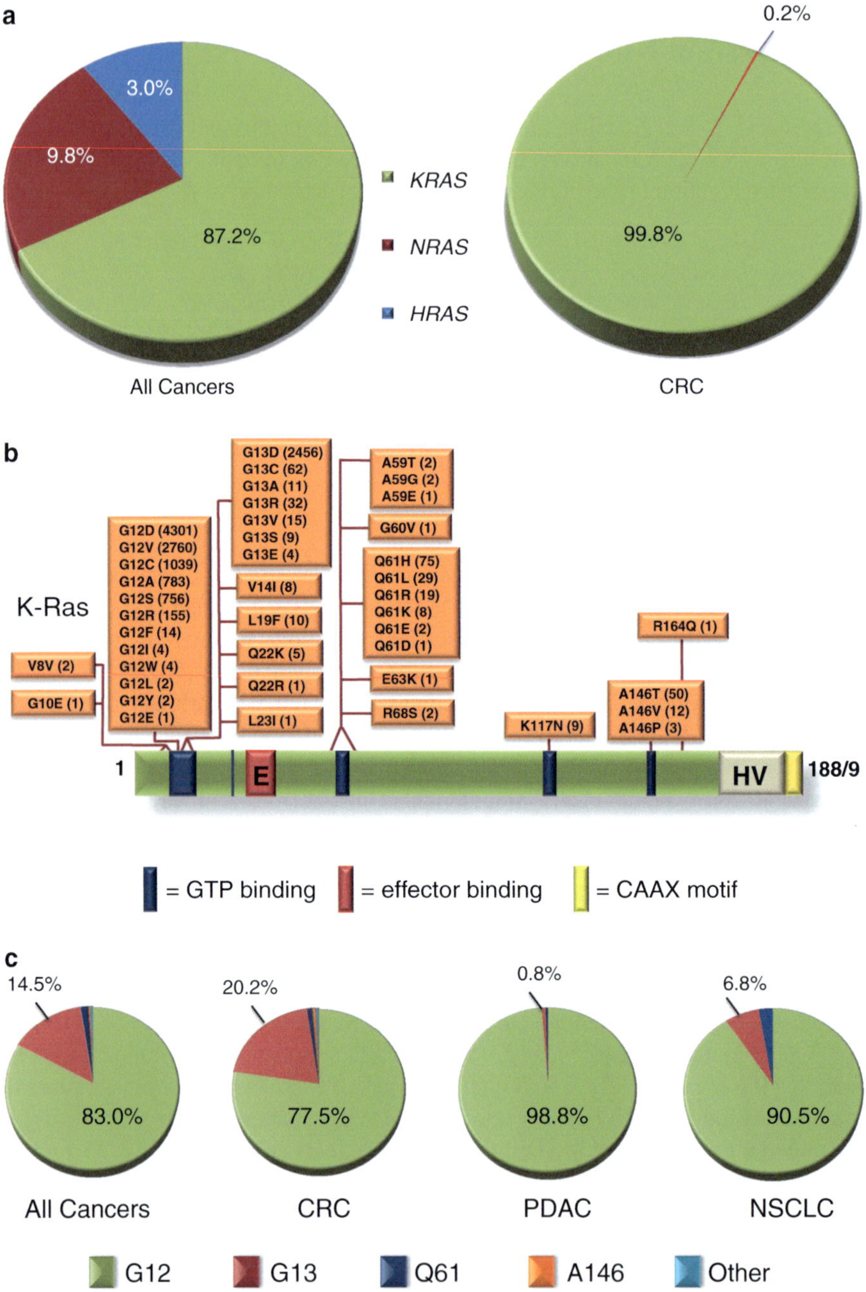

	Position (%)					
Cancer	**G12**	**G13**	**Q61**	**A146**	**Other**	**Total mutations**
All	83.0	14.5	1.5	0.4	0.54	24,289
CRC	77.5	20.2	1.0	0.7	0.7	2,728
PDAC	98.8	0.8	0.4	0	<0.1	2,968
NSCLC	90.5	6.8	2.6	0	0.1	887

NRAS and *KRAS* mutations are mutually exclusive. Due to the low frequency of *NRAS* mutation, no clinical or pathological characteristics have shown significant association with mutant *NRAS* alleles other than a trend for mutant *NRAS* to be observed in left-sided, MSS tumors found in women (Irahara et al. 2010). The genetic basis for mutation distribution among Ras isoforms is not understood, but functional studies in genetically engineered mice support the idea that K-Ras and N-Ras have different roles in CRC tumor biology. In particular, expression of mutationally activated K-Ras promoted progression of colonic adenocarcinoma in an *APC*-deficient mouse model, whereas N-Ras did not (Haigis et al. 2008).

Evidence for the emergence of *KRAS* mutation in the early stages of CRC tumorigenesis stems largely from sequencing efforts of representative patient specimens spanning CRC progression. Such analyses have shown that approximately half of all CRCs, as well as half of all larger precursor adenomas (>1 cm), harbor mutations in *KRAS* (Bos et al. 1987; Forrester et al. 1987; Vogelstein et al. 1988). By contrast, less than 10 % of adenomas smaller than one cm display similar mutations, suggesting that *KRAS* mutation, unlike loss of the APC tumor suppressor, is not an initiating event, but rather an early-stage oncogenic lesion that promotes colorectal tumor progression (Farr et al. 1988; Fearon and Vogelstein 1990; Vogelstein et al. 1988). A possible exception to this pattern is the role of K-Ras in aberrant crypt foci (ACF), naturally occurring histologic lesions which putatively give rise to a subset of CRCs (Pretlow and Pretlow 2005); *KRAS* mutations are found with high frequency in dysplastic ACF among patients with sporadic CRC, while *APC* lesions are absent (Takayama et al. 2001).

5.4.2 Validation of the Role of Mutant KRAS *in CRC*

Despite the early onset of *KRAS* mutations in CRC development, studies with *KRAS*-mutant CRC cell lines demonstrate that continued mutant *KRAS* expression is required for maintenance of the tumorigenic growth of advanced CRCs that harbor multiple genetic alterations. Genetic disruption of the mutant, but not wild-type, *KRAS* allele in DLD-1 and HCT 116 CRC cell lines via homologous recombination altered cell morphology, abrogated the capacity for CRC cells to grow under anchorage-independent conditions, and reduced the rate of tumorigenic growth in nude mice (Shirasawa et al. 1993). Induction of shRNA targeting *KRAS* caused regression of nude mouse tumors formed by SW480 CRC cells (Lim and Counter 2005). Similar shRNA studies also observed CRC growth suppression in vitro upon

Fig. 5.7 *RAS* mutations in CRC. (**a**) Frequency of mutation of the three *RAS* genes in all cancers compared with CRC. (**b**) Specific point mutations in K-Ras and N-Ras in CRC. The consensus GTP-binding motifs are indicated by the *dark blue boxes*, with the core effector domain (E; residues 32–40) indicated by the *red box*. The C-terminal hypervariable (HV) and adjacent CAAX motif (*yellow box*) are indicated. Ras residue numbers are indicated. The number of mutations for each amino acid substitution is in parentheses. (**c**) Distinct spectrum of *KRAS* mutations in different human cancers. Data compiled from COSMIC

suppression of mutant *KRAS* expression (Luo et al. 2009). These observations suggest that pharmacologic inhibition of mutant K-Ras function should have a significant clinical impact on metastatic CRC tumor growth.

5.4.3 KRAS4A *and CRC*

Whereas some normal tissues express undetectable levels of *KRAS4A*, *KRAS4A* is found in normal colon, pancreas and lung, tissues with frequent *KRAS* mutations (Plowman et al. 2003, 2006). For intestinal mouse tissue, there is a 1:1 ratio was seen for 4A:4B (Patek et al. 2008; Luo et al. 2010). A decrease in the ratio of 4A to 4B has been seen in human CRC cell lines (Plowman et al. 2006) and *K-Ras4A-deficient* mice showed enhanced carcinogen-induced colonic adenoma formation (Luo et al. 2010). These observations suggest a role for *KRAS4A* as a tumor suppressor. However, in a second study, a *K-Ras4A* deficiency did not alter small intestine tumorigenesis in the *Apc*$^{Min/+}$ mouse model of colon adenoma development (Patek et al. 2008), arguing against a tumor suppressive function. Since both models did not involve mutant K-Ras, the relative importance of wild-type and mutant K-Ras4A expression for mutant K-Ras4B-driven oncogenesis remains unclear. Thus, presently, the majority of studies evaluating K-Ras function are centered on K-Ras4B only.

5.4.4 BRAF *Mutation and CRC*

BRAF is mutationally activated in 11 % of CRC tumors (COSMIC). The majority of *BRAF* lesions found in human cancers occur within the kinase activation domain (89 %); 92 % of these are acidic substitutions at position V600 (initially described as V599 in earlier publications), with the V600E lesion being most prevalent (Fig. 5.1b). These changes are thought to mimic proximal phosphorylation events that are required for maximal kinase activity. Mutations in the highly conserved glycine-rich loop (GXGXXG motif) of the kinase domain account for the remaining 11 % of identified *BRAF* lesions and are thought to promote kinase domain structure for proper ATP interaction or orientation (Davies et al. 2002). In one study of 32 *BRAF*-mutant primary CRC tumors, 87.5 % harbored the V600E mutation (Rajagopalan et al. 2002). Other mutations identified at low frequency were R461I, I462S, and G463E.

Mutations in both the activation segment and the G-loop elevate basal kinase activity relative to wild-type B-Raf and increase ERK MAPK pathway phosphorylation. Representative mutants also are transforming in NIH 3T3 mouse fibroblasts (70- to 180-fold over wild-type) but still have 50-fold lower transforming activity than wild-type B-Raf activated by H-Ras12V (Davies et al. 2002). In general, the B-Raf (V600E) mutant shows greater potency in cell-based assays than the less common mutations and seems to be maximally activated in a Ras-independent fashion; other *BRAF*-mutant proteins can be further activated by Ras in cell culture studies (Davies et al. 2002). Among the three Raf family members (A-Raf, B-Raf, and C-Raf/Raf-1), B-Raf is the only one found mutated in patient tumors. Although the reason is unknown, B-Raf may be a preferred mutational target because, compared with Raf-1, it has higher basal kinase and transforming potential in vitro and requires fewer posttranslational phosphorylation events for activity (Mason et al. 1999). Additionally, wild-type B-Raf is more sensitive to activation by oncogenic Ras than are the other two Raf proteins, which, unlike B-Raf, respond strongly to stimulation by Src (Marais et al. 1997).

In general, *BRAF* mutations occur in a mutually exclusive manner with *RAS* lesions (Davies et al. 2002; Rajagopalan et al. 2002), indicating that a single oncogenic insult to the ERK MAPK pathway is sufficient for promoting tumorigenic activity. This pattern has been observed for cancer types, such as CRC and malignant melanoma, that seem to exhibit dependence on the ERK MAPK pathway by virtue of high *RAS/RAF* mutation frequency. Both oncogenes are present primarily in larger colorectal adenomas (>1 cm), so they likely are acquired after tumor initiation but before malignancy is determined (Rajagopalan et al. 2002). In contrast to *KRAS*, *BRAF* mutations more frequently appear in a genetic background with microsatellite instability and deficiencies in DNA MMR (Roth et al. 2010; Rajagopalan et al. 2002). MMR-deficient colorectal tumors are characterized by germ-line mutations in mismatch-repair genes, promoter hypermethylation, and microsatellite instability in coding and noncoding sequences. Together with *BRAF* mutation, these genetic anomalies also are found with high incidence in serrated polyps of the large intestine and are thought by some investigators to be precursor lesions to *BRAF-mutant* microsatellite-unstable carcinomas (Vakiani and Yantiss 2009). If validated, *BRAF* may represent a marker or pharmacotherapy target of serrated, pre-dysplastic CRC neoplasia.

5.4.5 **PIK3CA** *Mutation in CRC*

PIK3CA mutations are found in 11 % of CRC. The three predominant gain-of-function mutations (73 % of total) in p110α in CRC are located either in the helical (exon 9; E542K and E545K) or kinase (exon 20; H1047R) domain. There is limited evidence that exon 9 and exon 20 mutations have distinct functional properties. One study evaluated the transforming activities of different mutant *PIK3CA* allele in chicken embryo fibroblasts and determined the consequences of second site mutations that impaired either p85 PI3K regulatory subunit or Ras binding (Zhao and Vogt 2008). Impaired Ras, but not p85, binding reduced the transforming activities of the E542K and E545K helical domain mutants. In contrast, impaired p85, but not Ras, binding reduced the transforming activity of the H1047R kinase domain mutant. In a second study, where different *PIK3CA* mutants were ectopically expressed in the *KRAS*-mutant MDA-MB-231 breast carcinoma cells, E545K caused a more severe metastatic phenotype than the H1047R mutant (Prasad and

Baillie 1989). While the data are limited, there is evidence that the biological activities and consequences of different p110α mutant proteins are distinct.

In contrast to *BRAF* mutations, *PIK3CA* mutations can co-occur with *KRAS* mutations. In one retrospective analyses of 1022 cetuximab plus chemotherapy-refractory CRC tumors, 14.5 % of *KRAS*-mutant tumors also harbored exon 9 mutations, whereas only 3.8 % of *KRAS*-mutant tumors harbored exon 20 mutations (De Roock et al. 2010b). However, the fraction of exon 9 and 20 mutations found with *KRAS* mutations was essentially the same (62 % vs. 55 %), contrasting with the functional studies that suggested a need for activated Ras for exon 9 mutations. The remaining exon 9 or 20 mutations (39 %) were associated with *KRAS*-wild-type tumors.

PTEN loss, as assessed by immunocytochemical staining for protein expression, is also seen in *KRAS*-wild-type and mutant CRC (20–40 %) (Dienstmann et al. 2011). Together with *PIK3CA* mutations, this increases the percentage of CRC where there may be hyperactivation of PI3K-Akt signaling. However, *PTEN* loss and *PIK3CA* mutational activation are not likely to be functionally equivalent in biology or in therapeutic response, since the former mutation will promote activities of all class I PI3K isoforms. Finally, since PI3K is a known effector of Ras, it is somewhat surprising to find a concurrent mechanism for PI3K activation in some *KRAS*-mutant CRC tumors. This has been speculated to suggest that mutant K-Ras may not be a robust activator of PI3K in CRC. This has been seen in studies with model cell systems (McFall et al. 2001; Li et al. 2004). Additionally, studies in CRC found that mutant K-RAS was not required for PI3K activation and instead involved activated receptor tyrosine kinases (Ebi et al. 2011).

5.4.6 Other Ras Effectors and CRC

Other Ras effectors that have been shown to be involved in CRC biology in cell and mouse models include the guanine nucleotide exchange factors RalGDS and Tiam1, and the multifunctional phospholipase enzyme PLCε. RalGDS belongs to a family of four GEFs that act selectively on to closely related members of the Ras family branch of the Ras superfamily that share ~50 % amino acid identity with Ras (Neel et al. 2011), RalA and RalB, both of which are activated in CRC cell lines and patient-derived tumors (Martin et al. 2011). Consistent with findings reported in *KRAS*-dependent pancreatic cancer, RalA, signaling through its effectors RalBP1 and the exocyst component Exo84, was found to be required for anchorage-independent growth of CRC cells in soft agar, although the phenotype was not limited to *KRAS*-mutant cells (Lim et al. 2006; Martin et al. 2011). The Ras effector Tiam1 is a GEF for the Rac small GTPase. Tiam1 has been shown to play a role in polyp growth and formation and tumor metastasis in $Apc^{Min/+}$ mice, a genetically engineered model of CRC that carries a mutant allele of the Apc tumor suppressor (Malliri et al. 2006). Of note, Tiam1 represents another point of pathway cross-talk in CRC, as it is directly stimulated by Ras-GTP and is upregulated in response to Wnt signaling, a pathway synergistically activated in an *Apc*/*KRAS*-double mutant mouse model of CRC (Janssen et al. 2006). The second messenger signaling enzyme and Ras effector PLCε also contributes to the growth and progression of adenomas in $Apc^{Min/+}$ mice; depending on adenoma stage, tumors lacking PLCε undergo increased apoptosis, reduced proliferation, reduced expression of vascular endothelial growth factor (VEGF), and attenuated angiogenesis (low-grade adenomas) or reduced tumor-associated inflammation (high-grade adenomas) (Li et al. 2009).

5.5 *KRAS* and CRC Treatment

For resectable metastatic CRC, adjuvant chemotherapy is usually recommended. For unresectable CRC, where surgery is not possible, chemotherapy is recommended to reduce symptoms and prolong survival. Currently, first-line treatment involves 5-fluorouracil in combination with oxaliplatin (FOLFOX) or irinotecan (FOLFIRI) or the orally-available 5-fluorouracil prodrug capecitabine together with oxaliplatin (XELOX), which have limited response rates of 40–50 % and impact on overall survival (Giacchetti et al. 2000; Douillard et al. 2000). Three targeted therapeutics are approved for metastatic CRC, bevacizumab, a monoclonal antibody (mAb) inhibitor of VEGF, and cetuximab and panitumumab, mAb inhibitors of the epidermal growth factor receptor (EGFR/HER1). Bevacizumab is used in combination with FOLFOX, FOLFIRI, or XELOX for first- and second-line therapy. Cetuximab can be used alone or with irinotecan for second-line treatment, whereas panitumumab is approved for mCRC treatment after all other drugs have failed. In this section, we discuss the implications of *KRAS* and effector mutations for CRC treatment.

5.5.1 Prognostic and Predictive Value of Mutant **KRAS** *and* **BRAF**

The prognostic value of *KRAS* mutation in CRC remains unclear. Early retrospective analyses of patient-derived CRC tumors indicated that the G12V mutation was associated with significantly decreased failure-free survival and overall survival independent of treatment, particularly in Duke's C tumor types (Andreyev et al. 2001). However, more recent studies on the prognostic impact of *KRAS* mutations in CRC have yielded conflicting data. The prospective Phase III trials PETACC-3 and AGITG Max both showed that *KRAS* mutation status, including the G12V lesion and other codon 12 and 13 mutations, was not associated with disease outcome compared with wild-type *KRAS* tumors (Price et al. 2011; Roth et al. 2010). The less common exon 4 mutations (K117N and A146T/V), however, were shown in one study to be associated with significantly improved clinical outcome in 186 cases of stage I to III CRC (Janakiraman et al. 2010). Whether this is due to lower potency of the activating mutations compared with the more common *KRAS* lesions is unclear.

By contrast, independent studies consistently show that the primary activating mutation in *BRAF*, V600E, is a strong negative prognostic indicator of survival, suggesting that mutant B-Raf and K-Ras likely exert independent, as well as overlapping, functional roles in CRC biology (Price et al. 2011; Roth et al. 2010; Samowitz et al. 2005; Saridaki et al. 2010; Souglakos et al. 2009; Tol et al. 2009; Van Cutsem et al. 2011; Wong and Cunningham 2008).

5.5.2 **KRAS** *Mutation Status and EGFR Inhibitor Treatment*

While *KRAS* mutation status may not provide a clear prognostic marker of disease outcome, it does have well-established predictive value for therapeutic response to EGFR-targeted drugs. The observations that anti-EGFR therapy displayed only 10–20 % response with metastatic CRC (Saltz et al. 2004; Cunningham et al. 2004) prompted the search for molecular determinants that correlated with response or nonresponse. A 2009 meta-analysis of ten retrospective studies (Allegra et al. 2009) confirmed the seminal finding that *KRAS* mutation is associated with resistance to anti-EGFR mAb therapy with cetuximab and panitumumab (Allegra et al. 2009; Lievre et al. 2006, 2008; Amado et al. 2008; Karapetis et al. 2008). Based on these data, in 2009 the FDA recommended genotyping for mutational status of *KRAS* codons 12 and 13 in metastatic CRC and recommended against EGFR monoclonal antibody therapy in the presence of such mutations. The capacity for constitutively active K-Ras to render insensitivity to upstream inhibition of EGFR reinforces the understanding that Ras activation is a primary consequence of EGFR signaling.

However, since 50–65 % of patients with *KRAS*-wild-type advanced CRC also are resistant to EGFR-targeted biologics, *KRAS* mutation alone is not the sole determinant of insensitivity to this line of therapy (Allegra et al. 2009; De Roock et al. 2010a). Mounting evidence indicates that *BRAF*-mutant chemotherapy-refractory CRC tumors also exhibit resistance to EGFR therapies (Di Nicolantonio et al. 2008b; Laurent-Puig et al. 2009). Since *BRAF* mutations (~15 %) are mutually exclusive of *KRAS* mutations in CRC, this brings the fraction of nonresponders to ~55 %.

Recent retrospective studies suggest that, in addition to *BRAF* mutations, other genetic mutations may also correlate with EGFR inhibitor refractory CRCs. For example, mutations in *NRAS* and *PIK3CA* (exon 20 but not exon 9) had significantly lower response rates to treatment with cetuximab with chemotherapy (De Roock et al. 2011). This study also showed that exon 4 mutations of *KRAS* (117 and 146) had no effect on treatment outcome, but that, in contrast to other work, these mutations do coexist with the more common *KRAS* lesions that are associated with resistance.

Additional evidence indicates that *KRAS* mutation status alone may not solely dictate response to EGFR therapy. One recent retrospective study compared the response profiles of patients with metastatic CRC tumors with different mutant alleles of *KRAS* after treatment with cetuximab and found those with G13D-mutant disease had better progression-free and overall survival compared with those with

G12V-mutant disease (De Roock et al. 2010c). Prior studies have shown that K-Ras(G13D) had lower transforming activity in cell culture than the more prevalent K-Ras(G12V) mutant (Guerrero et al. 2000), and *KRAS* G12 but not G13 mutations were associated with inferior survival in *BRAF* WT CRC (Imamura et al. 2012).

The role of feedback impingement on ERK MAPK pathway activity also likely contributes to drug sensitivity in CRC tumors dependent on ERK MAPK output. Transcriptional analyses have shown that, due to down-regulation of transcripts including dual-specificity phosphatase (DUSP) and Sprouty family members, *BRAF*-mutant cancer cell lines are impacted less by intrinsic feedback inhibition mechanisms than are receptor tyrosine kinase-mutant lines (Pratilas et al. 2009). Related to these findings, published preliminary data indicate that high DUSP4 expression levels in *KRAS*-mutant CRC tumors are associated with improved survival with cetuximab treatment (De Roock et al. 2009, Abstract 289). Collectively, these data suggest that *KRAS* mutation status alone, as currently defined, is insufficient to accurately predict treatment response and that more extensive genotyping likely will be warranted before stratifying patients for or against EGFR therapy in the future.

5.6 Inhibitors of Mutant K-Ras Function: Drugging the "Undruggable"

The frequent mutational activation of *KRAS*, together with experimental evidence supporting the key role of mutant *KRAS* in tumor maintenance, argues that anti-KRAS-targeted therapies will be an effective treatment approach for a major fraction of CRCs. However, despite over 3 decades of intensive effort, the development of effective small molecule antagonists of mutant Ras has proven to be difficult and elusive. In this section, we summarize past and ongoing efforts to develop pharmacologic inhibitors of mutant Ras to provide a sense of why this goal has been so difficult to achieve.

5.6.1 Targeting Mutant Ras Itself: Is Mutant Ras Undruggable?

The fundamental biochemical difference between mutant and wild-type Ras proteins is impaired intrinsic and GAP-stimulated GTP hydrolysis, rendering Ras persistently GTP-bound and active. These differences have prompted efforts to directly antagonize mutant Ras function. Efforts to develop a GAP-mimetic small molecule that can reactivate the intrinsic GTPase activity of mutant Ras were unsuccessful. By analogy to ATP-competitive inhibitors of protein kinases (e.g., imatinib, gefitinib erlotinib), small molecule competitive inhibition of GTP binding has also been considered. However, in contrast to the success of small molecule ATP-binding competitive inhibitors of protein kinases, the analogous approach has not succeeded for GTP binding to Ras. Protein kinase inhibitors with low nanomolar affinities can be developed to block the low micromolar binding affinity of protein kinases for ATP. In contrast, GTP-binding competitive inhibitors are not feasible, due to the high micromolar intracellular GTP concentrations and low picomolar affinity levels of Ras for GTP (Cox and Der 2010; Gysin et al. 2011). While it remains possible that direct antagonists of mutant Ras can be developed, for example allosteric inhibitors of Ras association with other proteins, the perception that mutant Ras itself is "undruggable" prompted efforts to target Ras indirectly, to cripple two activities essential for mutant Ras-dependent cancer growth, membrane association or effector signaling.

5.6.2 Farnesyltransferase Inhibitors: Targeting the Wrong Ras Isoform

As described above, Ras function is critically dependent on CAAX box-signaled posttranslational modifications that facilitate Ras membrane attachment. The membrane association and oncogenic activities of mutant Ras proteins are completely abolished by mutation of the CAAX sequence to prevent farnesylation. These observations prompted efforts to therapeutically block Ras function by interfering with membrane targeting.

The most heavily pursued drug candidates have been inhibitors of farnesyltransferase (FTIs). Preclinical cell culture and mouse model evaluation of FTIs showed very promising antitumor activities. However, most of the models used involved mutant H-Ras-dependent transformation. It was determined subsequently that K-Ras4B and N-Ras can undergo alternative prenylation with a geranylgeranyl isoprenoid group when farnesyltransferase activity is blocked by FTI treatment (Cox and Der 2010; Gysin et al. 2011). This was an unexpected finding that provided an explanation for the failure of FTIs in clinical trials with *KRAS*-mutant CRC and other cancers. One positive outcome of this otherwise very disappointing era in anti-Ras drug discovery was the focusing of future efforts on K-Ras4B.

Some efforts to find other pharmacologic approaches to block K-Ras membrane association are ongoing. First, in light of the alternative modification of K-Ras4B caused by GGTase-I when FTase activity is blocked, concurrent inhibition of both FTase and GGTase-I may be an effective option (Berndt et al. 2011). Support for this is provided by mouse model studies where genetic ablation of *GGTASEI* impaired mutant *KRAS*-driven lung carcinoma development (Sjogren et al. 2007; Liu et al. 2010). However, concurrent inhibition of FTaseI and GGTase-I may also be limited by normal cell toxicity due to the large numbers of proteins believed essential for normal cell function (e.g., Rho GTPases) that depend on FTase and GGTase-I activity. Alternatively, because a number of GGTase-I substrates also contribute to Ras-dependent cancer growth, the increased antitumor activity may offset the increased non-Ras activities of duel FTase and GGTase-I inhibition.

Second, there is genetic evidence that the other two CAAX-signaled modifications may be useful targets for drug discovery. Genetic inhibition of *RCE1* and *ICMT* demonstrated that membrane displacement of K-Ras can affect tumorigenesis in cell culture and mouse models (Bergo et al. 2002, 2004; Michaelson et al. 2005; Wahlstrom et al. 2008). So this avenue of research may yet yield promising therapeutic options for the inhibition of Ras activity. However, uncertainties for this direction include the existence of up to 300 CXXX motif-terminating proteins that may depend on Rce1 and/or Icmt activity (Berndt et al. 2011), and whether potent and selective pharmacologic inhibitors of these two enzymes can be developed.

Finally, small molecule mimics of the farnesyl group have been considered. There is evidence that the membrane docking of farnesylated Ras is dependent on membrane-associated docking proteins, such as galectins (Paz et al. 2001; Shalom-Feuerstein et al. 2005). One such mimic, salirasib (farnesylthiosalicylic acid), is proposed to disrupt Ras association with their membrane docking proteins, thus disrupting Ras membrane stability and/or signaling (Blum et al. 2008). Studies in cell culture and mouse models support the anti-Ras and antitumor activities of salirasib and Phase I clinical trial analyses are ongoing.

5.7 Inhibition of K-Ras Effector Signaling

Although multiple effectors have been implicated in mutant RAS-dependent tumor growth, current efforts to develop inhibitors of effector signaling have focused primarily on the MAPK signaling cascade and, more recently, the PI3K effector signaling network. This focus is based largely on the fact that mutational activation of these effector signaling networks is seen in human cancers and in part because components of these pathway include protein kinases, tractable targets for inhibitor development.

5.7.1 Inhibitors of Raf-MEK-ERK *Mitogen-Activated Protein Kinase Cascade*

Inhibitors of this pathway have focused primarily on Raf or MEK, with two Raf inhibitors approved for cancer treatment (sorafenib and vemurafenib), although the significance of the anti-Raf activity of sorafenib for its clinical activities is unclear. Recently, one inhibitor of ERK1/2 has been described (Hatzivassiliou et al. 2012). Despite the apparent linear nature of the Raf-MEK-ERK cascade, inhibitors at each level are not likely to have equivalent consequences and there is evidence that concurrent inhibition at multiple levels may have synergistic value (Flaherty et al. 2012).

5.7.1.1 Raf Inhibitors

As a primary effector of Ras and a proto-oncogene in its own right, B-Raf is an attractive molecular target, and several ATP-competitive small molecules have advanced through the clinic and FDA approval for cancer treatment. The multikinase inhibitor sorafenib can block the activity of wild-type and mutant B-Raf, in addition to kinases including RAF-1, vascular endothelial growth factor receptors 1 and 2 (VEGFR1 and VEGFR2), platelet-derived growth factor receptor-β (PDGFR-β), and stem-cell growth factor receptor (SCFR). Based more on its anti-angiogenesis rather than anti-Raf activity, sorafenib was FDA-approved for the treatment of renal cell carcinoma and unresectable hepatocellular carcinoma, cancers with infrequent *RAS* or *BRAF* mutations. Cell culture studies with B-Raf(V600E)-expressing CRC lines showed that addition of sorafenib to cetuximab treatment synergistically decreased proliferation and viability of cells which were insensitive to cetuximab alone, suggesting that patients with *BRAF*-mutant CRC tumors may derive clinical benefit from combination EGFR/B-Raf-targeted therapies (Di Nicolantonio et al. 2008a). This has now been validated, where EGFR activation was shown to be a basis for vemurafenib insensitivity in *BRAF*-mutant CRC (Prahallad et al. 2012; Corcoran et al. 2012). Recruitment is active for a Phase II trial to investigate this hypothesis among patients with recurrent or refractory metastatic EGFR-expressing CRC (NCT00326495).

The Raf selective inhibitor vemurafenib (formerly PLX4032) was fast-tracked for FDA approval in 2011 for the treatment of *BRAF* V600E mutant metastatic or unresectable melanoma. This approval was based on results from a Phase III study where significant improvements in both progression-free and overall survival was seen among patients treated with vemurafenib compared with dacarbazine, the long-standing standard of care (Chapman et al. 2011). Earlier preclinical data also suggested that vemurafenib will show antitumor efficacy in *BRAF*-mutant CRC, where vemurafenib showed tenfold increased selectivity and inhibition of V600E over wild-type B-Raf CRC cancer lines (Tsai et al. 2008). However, despite potent antitumorigenic properties of vemurafenib in CRC xenograft models (Tsai et al. 2008), these results have not translated to the clinic. Findings from one small Phase I study showed only a modest 5 % response rate with vemurafenib in patients with *BRAF*-mutant metastatic CRC (Kopetz et al. 2010). The difference consequences of vemurafenib on *BRAF*-mutant CRC and melanomas may reflect tissue-specific, distinct roles of mutant B-Raf. Some evidence for this is suggested by the different impact of shRNA suppression of mutant B-Raf(V600E) expression in human tumor cell lines (Hao et al. 2007). Whereas suppression caused an acute response for melanoma cells, with a rapid and dramatic reduction in anchorage-dependent proliferation, no significant growth reduction was seen for *BRAF* V600E mutant CRC cell lines. Despite this difference, nevertheless, mutant BRAF suppression still reduced CRC growth under anchorage-independent conditions in cell culture and in mouse tumor xenografts, suggesting that vemurafenib-based therapy should still have some clinical benefit (Hao et al. 2007). Preclinical studies suggest that vemurafenib treatment in combination with other signaling inhibitors may be needed for effective antitumor activity for *BRAF*-mutant CRC.

Another compound that has shown a high degree of specificity for mutant B-Raf in preclinical studies is GDC-0879. *BRAF*-mutant CRC cells were highly sensitive to treatment with GDC-0879, while *KRAS*-mutant lines, in general, were resistant (Hoeflich et al. 2009). Important information derived from these and other preclinical studies with Raf inhibitors provided insight into innate resistance mechanisms in *KRAS*-mutant tumors. In the presence of mutant Ras, compounds targeting B-Raf induce heterodimerization with Raf-1, which is not inhibited, leading to a paradoxical activation of Raf-1 and ERK signaling (Hatzivassiliou et al. 2010; Poulikakos et al. 2010; Heidorn et al. 2010). The recent finding that vemurafenib treatment accelerated the development of preexisting *RAS* mutant skin lesions suggests that this mechanism is indeed a clinically relative limitation of Raf inhibitors (Su et al. 2012; Oberholzer et al. 2012). As a result, cancers such as CRC that have significant subsets of *KRAS*-and *BRAF*-mutant tumors could be differentially growth-enhanced or growth-restricted in response to pharmacologic targeting of mutant B-Raf depending on the tumor genotype. However, analysis in a mouse model of mutant *HRAS*-driven skin tumor formation suggests that concurrent treatment with a MEK1/2-selective inhibitor may overcome this limitation of Raf inhibitors (Su et al. 2012).

MEK Inhibitors

Inhibitors of the Raf effectors MEK1 and MEK2 have been pursued heavily in clinical trials for various Ras-dependent cancers in recent years, but to date none have received approval from the FDA or the European Medicines Agency. In general, MEK1/2 inhibitors are highly selective due to their non-ATP-competitive mode of action, but several promising candidates have been withdrawn from clinical development due to failed response in patients with CRC and other cancers with constitutively activated ERK MAPK signaling (Montagut and Settleman 2009). More recent preclinical analyses, however, have revealed that response patterns to MEK inhibitor treatment can be differentially dependent on the specific mechanism of ERK MAPK activation in a given tumor or cell type. One of the first examples of this was a study involving responses of *NRAS*-mutant, *BRAF*-mutant, or *RAS/RAF-wild-type* melanoma cell lines to treatment with the MEK1/2 inhibitor CI-1040, a compound which had failed to yield clinical response in a phase II trial for treatment of CRC (not selected for BRAF mutations), breast, PDAC, and NSCLC (Rinehart et al. 2004; Solit et al. 2006). These studies showed that B-Raf(V600E) cells were selectively and highly sensitized to MEK inhibition in in vitro growth assays and tumor xenograft experiments, whereas *NRAS*-mutant cells were only partially sensitive, and *RAS/RAF*-wild-type cells were insensitive. *BRAF*-dependent sensitivity correlated with down-regulation of cyclin D1 and a G1 arrest phenotype (Solit et al. 2006). Independent studies suggest that a similar dependence on mutant *BRAF* may extend to CRC. While three studies found that both *KRAS*- and *BRAF*-mutant cell lines exhibited anchorage-dependent and -independent growth inhibition with anti-MEK inhibitor treatment, a fourth study showed that *BRAF*-mutant, but not *KRAS-mutant*, CRC cells are growth inhibited under anchorage-independent conditions in response to the MEK inhibitor AZD6244 (ARRY-142886) (Davies et al. 2007; Martin et al. 2011; Yeh et al. 2009; Balmanno et al. 2009). Furthermore, in a mouse model of *KRAS*-driven CRC, CI-1040 treatment had no effect on tumor cell proliferation

(Haigis et al. 2008), and separate studies suggest that *KRAS* mutation status may represent a mechanism of resistance in CRC tumors unresponsive to CI-1040 treatment (Wang et al. 2005).

An important point realized by these studies is that elevated phospho-ERK levels may not necessarily correlate with response to MEK inhibitor sensitivity. Among CRC patient tumors, phospho-ERK1/2 levels correlated with *BRAF*-mutation status, but in cell lines, ERK activity showed no association with mutation status or MEK inhibition of anchorage-independent growth of either *KRAS*- or *BRAF*-mutant cells (Martin et al. 2011; Yeh et al. 2009). As mentioned previously, negative feedback regulators of ERK signaling such as DUSP6 are downregulated in *BRAF-mutant* cells and consequently, with *BRAF*-mutation status, may be superior to ERK1/2 activation as a biomarker of ERK MAPK-dependent transformation (Pratilas et al. 2009). Together, the current data indicate that MEK inhibitor therapy may be of particular benefit in patients with *BRAF*-mutant CRC, but clinical data addressing this hypothesis are not yet available. Currently a phase II study with AZD6244 is underway for patients with *BRAF*-mutant cancers (NCT00888134) and a phase I study with another MEK inhibitor, ARRY-438162, is recruiting patients with *KRAS*- or *BRAF*-mutant cancers including metastatic CRC (NCT00959127).

5.7.2 Inhibitors of PI3K *Effector Signaling*

Activated Akt is present at elevated levels in *KRAS*-mutant CRC cell lines independent of *PIK3CA* mutational status (Martin et al. 2011). However, a subset of CRCs co-segregate mutant *KRAS* and PI3K pathways lesions, suggesting that simultaneous activation of both signaling cascades is necessary for tumorigenic growth (Fearon 2011b). Another point of ERK MAPK-PI3K pathway cross-talk in CRC is the serine/threonine kinase mammalian target of rapamycin (mTOR), which directly regulates protein translation in response to upstream signaling and intracellular metabolic cues, leading to effects on transformation, motility, invasiveness, and angiogenesis (Mamane et al. 2006).

A likely mechanism for K-Ras-mediated drug resistance is alternative effector signaling output. Therapeutic intervention at the PI3K signaling axis is of interest because this pathway is known to be engaged in Ras-mediated transformation. In addition, PI3K can be activated directly by upstream receptor tyrosine kinases independently of Ras. *PIK3CA* mutations can occur independent of or in combination with mutant *RAS* or *RAF*. Sequencing studies have shown that *PIK3CA* mutations coexist with *KRAS* or *BRAF* mutations in 22 % of CRCs (Parsons et al. 2005; Velho et al. 2005). As a result, multiple compounds targeting PI3K and the downstream serine/threonine kinases AKT and mTOR are being tested in the clinic and will be discussed in more detail in the following chapters of this book.

Several lines of evidence indicate that elevated PI3K pathway activity contributes to MEK inhibitor resistance in CRC. First, while melanoma and CRC cells were sensitive to the Raf inhibitor GDC-0879, response was dependent on endogenous levels of PTEN. RNAi suppression of PTEN expression abrogated this response in otherwise sensitive melanoma cells (Hoeflich et al. 2009). Second, elevated phospho-AKT levels have been documented in *KRAS*-mutant CRC cell lines, and in separate studies, high pAKT correlated to varying degrees with resistance to the MEK inhibitor AZD6244 (Balmanno et al. 2009; Martin et al. 2011). Third, *KRAS*-mutant cancer cell lines expressing constitutively active PI3K are less sensitive to the growth effects of MEK inhibition than are *KRAS*-mutant lines with wild-type PI3K, and *KRAS*-mutant *PTEN*-null cells are completely resistant. Furthermore, *KRAS*/*PIK3CA*-double mutant colorectal tumors in mice required targeted inhibition of both pathways to achieve tumor stasis (Wee et al. 2009). Other work has shown that concurrent activation of the MAPK and PI3K pathways contributes to EGFR therapy insensitivity as well; dual lesions in the two signaling cascades confer resistance to cetuximab therapy in CRC cell lines (Jhawer et al. 2008).

The cumulative clinical and preclinical data demonstrate that single-agent treatment is insufficient to achieve sustained responses in patients with *KRAS*- or *BRAF-mutant* CRC. Complex intrinsic mechanisms including independent oncogenic genetic events, loss of tumor suppressor genes, pathway cross-talk, and feedback promotion of alternative signaling pathways all contribute to tumorigenic behavior and drug resistance. In vitro studies provide evidence that combination drug treatment of CRC and other cancer cell lines may help overcome these issues. Simultaneous inhibition of the ERK MAPK pathway and PI3K, mTOR, or both, has been shown to be growth inhibitory and proapoptotic in colorectal and breast cancer cells (Mirzoeva et al. 2009; Zhang et al. 2009), and causes tumor regression in a mouse model of *KRAS*-mutant lung adenocarcinoma (Engelman et al. 2008). As mentioned previously, combination treatment with cetuximab and sorafenib enhanced growth inhibition of *BRAF*-mutant CRC cell lines (Di Nicolantonio et al. 2008a). Together these findings offer support for clinical investigation of multi-agent targeted therapy for the treatment of *KRAS*- and *BRAF*-mutant CRC. However, in a recently completed Phase I clinical trial analysis of 254 chemorefractory CRC patients, where *KRAS*, *BRAF*, and *PIK3CA* mutation status was used to determine matched treatment with a MEK and/or PI3K inhibitor nevertheless did not result in a significant clinical benefit (Dienstmann et al. 2012). Thus, more complex combinations beyond inhibition of the two canonical K-Ras effector pathways may be needed for effective inhibition of mutant K-Ras-dependent cancer growth.

5.8 Synthetic Lethal Partners of Mutant *KRAS*: Novel Approaches for Anti-K-Ras Therapeutics?

Recently, in an effort to understand, and potentially exploit, molecular vulnerabilities in *KRAS*-mutant cancer cells, multiple research groups have utilized RNA interference-based screens to identify other proteins on which Ras-mutant cells, including CRC cells, are dependent for viability. The underlying concept is that targeted therapies can induce selective lethality in *KRAS*-mutant cells by inhibiting protein function required by tumor cells harboring mutant, but not normal, K-Ras. Such "synthetic lethal" partners of mutant *KRAS* were searched for using functional screens with high-throughput RNA interference libraries. Two such studies applied high-throughput RNAi largely directed against kinases and phosphatases to bias hits toward tractable pharmaceutical drug targets (Scholl et al. 2009; Barbie et al. 2009). In one study, the anti-apoptotic serine/threonine kinase STK33 was identified, whereas a second study identified the pro-survival NF-κB-activating serine/threonine kinase TBK1. While K-Ras dependence on both STK33 and TBK1 in these studies seemed to be independent of ERK MAPK or PI3K effector signaling, RalB was identified as a possible linkage point between K-Ras and TBK1 activation, as depletion of RalB resulted in significant lethality in *KRAS*-mutant lung cancer cells (Barbie et al. 2009). This linkage verified an earlier finding that TBK1 functioned downstream of RalB-dependent tumor cell signaling.

Third, genome-wide RNAi screen revealed that mutant K-Ras (K-RasG13D) in the isogenic DLD-1 CRC cell lines required partnership with numerous pro-mitotic proteins, including the mitotic kinase polo-like kinase 1 (PLK1) and members of the anaphase-promoting complex/cyclosome (APC/C), for viability (Luo et al. 2009). In addition, *KRAS*-mutant CRC cells were more sensitive than *KRAS*-wild-type cells to direct inhibition of both PLK1 and the proteasome that is required for APC/C function. Finally, a fourth study that also used the same set of isogenic paired WT and mutant *KRAS* CRC cell lines identified the Snail2 transcriptional repressor as a synthetic lethal partner of mutant *KRAS* (Wang et al. 2010).

Although these studies hold great promise for the identification of therapeutic targets for the treatment of *KRAS*-mutant CRC, more rigorous validation of the identified genes is needed to establish the significance of identified molecules to patient tumors. For example, the relevance of STK33 (Scholl et al. 2009; Barbie et al. 2009) has been questioned in an independent study where both pharmacologic and genetic ablation of STK33 failed to identify it as a synthetic lethal partner of mutant *KRAS* (Babij et al. 2011). One limitation of these studies, introduced by the need for a simple two-dimensional cell culture biological screen amenable to high-throughput analyses, is that this cellular setting may not accurately model the biology of *KRAS*-mutant tumor cells in the context of a three-dimensional tumor and in the presence of nontumor stromal cells. Future studies with more advanced biological assays will help overcome this concern.

5.9 Conclusions

While the restriction of EGFR inhibitor treatment to *KRAS*-wild-type patients has improved the response for this subset of CRC patients, it also emphasizes the need for targeted therapies for the 40 % of CRC patients with mutant *KRAS*. Despite nearly 3 decades of intensive effort that has been met with limited success, there remains strong optimize that effective anti-K-Ras inhibitors will hold great clinical benefit for *KRAS*-mutant CRC patients. An outcome of the failed past efforts is a better understanding of the complex and dynamic nature of K-Ras-dependent signaling networks. The realization of Ras isoform differences has prompted a focus on the K-Ras4B isoform and mutation-specific differences in K-Ras function will likely emerge. Presently, the best hope for anti-K-Ras therapy involves inhibition of K-Ras effector signaling. Of the multitude of effectors, the mutational activation of B-Raf argues that this is the Raf-MEK-ERK protein kinase cascade will be a key signaling pathway for blocking K-Ras function. However, lessons learned from preclinical and clinical evaluation of Raf and MEK inhibitors reveal mechanisms of resistance to Raf-MEK-ERK inhibition need to be better understood. Concurrent inhibition of non-Raf effectors, in particular the P3K-Akt pathway, will likely be needed for effective inhibition of K-Ras signaling. Cocktails of inhibitors that concurrently disrupt K-Ras signaling at multiple points will need to be defined. It remains possible that mutant K-Ras itself is druggable. Improved genome-wide screens for more robust and physiologically relevant synthetic lethal partners of mutant K-Ras still hold great promise for unanticipated approaches for antagonism of mutant K-Ras.

Experimental observations have shown that the cellular response to mutant Ras is dictated by the existence of other genetic lesions. The 40 % of *KRAS*-mutant CRCs are not uniform and are characterized by significant heterogeneity, each with a distinct set of "hills." Furthermore, there is emerging evidence that the different activating mutations of K-Ras will have different consequences for drug response. Clearly, *KRAS*-mutant CRCs cannot be considered a homogeneous subset of this cancer, with one simple therapeutic option. Furthermore, there is considerable genetic heterogeneity within one cancer, where the fraction of tumor cells with specific mutations is highly variable. Additional genetic and biochemical profiling will be needed to divide *KRAS*-mutant CRCs into subsets responsive to different cocktails of K-Ras-directed therapies.

References

Ahearn IM, Haigis K, Bar-Sagi D et al (2011) Regulating the regulator: post-translational modification of RAS. Nat Rev Mol Cell Biol 13(1):39–51. doi:10.1038/nrm3255

Allegra CJ, Jessup JM, Somerfield MR et al (2009) American Society of Clinical Oncology provisional clinical opinion: testing for KRAS gene mutations in patients with metastatic colorectal carcinoma to predict response to anti-epidermal growth factor receptor monoclonal antibody therapy. J Clin Oncol 27(12):2091–2096

Amado RG, Wolf M, Peeters M et al (2008) Wild-type KRAS is required for panitumumab efficacy in patients with metastatic colorectal cancer. J Clin Oncol 26(10):1626–1634. doi:10.1200/JCO.2007.14.7116

Andreyev HJ, Norman AR, Cunningham D et al (2001) Kirsten ras mutations in patients with colorectal cancer: the 'RASCAL II' study. Br J Cancer 85(5):692–696

Babij C, Zhang Y, Kurzeja RJ et al (2011) STK33 kinase activity is nonessential in KRAS-dependent cancer cells. Cancer Res 71(17):5818–5826

Balmanno K, Chell SD, Gillings AS et al (2009) Intrinsic resistance to the MEK1/2 inhibitor AZD6244 (ARRY-142886) is associated with weak ERK1/2 signalling and/or strong PI3K signalling in colorectal cancer cell lines. Int J Cancer 125(10):2332–2341

Bandyopadhyay S, Chiang CY, Srivastava J et al (2010) A human MAP kinase interactome. Nat Methods 7(10):801–805
Barbie DA, Tamayo P, Boehm JS et al (2009) Systematic RNA interference reveals that oncogenic KRAS-driven cancers require TBK1. Nature 462(7269):108–112
Bergo MO, Ambroziak P, Gregory C et al (2002) Absence of the CAAX endoprotease Rce1: effects on cell growth and transformation. Mol Cell Biol 22(1):171–181
Bergo MO, Gavino BJ, Hong C et al (2004) Inactivation of Icmt inhibits transformation by oncogenic K-Ras and B-Raf. J Clin Invest 113(4):539–550. doi:10.1172/JCI18829
Bermudez O, Pages G, Gimond C (2010) The dual-specificity MAP kinase phosphatases: critical roles in development and cancer. Am J Physiol Cell Physiol 299(2):C189–C202. doi:10.1152/ajpcell.00347.2009
Berndt N, Hamilton AD, Sebti SM (2011) Targeting protein prenylation for cancer therapy. Nat Rev Cancer 11(11):775–791. doi:10.1038/nrc3151
Bivona TG, Quatela SE, Bodemann BO et al (2006) PKC regulates a farnesyl-electrostatic switch on K-Ras that promotes its association with Bcl-XL on mitochondria and induces apoptosis. Mol Cell 21(4):481–493. doi:10.1016/j.molcel.2006.01.012
Blum R, Cox AD, Kloog Y (2008) Inhibitors of chronically active ras: potential for treatment of human malignancies. Recent Pat Anticancer Drug Discov 3(1):31–47
Bodemann BO, White MA (2008) Ral GTPases and cancer: linchpin support of the tumorigenic platform. Nat Rev Cancer 8(2):133–140. doi:10.1038/nrc2296
Bos JL, Fearon ER, Hamilton SR et al (1987) Prevalence of ras gene mutations in human colorectal cancers. Nature 327(6120):293–297
Brown MD, Sacks DB (2009) Protein scaffolds in MAP kinase signalling. Cell Signal 21(4):462–469. doi:10.1016/j.cellsig.2008.11.013
Cancer Genome Atlas Network (2012) Comprehensive molecular characterization of human colon and rectal cancer. Nature 487(7407):330–337. doi:10.1038/nature11252
Cejas P, Lopez-Gomez M, Aguayo C et al (2009) KRAS mutations in primary colorectal cancer tumors and related metastases: a potential role in prediction of lung metastasis. PLoS One 4(12):e8199
Chan TL, Zhao W, Leung SY et al (2003) BRAF and KRAS mutations in colorectal hyperplastic polyps and serrated adenomas. Cancer Res 63(16):4878–4881
Chapman PB, Hauschild A, Robert C et al (2011) Improved survival with vemurafenib in melanoma with BRAF V600E mutation. N Engl J Med 364(26):2507–2516
Corcoran RB, Ebi H, Turke AB et al (2012) EGFR-mediated re-activation of MAPK signaling contributes to insensitivity of BRAF mutant colorectal cancers to RAF inhibition with vemurafenib. Cancer Discov 2(3):227–235. doi:10.1158/2159-8290.CD-11-0341
Cox AD, Der CJ (2010) Ras history: the saga continues. Small GTPases 1(1):2–27
Cunningham D, Humblet Y, Siena S et al (2004) Cetuximab monotherapy and cetuximab plus irinotecan in irinotecan-refractory metastatic colorectal cancer. N Engl J Med 351(4):337–345. doi:10.1056/NEJMoa033025
Davies H, Bignell GR, Cox C et al (2002) Mutations of the BRAF gene in human cancer. Nature 417(6892):949–954
Davies BR, Logie A, McKay JS et al (2007) AZD6244 (ARRY-142886), a potent inhibitor of mitogen-activated protein kinase/extracellular signal-regulated kinase kinase 1/2 kinases: mechanism of action in vivo, pharmacokinetic/pharmacodynamic relationship, and potential for combination in preclinical models. Mol Cancer Ther 6(8):2209–2219
de la Vega M, Burrows JF, Johnston JA (2011) Ubiquitination: added complexity in Ras and Rho family GTPase function. Small GTPases 2(4):192–201. doi:10.4161/sgtp.2.4.16707
De Roock W, Fieuws S, Biesmans B et al (2009, Abstract 289) DUSP expression as a predictor of outcome after cetuximab treatment in Kras wild type and mutant colorectal tumors. Gastrointestinal cancers symposium of the American Society of Clinical Oncology, San Francisco, 22–24 Jan 2009
De Roock W, Claes B, Bernasconi D et al (2010a) Effects of KRAS, BRAF, NRAS, and PIK3CA mutations on the efficacy of cetuximab plus chemotherapy in chemotherapy-refractory metastatic colorectal cancer: a retrospective consortium analysis. Lancet Oncol 11(8):753–762

De Roock W, Claes B, Bernasconi D et al (2010b) Effects of KRAS, BRAF, NRAS, and PIK3CA mutations on the efficacy of cetuximab plus chemotherapy in chemotherapy-refractory metastatic colorectal cancer: a retrospective consortium analysis. Lancet Oncol 11(8):753–762. doi:10.1016/S1470-2045(10)70130-3

De Roock W, Jonker DJ, Di Nicolantonio F et al (2010c) Association of KRAS p.G13D mutation with outcome in patients with chemotherapy-refractory metastatic colorectal cancer treated with cetuximab. JAMA 304(16):1812–1820

De Roock W, De Vriendt V, Normanno N et al (2011) KRAS, BRAF, PIK3CA, and PTEN mutations: implications for targeted therapies in metastatic colorectal cancer. Lancet Oncol 12(6):594–603

Di Nicolantonio F, Martini M, Molinari F et al (2008a) Wild-type BRAF is required for response to panitumumab or cetuximab in metastatic colorectal cancer. J Clin Oncol 26(35):5705–5712

Di Nicolantonio F, Martini M, Molinari F et al (2008b) Wild-type BRAF is required for response to panitumumab or cetuximab in metastatic colorectal cancer. J Clin Oncol 26(35):5705–5712. doi:10.1200/JCO.2008.18.0786

Dienstmann R, Vilar E, Tabernero J (2011) Molecular predictors of response to chemotherapy in colorectal cancer. Cancer J 17(2):114–126. doi:10.1097/PPO.0b013e318212f844

Dienstmann R, Serpico D, Rodon J et al (2012) Molecular profiling of patients with colorectal cancer and matched targeted therapy in phase I clinical trials. Mol Cancer Ther 11(9):2062–2071. doi:10.1158/1535-7163.MCT-12-0290

Douillard JY, Cunningham D, Roth AD et al (2000) Irinotecan combined with fluorouracil compared with fluorouracil alone as first-line treatment for metastatic colorectal cancer: a multicentre randomised trial. Lancet 355(9209):1041–1047

Ebi H, Corcoran RB, Singh A et al (2011) Receptor tyrosine kinases exert dominant control over PI3K signaling in human KRAS mutant colorectal cancers. J Clin Invest 121(11):4311–4321. doi:10.1172/JCI57909

Edkins S, O'Meara S, Parker A et al (2006) Recurrent KRAS codon 146 mutations in human colorectal cancer. Cancer Biol Ther 5(8):928–932

Engelman JA, Chen L, Tan X et al (2008) Effective use of PI3K and MEK inhibitors to treat mutant Kras G12D and PIK3CA H1047R murine lung cancers. Nat Med 14(12):1351–1356

Esteban LM, Vicario-Abejon C, Fernandez-Salguero P et al (2001) Targeted genomic disruption of H-ras and N-ras, individually or in combination, reveals the dispensability of both loci for mouse growth and development. Mol Cell Biol 21(5):1444–1452. doi:10.1128/MCB.21.5.1444-1452.2001

Farr CJ, Marshall CJ, Easty DJ et al (1988) A study of ras gene mutations in colonic adenomas from familial polyposis coli patients. Oncogene 3(6):673–678

Fearon ER (2011a) Molecular genetics of colorectal cancer. Annu Rev Pathol 6:479–507. doi:10.1146/annurev-pathol-011110-130235

Fearon ER (2011b) Molecular genetics of colorectal cancer. Annu Rev Pathol 6:479–507

Fearon ER, Vogelstein B (1990) A genetic model for colorectal tumorigenesis. Cell 61(5):759–767

Fehrenbacher N, Bar-Sagi D, Philips M (2009) Ras/MAPK signaling from endomembranes. Mol Oncol 3(4):297–307

Flaherty KT, Infante JR, Daud A et al (2012) Combined BRAF and MEK inhibition in melanoma with BRAF V600 mutations. N Engl J Med. doi:10.1056/NEJMoa1210093

Forrester K, Almoguera C, Han K et al (1987) Detection of high incidence of K-ras oncogenes during human colon tumorigenesis. Nature 327(6120):298–303

Giacchetti S, Perpoint B, Zidani R et al (2000) Phase III multicenter randomized trial of oxaliplatin added to chronomodulated fluorouracil-leucovorin as first-line treatment of metastatic colorectal cancer. J Clin Oncol 18(1):136–147

Gonzalez-Garcia A, Pritchard CA, Paterson HF et al (2005) RalGDS is required for tumor formation in a model of skin carcinogenesis. Cancer Cell 7(3):219–226. doi:10.1016/j.ccr.2005.01.029

Guerrero S, Casanova I, Farre L et al (2000) K-ras codon 12 mutation induces higher level of resistance to apoptosis and predisposition to anchorage-independent growth than codon 13 mutation or proto-oncogene overexpression. Cancer Res 60(23):6750–6756

Gupta S, Ramjaun AR, Haiko P et al (2007) Binding of ras to phosphoinositide 3-kinase p110alpha is required for ras-driven tumorigenesis in mice. Cell 129(5):957–968. doi:10.1016/j.cell.2007.03.051

Gysin S, Salt M, Young A et al (2011) Therapeutic strategies for targeting ras proteins. Genes Cancer 2(3):359–372
Haigis KM, Kendall KR, Wang Y et al (2008) Differential effects of oncogenic K-Ras and N-Ras on proliferation, differentiation and tumor progression in the colon. Nat Genet 40(5):600–608
Hao H, Muniz-Medina VM, Mehta H et al (2007) Context-dependent roles of mutant B-Raf signaling in melanoma and colorectal carcinoma cell growth. Mol Cancer Ther 6(8):2220–2229. doi:10.1158/1535-7163.MCT-06-0728
Hatzivassiliou G, Song K, Yen I et al (2010) RAF inhibitors prime wild-type RAF to activate the MAPK pathway and enhance growth. Nature 464(7287):431–435
Hatzivassiliou G, Liu B, O'Brien C et al (2012) ERK inhibition overcomes acquired resistance to MEK inhibitors. Mol Cancer Ther. doi:10.1158/1535-7163.MCT-11-1010
Heid I, Lubeseder-Martellato C, Sipos B et al (2011) Early requirement of Rac1 in a mouse model of pancreatic cancer. Gastroenterology 141(2):719–730, 730. e1–7. doi: 10.1053/j.gastro.2011.04.043
Heidorn SJ, Milagre C, Whittaker S et al (2010) Kinase-dead BRAF and oncogenic RAS cooperate to drive tumor progression through CRAF. Cell 140(2):209–221. doi:10.1016/j.cell.2009.12.040
Hoeflich KP, Herter S, Tien J et al (2009) Antitumor efficacy of the novel RAF inhibitor GDC-0879 is predicted by BRAFV600E mutational status and sustained extracellular signal-regulated kinase/mitogen-activated protein kinase pathway suppression. Cancer Res 69(7):3042–3051
Ikuta S, Edamatsu H, Li M et al (2008) Crucial role of phospholipase C epsilon in skin inflammation induced by tumor-promoting phorbol ester. Cancer Res 68(1):64–72. doi:10.1158/0008-5472.CAN-07-3245
Imamura Y, Morikawa T, Liao X et al (2012) Specific mutations in KRAS codons 12 and 13, and patient prognosis in 1075 BRAF wild-type colorectal cancers. Clin Cancer Res 18(17):4753–4763. doi:10.1158/1078-0432.CCR-11-3210
Irahara N, Baba Y, Nosho K et al (2010) NRAS mutations are rare in colorectal cancer. Diagn Mol Pathol 19(3):157–163
Janakiraman M, Vakiani E, Zeng Z et al (2010) Genomic and biological characterization of exon 4 KRAS mutations in human cancer. Cancer Res 70(14):5901–5911
Janssen KP, Alberici P, Fsihi H et al (2006) APC and oncogenic KRAS are synergistic in enhancing Wnt signaling in intestinal tumor formation and progression. Gastroenterology 131(4):1096–1109
Jemal A, Bray F, Center MM et al (2011) Global cancer statistics. CA Cancer J Clin 61(2):69–90. doi:10.3322/caac.20107
Jhawer M, Goel S, Wilson AJ et al (2008) PIK3CA mutation/PTEN expression status predicts response of colon cancer cells to the epidermal growth factor receptor inhibitor cetuximab. Cancer Res 68(6):1953–1961
Johnson L, Greenbaum D, Cichowski K et al (1997) K-ras is an essential gene in the mouse with partial functional overlap with N-ras. Genes Dev 11(19):2468–2481
Karapetis CS, Khambata-Ford S, Jonker DJ et al (2008) K-ras mutations and benefit from cetuximab in advanced colorectal cancer. N Engl J Med 359(17):1757–1765. doi:10.1056/NEJMoa0804385
Kissil JL, Walmsley MJ, Hanlon L et al (2007) Requirement for Rac1 in a K-ras induced lung cancer in the mouse. Cancer Res 67(17):8089–8094. doi:10.1158/0008-5472.CAN-07-2300
Koera K, Nakamura K, Nakao K et al (1997) K-ras is essential for the development of the mouse embryo. Oncogene 15(10):1151–1159. doi:10.1038/sj.onc.1201284
Kopetz S, Desai J, Chan E et al (2010) PLX4032 in metastatic colorectal cancer patients with mutant BRAF tumors. ASCO Meeting Abstracts no 28(15 suppl):3534
Laurent-Puig P, Cayre A, Manceau G et al (2009) Analysis of PTEN, BRAF, and EGFR status in determining benefit from cetuximab therapy in wild-type KRAS metastatic colon cancer. J Clin Oncol 27(35):5924–5930. doi:10.1200/JCO.2008.21.6796
Li W, Zhu T, Guan KL (2004) Transformation potential of Ras isoforms correlates with activation of phosphatidylinositol 3-kinase but not ERK. J Biol Chem 279(36):37398–37406. doi:10.1074/jbc.M405730200

Li M, Edamatsu H, Kitazawa R et al (2009) Phospholipase Cepsilon promotes intestinal tumorigenesis of Apc(Min/+) mice through augmentation of inflammation and angiogenesis. Carcinogenesis 30(8):1424–1432
Lievre A, Bachet JB, Le Corre D et al (2006) KRAS mutation status is predictive of response to cetuximab therapy in colorectal cancer. Cancer Res 66(8):3992–3995
Lievre A, Bachet JB, Boige V et al (2008) KRAS mutations as an independent prognostic factor in patients with advanced colorectal cancer treated with cetuximab. J Clin Oncol 26(3):374–379
Lim KH, Counter CM (2005) Reduction in the requirement of oncogenic Ras signaling to activation of PI3K/AKT pathway during tumor maintenance. Cancer Cell 8(5):381–392. doi:10.1016/j.ccr.2005.10.014
Lim KH, O'Hayer K, Adam SJ et al (2006) Divergent roles for RalA and RalB in malignant growth of human pancreatic carcinoma cells. Curr Biol 16(24):2385–2394
Liu M, Sjogren AK, Karlsson C et al (2010) Targeting the protein prenyltransferases efficiently reduces tumor development in mice with K-RAS-induced lung cancer. Proc Natl Acad Sci U S A 107(14):6471–6476. doi:10.1073/pnas.0908396107
Luo J, Emanuele MJ, Li D et al (2009) A genome-wide RNAi screen identifies multiple synthetic lethal interactions with the Ras oncogene. Cell 137(5):835–848
Luo F, Ye H, Hamoudi R et al (2010) K-ras exon 4A has a tumour suppressor effect on carcinogen-induced murine colonic adenoma formation. J Pathol 220(5):542–550. doi:10.1002/path.2672
Malliri A, van der Kammen RA, Clark K et al (2002) Mice deficient in the Rac activator Tiam1 are resistant to Ras-induced skin tumours. Nature 417(6891):867–871. doi:10.1038/nature00848
Malliri A, Rygiel TP, van der Kammen RA et al (2006) The rac activator Tiam1 is a Wnt-responsive gene that modifies intestinal tumor development. J Biol Chem 281(1):543–548
Mamane Y, Petroulakis E, LeBacquer O et al (2006) mTOR, translation initiation and cancer. Oncogene 25(48):6416–6422
Marais R, Light Y, Paterson HF et al (1997) Differential regulation of Raf-1, A-Raf, and B-Raf by oncogenic ras and tyrosine kinases. J Biol Chem 272(7):4378–4383
Markowitz SD, Bertagnolli MM (2009) Molecular origins of cancer: molecular basis of colorectal cancer. N Engl J Med 361(25):2449–2460. doi:10.1056/NEJMra0804588
Martin TD, Samuel JC, Routh ED et al (2011) Activation and involvement of Ral GTPases in colorectal cancer. Cancer Res 71(1):206–215
Mason CS, Springer CJ, Cooper RG et al (1999) Serine and tyrosine phosphorylations cooperate in Raf-1, but not B-Raf activation. EMBO J 18(8):2137–2148
Matallanas D, Birtwistle M, Romano D et al (2011) Raf family kinases: old dogs have learned new tricks. Genes Cancer 2(3):232–260. doi:10.1177/1947601911407323
McFall A, Ulku A, Lambert QT et al (2001) Oncogenic Ras blocks anoikis by activation of a novel effector pathway independent of phosphatidylinositol 3-kinase. Mol Cell Biol 21(16):5488–5499. doi:10.1128/MCB.21.16.5488-5499.2001
Michaelson JS, Cho S, Browning B et al (2005) Tweak induces mammary epithelial branching morphogenesis. Oncogene 24(16):2613–2624. doi:10.1038/sj.onc.1208208
Mirzoeva OK, Das D, Heiser LM et al (2009) Basal subtype and MAPK/ERK kinase (MEK)-phosphoinositide 3-kinase feedback signaling determine susceptibility of breast cancer cells to MEK inhibition. Cancer Res 69(2):565–572
Montagut C, Settleman J (2009) Targeting the RAF-MEK-ERK pathway in cancer therapy. Cancer Lett 283(2):125–134
Neel NF, Martin TD, Stratford JK et al (2011) The RalGEF-Ral effector signaling network: the road less traveled for anti-Ras drug discovery. Genes Cancer 2(3):275–287. doi:10.1177/1947601911407329
Oberholzer PA, Kee D, Dziunycz P et al (2012) RAS mutations are associated with the development of cutaneous squamous cell tumors in patients treated with RAF inhibitors. J Clin Oncol 30(3):316–321. doi:10.1200/JCO.2011.36.7680
Ong CC, Jubb AM, Haverty PM et al (2011) Targeting p21-activated kinase 1 (PAK1) to induce apoptosis of tumor cells. Proc Natl Acad Sci U S A 108(17):7177–7182. doi:10.1073/pnas.1103350108

Parsons DW, Wang TL, Samuels Y et al (2005) Colorectal cancer: mutations in a signalling pathway. Nature 436(7052):792
Patek CE, Arends MJ, Rose L et al (2008) The pro-apoptotic K-Ras 4A proto-oncoprotein does not affect tumorigenesis in the ApcMin/+ mouse small intestine. BMC Gastroenterol 8:24. doi:10.1186/1471-230X-8-24
Paz A, Haklai R, Elad-Sfadia G et al (2001) Galectin-1 binds oncogenic H-Ras to mediate Ras membrane anchorage and cell transformation. Oncogene 20(51):7486–7493. doi:10.1038/sj.onc.1204950
Plowman SJ, Williamson DJ, O'Sullivan MJ et al (2003) While K-ras is essential for mouse development, expression of the K-ras 4A splice variant is dispensable. Mol Cell Biol 23(24):9245–9250
Plowman SJ, Berry RL, Bader SA et al (2006) K-ras 4A and 4B are co-expressed widely in human tissues, and their ratio is altered in sporadic colorectal cancer. J Exp Clin Cancer Res 25(2):259–267
Potenza N, Vecchione C, Notte A et al (2005) Replacement of K-Ras with H-Ras supports normal embryonic development despite inducing cardiovascular pathology in adult mice. EMBO Rep 6(5):432–437. doi:10.1038/sj.embor.7400397
Poulikakos PI, Zhang C, Bollag G et al (2010) RAF inhibitors transactivate RAF dimers and ERK signalling in cells with wild-type BRAF. Nature 464(7287):427–430
Prahallad A, Sun C, Huang S et al (2012) Unresponsiveness of colon cancer to BRAF(V600E) inhibition through feedback activation of EGFR. Nature 483(7387):100–103. doi:10.1038/nature10868
Prasad SS, Baillie DL (1989) Evolutionarily conserved coding sequences in the dpy-20-unc-22 region of Caenorhabditis elegans. Genomics 5(2):185–198
Pratilas CA, Taylor BS, Ye Q et al (2009) (V600E)BRAF is associated with disabled feedback inhibition of RAF-MEK signaling and elevated transcriptional output of the pathway. Proc Natl Acad Sci U S A 106(11):4519–4524
Pretlow TP, Pretlow TG (2005) Mutant KRAS in aberrant crypt foci (ACF): initiation of colorectal cancer? Biochim Biophys Acta 1756(2):83–96
Price TJ, Hardingham JE, Lee CK et al (2011) Impact of KRAS and BRAF gene mutation status on outcomes from the phase III AGITG MAX trial of capecitabine alone or in combination with bevacizumab and mitomycin in advanced colorectal cancer. J Clin Oncol 29(19):2675–2682
Rajagopalan H, Bardelli A, Lengauer C et al (2002) Tumorigenesis: RAF/RAS oncogenes and mismatch-repair status. Nature 418(6901):934. doi:10.1038/418934a
Rinehart J, Adjei AA, Lorusso PM et al (2004) Multicenter phase II study of the oral MEK inhibitor, CI-1040, in patients with advanced non-small-cell lung, breast, colon, and pancreatic cancer. J Clin Oncol 22(22):4456–4462
Roth AD, Tejpar S, Delorenzi M et al (2010) Prognostic role of KRAS and BRAF in stage II and III resected colon cancer: results of the translational study on the PETACC-3, EORTC 40993, SAKK 60–00 trial. J Clin Oncol 28(3):466–474
Saltz LB, Meropol NJ, Loehrer PJ Sr et al (2004) Phase II trial of cetuximab in patients with refractory colorectal cancer that expresses the epidermal growth factor receptor. J Clin Oncol 22(7):1201–1208. doi:10.1200/JCO.2004.10.182
Samowitz WS, Sweeney C, Herrick J et al (2005) Poor survival associated with the BRAF V600E mutation in microsatellite-stable colon cancers. Cancer Res 65(14):6063–6069
Samuels Y, Wang Z, Bardelli A et al (2004) High frequency of mutations of the PIK3CA gene in human cancers. Science 304(5670):554. doi:10.1126/science.1096502
Santini D, Loupakis F, Vincenzi B et al (2008) High concordance of KRAS status between primary colorectal tumors and related metastatic sites: implications for clinical practice. Oncologist 13(12):1270–1275
Saridaki Z, Papadatos-Pastos D, Tzardi M et al (2010) BRAF mutations, microsatellite instability status and cyclin D1 expression predict metastatic colorectal patients' outcome. Br J Cancer 102(12):1762–1768

Scholl C, Frohling S, Dunn IF et al (2009) Synthetic lethal interaction between oncogenic KRAS dependency and STK33 suppression in human cancer cells. Cell 137(5):821–834

Shalom-Feuerstein R, Cooks T, Raz A et al (2005) Galectin-3 regulates a molecular switch from N-Ras to K-Ras usage in human breast carcinoma cells. Cancer Res 65(16):7292–7300. doi:10.1158/0008-5472.CAN-05-0775

Shimizu K, Goldfarb M, Suard Y et al (1983) Three human transforming genes are related to the viral ras oncogenes. Proc Natl Acad Sci U S A 80(8):2112–2116

Shirasawa S, Furuse M, Yokoyama N et al (1993) Altered growth of human colon cancer cell lines disrupted at activated Ki-ras. Science 260(5104):85–88

Siegel R, Ward E, Brawley O et al (2011) Cancer statistics, 2011: the impact of eliminating socio-economic and racial disparities on premature cancer deaths. CA Cancer J Clin 61(4):212–236. doi:10.3322/caac.20121

Sjoblom T, Jones S, Wood LD et al (2006) The consensus coding sequences of human breast and colorectal cancers. Science 314(5797):268–274. doi:10.1126/science.1133427

Sjogren AK, Andersson KM, Liu M et al (2007) GGTase-I deficiency reduces tumor formation and improves survival in mice with K-RAS-induced lung cancer. J Clin Invest 117(5):1294–1304. doi:10.1172/JCI30868

Solit DB, Garraway LA, Pratilas CA et al (2006) BRAF mutation predicts sensitivity to MEK inhibition. Nature 439(7074):358–362

Souglakos J, Philips J, Wang R et al (2009) Prognostic and predictive value of common mutations for treatment response and survival in patients with metastatic colorectal cancer. Br J Cancer 101(3):465–472

Su F, Viros A, Milagre C et al (2012) RAS mutations in cutaneous squamous-cell carcinomas in patients treated with BRAF inhibitors. N Engl J Med 366(3):207–215. doi:10.1056/NEJMoa1105358

Takayama T, Ohi M, Hayashi T et al (2001) Analysis of K-ras, APC, and beta-catenin in aberrant crypt foci in sporadic adenoma, cancer, and familial adenomatous polyposis. Gastroenterology 121(3):599–611

Tol J, Nagtegaal ID, Punt CJ (2009) BRAF mutation in metastatic colorectal cancer. N Engl J Med 361(1):98–99

Tsai J, Lee JT, Wang W et al (2008) Discovery of a selective inhibitor of oncogenic B-Raf kinase with potent antimelanoma activity. Proc Natl Acad Sci U S A 105(8):3041–3046

Udell CM, Rajakulendran T, Sicheri F et al (2011) Mechanistic principles of RAF kinase signaling. Cell Mol Life Sci 68(4):553–565. doi:10.1007/s00018-010-0520-6

Vakiani E, Solit DB (2011) KRAS and BRAF: drug targets and predictive biomarkers. J Pathol 223(2):219–229

Vakiani E, Yantiss RK (2009) Pathologic features and biologic importance of colorectal serrated polyps. Adv Anat Pathol 16(2):79–91

Vakiani E, Janakiraman M, Shen R et al (2012) Comparative genomic analysis of primary versus metastatic colorectal carcinomas. J Clin Oncol 30(24):2956–2962. doi:10.1200/JCO.2011.38.2994

Van Cutsem E, Kohne CH, Lang I et al (2011) Cetuximab plus irinotecan, fluorouracil, and leucovorin as first-line treatment for metastatic colorectal cancer: updated analysis of overall survival according to tumor KRAS and BRAF mutation status. J Clin Oncol 29(15):2011–2019

Velho S, Oliveira C, Ferreira A et al (2005) The prevalence of PIK3CA mutations in gastric and colon cancer. Eur J Cancer 41(11):1649–1654

Vetter IR, Wittinghofer A (2001) The guanine nucleotide-binding switch in three dimensions. Science 294(5545):1299–1304

Vigil D, Cherfils J, Rossman KL et al (2010) Ras superfamily GEFs and GAPs: validated and tractable targets for cancer therapy? Nature reviews. Cancer 10(12):842–857. doi:10.1038/nrc2960

Vogelstein B, Fearon ER, Hamilton SR et al (1988) Genetic alterations during colorectal-tumor development. N Engl J Med 319(9):525–532

Wahlstrom AM, Cutts BA, Liu M et al (2008) Inactivating Icmt ameliorates K-RAS-induced myeloproliferative disease. Blood 112(4):1357–1365. doi:10.1182/blood-2007-06-094060

Wang Y, Van Becelaere K, Jiang P et al (2005) A role for K-ras in conferring resistance to the MEK inhibitor, CI-1040. Neoplasia 7(4):336–347

Wang Y, Ngo VN, Marani M et al (2010) Critical role for transcriptional repressor Snail2 in transformation by oncogenic RAS in colorectal carcinoma cells. Oncogene 29(33):4658–4670. doi:10.1038/onc.2010.218

Wee S, Jagani Z, Xiang KX et al (2009) PI3K pathway activation mediates resistance to MEK inhibitors in KRAS mutant cancers. Cancer Res 69(10):4286–4293

Wennerberg K, Rossman KL, Der CJ (2005) The Ras superfamily at a glance. J Cell Sci 118(Pt 5):843–846. doi:10.1242/jcs.01660

Wilson PM, Labonte MJ, Lenz HJ (2010) Molecular markers in the treatment of metastatic colorectal cancer. Cancer J 16(3):262–272. doi:10.1097/PPO.0b013e3181e07738

Wong R, Cunningham D (2008) Using predictive biomarkers to select patients with advanced colorectal cancer for treatment with epidermal growth factor receptor antibodies. J Clin Oncol 26(35):5668–5670

Wood LD, Parsons DW, Jones S et al (2007) The genomic landscapes of human breast and colorectal cancers. Science 318(5853):1108–1113. doi:10.1126/science.1145720

Yeh JJ, Routh ED, Rubinas T et al (2009) KRAS/BRAF mutation status and ERK1/2 activation as biomarkers for MEK1/2 inhibitor therapy in colorectal cancer. Mol Cancer Ther 8(4):834–843

Zhang YJ, Tian XQ, Sun DF et al (2009) Combined inhibition of MEK and mTOR signaling inhibits initiation and progression of colorectal cancer. Cancer Invest 27(3):273–285

Zhao L, Vogt PK (2008) Helical domain and kinase domain mutations in p110alpha of phosphatidylinositol 3-kinase induce gain of function by different mechanisms. Proc Natl Acad Sci U S A 105(7):2652–2657. doi:10.1073/pnas.0712169105

Chapter 6
The PI3K Pathway in Colorectal Cancers

Jihye Yun, George Poulogiannis, Evan T. Brower, Samuel Klempner, and Lewis L. Cantley

Abstract Phosphoinositide 3-kinase (PI3K) belongs to the intracellular lipid kinases family involved in diverse physiological processes, including proliferation, apoptosis, growth, and metabolism. Recent mutation analysis has shown that the PI3K pathway is one of the most frequently dysregulated pathways in human cancer, including colorectal cancer (CRC). Accordingly, significant effort has been made to develop pharmacological inhibitors targeting PI3K or key nodes in this pathway, such as AKT and mTOR. There are currently more than 20 unique compounds targeting the PI3K pathway being assessed in numerous cancer-related clinical trials. In addition, the mutation status of PI3K pathway in cancers may have

J. Yun • L.L. Cantley (✉)
Department of Medicine, Weill Cornell Medical College, New York, NY 10065, USA

Department of Medicine, Division of Signal Transduction, Beth Israel Deaconess Medical Center, Boston, MA 02115, USA

Department of Systems Biology, Harvard Medical School, Boston, MA 02115, USA
e-mail: lec2014@med.cornell.edu

G. Poulogiannis
Department of Medicine, Division of Signal Transduction, Beth Israel Deaconess Medical Center, Boston, MA 02115, USA

Department of Systems Biology, Harvard Medical School, Boston, MA 02115, USA

E.T. Brower
Ludwig Center for Cancer Genetics and Therapeutics, Sidney Kimmel Comprehensive Cancer Center, Johns Hopkins University School of Medicine, Baltimore, MD 21287, USA

S. Klempner
Division of Hematology-Oncology, Beth Israel Deaconess Medical Center, Boston, MA 02115, USA

K.M. Haigis (ed.), *Molecular Pathogenesis of Colorectal Cancer*,
DOI 10.1007/978-1-4614-8412-7_6, © Springer Science+Business Media New York 2013

predictive and prognostic implications. After 3 decades of the discovery of PI3K, we are now at an exciting intersection in translating our knowledge of the PI3K signaling pathway into developing effective therapeutics for the treatment of cancer. A comprehensive understanding of circuits and regulations of this pathway are essential to the rational development of such therapies. In this chapter, we will discuss recent advances in our understanding of the functions and mutations of PI3K signaling pathway in the pathogenesis of CRC. We will also review current drug-discovery efforts and challenges targeting PI3K signaling for the treatment of CRC.

6.1 Introduction

Colorectal cancer (CRC) is one of the most common cancers, with an estimated 143,460 new diagnoses and 51,690 deaths in the United States in 2012. CRC remains the third most frequent cancer in the United States, as well as the third most common cause of cancer-related death, accounting for roughly 9 % of US cancer deaths for both men and women. Despite advances in screening, nearly 20 % of patients present with metastatic disease, which carries a poor prognosis, with 5-year survival rates of 12 % (Grothey 2009). While cytotoxic fluorouracil (5-FU)-based regimens have been standard of care as adjuvant and first-line metastatic therapies for CRC treatment, advances have been slow and the efficacy of these agents has reached a plateau (Cook et al. 1969). As such, recent efforts have shifted toward the development of novel therapeutic agents that inhibit specific molecular pathways. Inhibiting the epidermal growth factor receptor (EGFR) with monoclonal antibodies, such as cetuximab and panitumumab, and blocking angiogenesis with antibodies against vascular endothelial growth factor receptor (VEGFR), such as bevacizumab, are successful examples of targeted therapies in CRC. However, the clinical benefits of these targeted therapies in most cases are short-term and often limited to subgroups of patients, indicating the need for evaluating novel biomarkers and identifying new targets for drug therapy of CRC.

The phosphoinositide 3-kinase (PI3K) pathway is one of the most deregulated pathways in human cancer. Several components of this pathway, including PI3K, the v-akt murine thymoma viral oncogene homolog (AKT), and the mammalian target of rapamycin (mTOR), are druggable and plausible targets for cancer therapy. Consequently, the development of therapeutics targeting this pathway has occurred at a rapid pace over the past 10 years, and preclinical and early clinical studies are beginning to suggest strategies to increase efficacy.

In this chapter we provide a comprehensive analysis of genetic alterations in the PI3K pathway detected in CRC and discuss their value as prognostic indicators and potential roles as predictive biomarkers for anti-EGFR therapy. In order to frame the discussion, we begin with a review of the current understanding of the PI3K signaling pathway and the effects that it confers on cellular growth, proliferation, survival, and metabolism. Finally, we discuss some of the current and emerging therapeutic approaches to targeting the PI3K pathway in CRC.

6.2 The Key Players of the PI3K Pathway

6.2.1 PI3Ks

PI3Ks are divided into three classes (I, II, and III) according to their structural characteristics and lipid substrate preferences (Fig. 6.1). Different classes of PI3K also have distinct roles in cellular signaling pathways (Engelman et al. 2006). Class I enzymes are the best characterized of the PI3K classes and are the major class known to be associated with cancer. Therefore, we will mainly focus on class I PI3K throughout this chapter. All PI3K classes catalyze the phosphorylation of inositol-containing lipids, known as phosphatidylinositols (PtdIns), at the 3′-position of the inositol ring. The primary substrate of class I PI3K is phosphatidylinositol (4,5)-bisphosphate (PIP_2), which is converted to phosphatidylinositol (3,4,5)-trisphosphate (PIP_3). PIP_3 is an important second messenger in cell physiology. Through direct interactions with pleckstrin homology (PH) domains on a wide variety of signaling proteins, including Tec family protein-tyrosine kinases, AKT family kinases, and PDK1 (3-phosphoinositide-dependent protein kinase-1) and with various guanine exchange factors (GEFs) and GTPase-activating proteins (GAPs) of low molecular weight GTP-binding proteins, PIP_3 orchestrates a set of events controlling cellular growth, metabolism, proliferation, and survival. In contrast to PIP_3, which is produced only from Class I PI3Ks, phosphatidylinositol-3-phosphate (PI3P) is produced by both Class II and Class III PI3Ks. This lipid binds to FYVE domains and PX domains in a variety of proteins to control intracellular membrane trafficking, especially trafficking through early endosomes. Also of interest in cancer is phosphatidylinositol-3,4-bisphosphate (PI3, $4P_2$), which is produced by Class II PI3Ks via phosphorylation of the 3′ position of phosphatidylinositol-4-phosphate or by SHIP family phosphatases via dephosphorylation of the 5′ position of PIP_3. PI3, $4P_2$, like PIP_3, binds to AKT and PDK1 to facilitate AKT activation, but fails to bind to most of the other PIP_3 targets.

6.2.1.1 Class IA PI3Ks and Cancer

Class IA PI3Ks have been most frequently associated with human cancer. These enzymes are heterodimers of p110 family catalytic subunits and p85 family regulatory subunits. PIK3R1 encodes p85α (and its alternative start site variants p55α and p50α). PIK3R2 and PIK3R3 encode the p85β and p55γ isoforms of the p85 regulatory subunit, respectively. This group of subunits is collectively called p85 (Engelman et al. 2006; Bader et al. 2005). The class IA p85 regulatory isoforms have a common structure composed of a p110-binding domain (also called the inter-SH2 domain, iSH2) flanked by two Src-homology 2 (SH2) domains and this core structure is conserved back to worms and flies. The two longer isoforms, p85α and p85β, have an extended N-terminal region containing a Src-homology 3 (SH3)

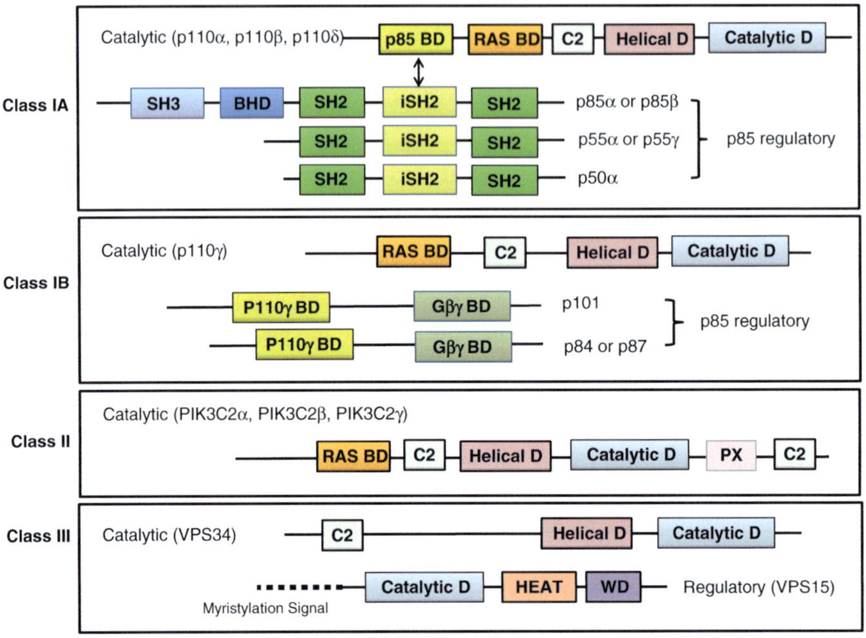

Fig. 6.1 Classification and domain structure of phosphatidylinositol 3-kinase (PI3K). PI3Ks are divided into three classes (I, II, and III) based on their structural characteristics and lipid substrate preference. Class I PI3Ks are further divided into two subfamilies, IA and IB, according to the receptors with which they interact. Class IA PI3Ks are heterodimers consisting of a p110 catalytic subunit and a p85 regulatory subunit. There are three p110 catalytic isoforms: p110**α**, p110**β**, and p110**δ**, which all have a p85-binding domain (p85BD) at the N-terminus, followed by a Ras-binding domain (RBD), a putative membrane-binding C2 domain (C2), and helical domain (Helical D), and a C-terminal catalytic domain (Catalytic D). The p85 regulatory subunits share a core structure consisting of a p110-binding domain called the iSH2 domain, flanked by SH2 domains. The two longer isoforms, p85**α** and p85**β**, have an SH3 domain and BCR homology domain (BHD) located in their extended N-terminal region. Class IB PI3K is a heterodimer composed of the catalytic subunit p110**γ** and the regulatory subunit p101. p110**γ** is activated by G protein-coupled receptors (GPCRs). Class II PI3Ks consist of only a p110-like catalytic subunit. There are three class II PI3K isoforms: PI3KC2**α**, PI3KC2**β**, and PI3KC2**γ**, each of which has an extended divergent N-terminus followed by a RBD, a C2 domain, a helical domain, a catalytic domain, PX (Phox homology), and C2 domains at the C-termini. The class III PI3K consists of a heterodimer of the catalytic subunit, VPS34 (homologue of the yeast vacuolar protein 34; also known as PIK3C3 in mammal) and a regulatory subunit, VPS15 (also known as PIK3R4 in mammal). Vps15 consists of a catalytic domain (which is thought to be inactive), HEAT domains (which might mediate protein–protein interactions), and WD repeats, which have structural and functional characteristics similar to a G**β** subunit

domain and a BCR homology domain (BHD) (Fruman et al. 1998). The SH3 and BHD are speculated to have a negative regulatory role toward the catalytic activity of the p110 subunit. This is consistent with the observation that the p55α and p50α subunits are more effective activators of p110 than is p85α (Vivanco and Sawyers 2002; Inukai et al. 1997; Ueki et al. 2000). Three genes, PIK3CA, PIK3CB, and

PIK3CD, encode the highly homologous p110 catalytic subunit isoforms p110α, p110β, and p110δ, respectively. These three p110 subunits are comprised of five domains; an N-terminal adaptor binding domain (ABD) that interacts with the p85 regulatory subunit, a Ras-binding domain (RBD) that mediates activation by the small GTPase Ras, a C2 domain that might be important for membrane anchoring, a helical domain, and a C-terminal catalytic domain. The last four domains have significant sequence homology among all isoforms.

The p85 regulatory subunit is essential for mediating class IA PI3K activation by receptor tyrosine kinases (RTKs), such as EGFR, platelet-derived growth factor receptor (PDGFR), and IGF-1R (insulin-like growth factor-1 receptor). The SH2 domains of p85 bind to phosphotyrosine residues arranged in a pYXXM (in which "pY" indicates a phosphorylated tyrosine) motif on activated RTKs. In some cases, the p85–RTK interactions occur indirectly through adaptor proteins, such as the insulin receptor substrates (IRS1 and IRS2) downstream of IGF-1R (Vivanco and Sawyers 2002; White 1998). Binding of p85 to RTKs or phosphoprotein intermediaries relieves the basal inhibition of p110 by p85 and recruits the p85-p110 heterodimer to the plasma membrane, where it phosphorylates the membrane lipid PIP_2 to produce PIP_3 (Yu et al. 1998a, b). This leads to activation of various cellular processes, such as proliferation, growth, survival, and metabolism. Interestingly, the p110β isoform is regulated not only by the p85 regulatory subunit but also by binding to Gβγ subunits of heterotrimeric G proteins, suggesting that p110β might integrate signals from GPCRs as well as RTKs (Kurosu et al. 1997; Roche et al. 1998). However, the implication of the p110β activation by GPCRs in cancer remains less well defined. p110α and p110β are expressed ubiquitously, whereas p110δ is predominantly expressed in lymphocytes. Although p110α, p110β, and p110δ have very similar structures and share the p85 regulatory subunits, numerous studies indicated that they may have distinct functions. For example, germline deletion of either p110α or p110β results in embryonic lethality (Bi et al. 1999; Foukas et al. 2006). Mice heterozygous for kinase dead mutation in p110α were viable and fertile, but showed severe defects in the insulin pathway such as hyperinsulinemia, glucose intolerance, and increased adiposity (Foukas et al. 2006). p110δ, although not essential, has an important role in the regulation of the immune compartment, especially B-cell growth (Fruman 2004). As we will be discussed in detail later, both PIK3CA (p110α) and PIK3R1 (p85α) are somatically mutated in various cancers including CRC (Ikenoue et al. 2005; Mizoguchi et al. 2004; Philp et al. 2001; Samuels et al. 2004).

6.2.1.2 Class IB PI3Ks

Similar to class IA PI3Ks, class IB PI3Ks are heterodimers composed of the catalytic subunit p110γ and the regulatory subunit p101 (Fig. 6.1). Although p110γ is highly homologous with the class IA p110 proteins, the p101 regulatory subunit is distinct from the p85 Class IA regulatory subunit. Recently, two additional regulatory subunits, p84 and p87PIKAP (PI3Kγ adaptor protein of 87 kDa), have been identified (Voigt et al. 2006; Suire et al. 2005). The regulatory subunits complexed with p110γ do not have SH2 domains, and thus do not interact with RTKs. Instead, p110γ is activated exclusively by GPCRs through direct interaction with the Gβγ subunit of trimeric G proteins. p110γ is primarily expressed in leukocytes but is also found in the heart, pancreas, liver, and skeletal muscle (Sasaki et al. 2000).

6.2.2 PIP_3 Phosphatases

The main consequence of class I PI3K activation is the generation of PIP_3 in the plasma membrane. PIP_3 functions as a second messenger to activate effector protein kinases such as AKT. Thus, in most tissues, PIP_3 has a pivotal role in the actions of insulin, growth factors, and cytokines, thereby mediating effects of diverse physiological processes including proliferation, apoptosis, growth, and metabolism. The cellular levels of PIP_3 are hardly detectable in mammalian cells under unstimulated growth conditions and are tightly regulated by the opposing activity of several PIP_3 phosphatases (PTEN, SHIP1, and SHIP2). PTEN (phosphatase and tensin homologue, deleted on chromosome ten), also called MMAC1 and TEP1, is an important tumor suppressor and is the PIP_3 phosphatase most clearly involved in cancer. Loss of PTEN function occurs through mutations, deletions, or epigenetic silencing in a variety of human cancers at high frequency, making PTEN the second most frequently mutated tumor suppressor gene after p53 (Stokoe 2001). PTEN functionally antagonizes PI3K activity through its intrinsic lipid phosphatase activity that decreases the cellular level of PIP_3 by converting PIP_3 back into PtdIns $(4,5)P_2$ ($PI4,5P_2$) (Fig. 6.2). Thus, loss of PTEN results in constitutively active signaling through the PI3K pathway, leading to tumor development (Cully et al. 2006).

The SHIP phosphatases also act on PIP_3, but remove phosphate from the 5-position instead of 3-position, generating PtdIns $(3,4)P_2$ ($PI3, 4P_2$) (Fig. 6.2). PI3, $4P_2$ can function as a second messenger to recruit certain PH-domain-containing proteins, such as AKT and PDK1 to the plasma membrane. But several PIP_3 binding proteins, such as TEC family protein-Tyr kinases, fail to bind to PI3, $4P_2$. Therefore, although both PTEN and SHIP reduce the level of PIP_3 in cells, PTEN terminates all downstream PI3K signaling, while SHIP only terminates a subset of downstream signals. To completely shut off downstream signaling, PI3, $4P_2$ is dephosphorylated by distinct 4-phosphatases called INPP4A and INPP4B. INPP4B has also been identified as a tumor suppressor in breast and ovarian cancers (Agoulnik et al. 2011). Targeted deletion of PTEN recapitulates many of the ramifications of PTEN loss in human cancers. Homozygous deletion of PTEN causes embryonic lethality, indicating an essential role of PTEN during embryonic development (Di Cristofano et al. 1998). Mice that are heterozygous for PTEN are viable, but have a high incidence of T-cell lymphomas, germline tumors and cancers in several epithelial tissues, including the intestine, endometrium, prostate, and mammary glands (Di Cristofano et al. 1998). Tissue-specific homozygous deletion of PTEN in the prostate epithelium leads to aggressive prostate carcinoma (Wang et al. 2003). Likewise, PTEN deletion in T cells and mammary glands causes aggressive lymphomas and breast tumors, respectively (Kishimoto et al. 2003).

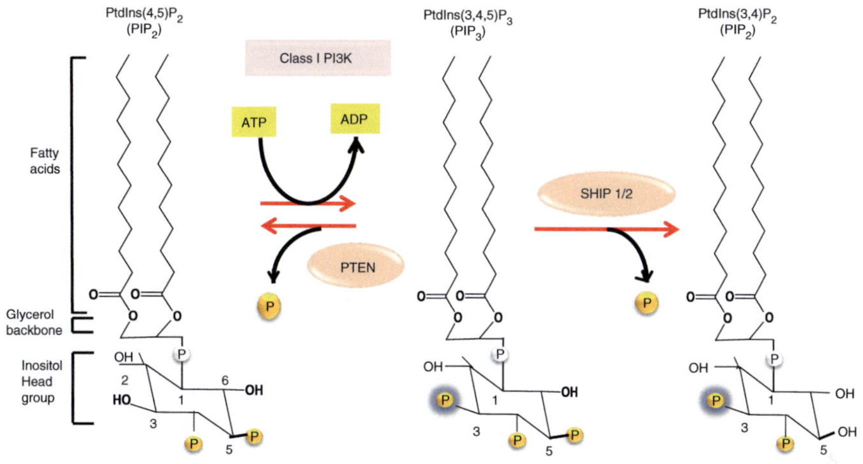

Fig. 6.2 Structure and generation of phosphatidylinositol-3,4,5-trisphosphate. Phosphatidylinositol-3,4,5-triphosphate (PtdIns(3,4,5)P_3) is an essential second messenger that regulates many cellular processes. Class I PI3Ks phosphorylate the inositol ring of phosphatidylinositol-4,5-triphosphate (PtdIns(4,5)P_2) at the 3-position, to generate PtdIns (3,4,5)P_3. PTEN (phosphatase and tensin homologues) is a lipid phosphatase that removes phosphate at the 3-position of PtdIns (3,4,5)P_3, converting it back to PtdIns (4,5)P_2. Additionally, PtdIns (3,4,5)P_3 can be dephosphorylated at the 5-position by SHIP1 or SHIP2 to generate PtdIns (3,4)P_2

6.2.3 AKT: Direct Effector of PIP_3

AKT, also known as protein kinase B (PKB), the human homologue of the retroviral oncogene v-Akt, is the main downstream executor of the PI3K signaling pathway (Fig. 6.3). This serine-threonine protein kinase has three isoforms—AKT1, AKT2, and AKT3. The three isoforms share a similar structure: an amino-terminal pleckstrin homoloy (PH) domain, a central catalytic domain, and a short carboxy-terminal regulatory domain. AKT is activated by a dual regulatory mechanism that requires both translocation to the plasma membrane and phosphorylation at Thr308 and Ser473 (Alessi et al. 1997; Stephens et al. 1998). The generation of PIP_3 at the plasma membrane by activated PI3K facilitates the recruitment of both AKT and a second protein kinase, PDK1, to the membrane due to the ability of the PH domains of these proteins to bind to PIP_3. PIP_3 binding induces a conformational change in AKT, resulting in the exposure of Thr308 for phosphorylation by PDK1. Full activation of AKT requires phosphorylation of Ser473 by TORC2, or in certain situations, another PIK-family protein kinase such as DNA-PK or ATM (Sarbassov et al. 2005). After dual-phosphorylation and activation, AKT phosphorylates an array of target proteins containing the amino acid sequence RxRxxS/T-B (where x represents any amino acid and B is any bulky hydrophobic residue) (Alessi et al. 1996). Currently, more than 100 different AKT substrates have been reported, although it is not clear that all of these are direct substrates in vivo (Manning and Cantley 2007). This variety of substrates indicates broad biological functions mediated by multiple downstream effectors.

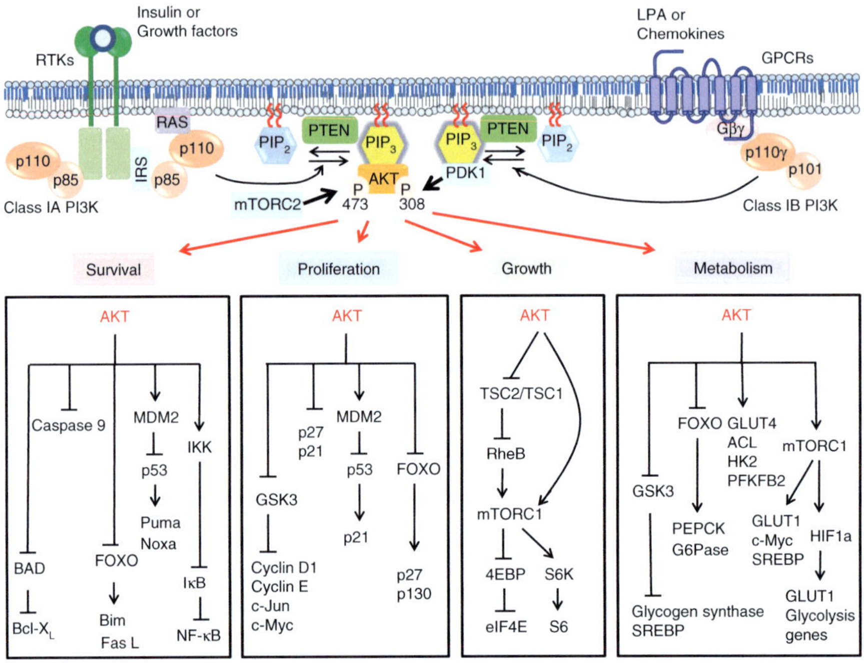

Fig. 6.3 Schematic of signaling through the PI3K/AKT pathway. PI3K signaling affects processes related to tumorigenesis such as cell survival, proliferation, growth, and metabolism. *Arrows* represent activation, while *bars* reflect inhibition

The three AKT isoforms are very similar in amino acid sequence and might have indistinguishable substrate specificity in vitro (Walker et al. 1998), yet many isoform-specific substrates may exist in vivo (Stambolic and Woodgett 2006). One possible scenario suggesting isoform-specific AKT substrates is that a different cellular localization of the three isoforms could determine their accessibility to a selective group of substrate proteins (Bhaskar and Hay 2007). Furthermore, it is possible that each AKT isoform possesses different functions, as demonstrated by germ line deletions in mice. AKT1 null mice show developmental defects especially in the thymus and testes (Chen et al. 2005; Cho et al. 2001a), AKT2 null mice have defects in glucose homeostasis (Cho et al. 2001b; Garofalo et al. 2003), and AKT3 null mice display defects in brain development (Easton et al. 2005; Tschopp et al. 2005). The relative expression of the three isoforms also differs in mammalian cells in that AKT1 is predominantly expressed in the majority of tissues, AKT2 is the predominant isoform in insulin-responsive tissues such as adipocytes and muscle tissues, while AKT3 is the predominant isoform in the brain and testis (Manning and Cantley 2007). All three AKT isoforms have been found to be mutated or amplified in in subsets of human cancers, albeit at a relatively low frequency.

6.3 Biological Effects of PI3K/AKT Activation

The primary biological effects of AKT activation can be classified into four categories—survival, proliferation (increased cell number), growth (increased cell size), and metabolism (Fig. 6.3). AKT has additional effects on tumor-induced angiogenesis that are mediated, in part, through hypoxia-inducible factor 1, alpha subunit (HIF1A), and vascular endothelial growth factor (VEGF) (Vivanco and Sawyers 2002).

6.3.1 Cell Survival

The balance of proliferation and apoptosis is critical for normal homeostasis. Increased proliferation and/or decreased apoptosis is the basis of tumorigenesis. Even before the relevant substrates were identified, several groups showed a critical role for AKT in promoting cell survival by preventing apoptosis. For example, dominant-negative alleles of AKT induce cell death (Dudek et al. 1997) and constitutively active AKT rescues PTEN-mediated apoptosis (Li et al. 1998). Later, numerous studies led to the discovery that many of the apoptosis-related proteins are directly or indirectly regulated by AKT. AKT protects cells from death by directly phosphorylating several downstream substrates that are involved in apoptosis. AKT negatively regulates the function or expression of several Bcl-2 homology domains (BH3)-only proteins, which exert their pro-apoptotic effects by binding to and inactivating pro-survival Bcl-2 family members. For instance, BAD, a BH3-only protein is a pro-apoptotic member of the Bcl-2 family of proteins that promote cell death by binding to the survival factor BCL-X_L, thereby blocking the function of BCL-X_L. Phosphorylation of BAD by AKT creates a binding site for 14-3-3 proteins, which triggers release of BAD from BCL-X_L (Datta et al. 1997, 2000; del Peso et al. 1997). The consequence is restoration of BCL-X_L's anti-apoptotic function. AKT also inhibits the expression of BH3-only proteins through effects on FOXO transcription factors. Phosphorylation of FOXO proteins by AKT allows 14-3-3 proteins to bind to FOXOs, resulting in their inactivation through sequestration in the cytoplasm (Tran et al. 2003). Through this mechanism, AKT blocks FOXO-mediated transcription of target genes that promote apoptosis, cell-cycle arrest, and metabolic processes (see below). Two major pro-apoptotic targets of FOXO are the BH3-only protein BIM and cytokine FAS ligand (Fas L) (Dijkers et al. 2002; Brunet et al. 1999).

AKT can also influence cell survival through indirect effects on two central regulators of cell death—nuclear factor of kB (NF-kB) and p53. MDM2 is a negative regulator of p53 that targets p53 for degradation by the proteasome through its E3 ubiquitin ligase activity. AKT phosphorylates MDM2, promoting its translocation to the nucleus where it binds to p53 to promote ubiquitination and degradation (Mayo and Donner 2001; Zhou et al. 2001). The BH3-only proteins Puma and Noxa are two transcriptional targets of p53 that seem to be the important targets in p53-induced apoptosis (Villunger et al. 2003). NF-κB is a transcription factor that can promote survival in response to several extracellular stimuli. When it forms a complex with IκB (inhibitor of NF-κB), it remains in the cytoplasm. AKT can have a positive effect on NF-κB function by phosphorylation and activation of IκB kinase (IKK), a kinase that induces degradation of IκB (Ozes et al. 1999; Huang and Chen 2009). Degradation of IκB releases NF-kB from the cytoplasm, and the free NF-κB enters the nucleus to activate its target genes related to increased cell survival.

6.3.2 Cell Proliferation (Cell Cycle)

AKT can stimulate proliferation through multiple downstream targets regulating cell-cycle machinery. The cell cycle is regulated by the coordinated action of cyclin-dependent kinase (CDK) complexes and CDK inhibitors (CKIs). Glycogen synthase kinase-3 (GSK3) phosphorylates cyclin D1 and cyclin E and transcription factors c-Jun and c-Myc, which all play an important role in the G1/S phase cell-cycle transition, targeting them for degradation by the proteasome (Diehl et al. 1998; Wei et al. 2005; Welcker et al. 2003; Yeh et al. 2004). AKT directly phosphorylates GSK3 and blocks its kinase activity, thereby enhancing the stability of these G1/S transition-related proteins. AKT can also negatively regulate the function of the CKI p21 (also known as CIP1 or WAF1). The expression of p21 can also be negatively regulated by activation of MDM2 by AKT (Mayo and Donner 2001; Zhou et al. 2001). Activated MDM2 subsequently down-regulates p53-mediated transcription of p21. Moreover, Akt inhibits p27 expression and retinoblastoma-related protein p130 through phosphorylation and inhibition of the FOXO transcription factors. p27 and p130 are known to cooperate to inhibit the cell cycle at the G1/S transition (Liang and Slingerland 2003).

6.3.3 Cell Growth

One of the well-characterized functions of AKT is its role in promoting cell growth (i.e., an increase in cell size). The major mechanism by which AKT regulates cell mass increase is through activation of the mTOR complex 1 (mTORC1), which is regulated by both nutrients and growth factor signaling. mTOR (mammalian target of Rapamycin), a catalytic subunit of mTORC1, is one of the best-studied downstream responders to AKT activation and belongs to a group of serine-threonine protein kinases of the PI3K superfamily, referred to as class IV PI3Ks, which also includes ATM, ATR, and DNA-PK. mTOR exists in two distinct complexes—mTORC1 and mTORC2. mTORC1 consists of the mTOR catalytic subunit, regulatory associated protein of mTOR (RAPTOR), proline-rich AKT substrate 40 kDa (PRAS40) and a common regulatory subunit called mLST8 (Wullschleger et al. 2006). mTORC2 consists of mTOR, rapamycin-insensitive companion of mTOR

(RICTOR), mammalian stress-activated protein kinase interacting protein 1 (SIN1) and mLST8 (Liu et al. 2009). The mTORC1 complex is strongly inhibited by rapamycin treatment, while mTORC2 is not affected by acute treatment and only chronic rapamycin treatment at high concentration inhibits its assembly and signaling capacity (Sarbassov et al. 2006).

mTOR has a dual role in PI3K/AKT signaling; when in the TORC2 complex it participates in activation of AKT via phosphorylation of AKT at Ser473 (as discussed above) and when in the TORC1 complex it is activated downstream of AKT. AKT activates mTORC1 multiple ways. The major mechanism appears to be through phosphorylation and inactivation of TSC2 (tuberous sclerosis complex 2, also called tuberin) (Inoki et al. 2003a; Manning et al. 2002). TSC2 shares homology with GAPs, and its heterodimerization with TSC1 is required to exert a GAP activity toward the small GTPase Rheb (Ras homolog enriched in brain) (Castro et al. 2003; Garami et al. 2003; Inoki et al. 2003b). The GTP-bound form of Rheb strongly activates mTORC1, through direct binding to this complex. Therefore, AKT activates mTORC1 indirectly by phosphorylating and inhibiting the Rheb-GAP activity of TSC2, thereby allowing Rheb-GTP to activate mTORC1 signaling (Manning and Cantley 2007). The most extensively characterized downstream targets of mTORC1 are ribosomal protein S6 kinase (S6K; also known as p70S6K) and eukaryotic translation-initiation factor 4E-binding protein 1 (4EBP1). mTORC1 mediates phosphorylation of S6K at a threonine residue (T381) in a hydrophobic motif at the C terminus of the kinase domain. Phosphorylation at this site allows the recruitment and subsequent phosphorylation and activation of S6K by PDK1 (Pullen and Thomas 1997; Pullen et al. 1998). Active S6K1 appears to play multiple roles in the initiation of protein synthesis through phosphorylation of S6 Ribosomal protein (S6, a component of the ribosome) and other components of the translational machinery, thereby enhancing the translation of mRNAs containing 5′ polypyrimidine tracts (Peterson and Schreiber 1998). Phosphorylation of 4EBP1 by mTORC1 suppresses its ability to bind and inhibit the translation-initiation factor eIF4E (Pause et al. 1994; Gingras et al. 1998). eIF4E that is not inhibited by 4EBP binds an mRNA 5′ cap structure and ultimately bring it to the ribosome, increasing translational efficiency of mRNAs (Pause et al. 1994).

6.3.4 Cellular Metabolism

More than 80 years ago, the biochemist Otto Warburg observed that cancer cells consume glucose fervently and produce more lactate, even in the presence of ample oxygen, as compared with normal cells (Warburg 1956; Vander Heiden et al. 2009). Research over the past few years reinforced his observation—a high rate of glycolysis in cancer, also called “Warburg effect”—and also revealed altered metabolism of lipids, amino acids, and nucleotides in cancer cells (Vander Heiden et al. 2009). Oncogenic events, most notably the uncontrolled activation of the PI3K/AKT pathway, have been found to be directly related to altered metabolisms in cancer.

Under normal conditions, AKT2, the primary isoform in insulin-responsive tissues, has been associated with glucose transporter 4 (GLUT4)-containing vesicles upon insulin stimulation, increasing glucose uptake in fat and muscle tissues (Calera et al. 1998; Kohn et al. 1996). GLUT1 is the main glucose transporter in most cell types. Unlike GLUT4, GLUT1 seems to be regulated mainly through alterations in expression levels. Activation of mTORC1, through AKT-mediated phosphorylation of TSC2, regulates both HIF1α-dependent transcription of glycolytic enzymes, including Glut1, and cap-dependent translation of Glut1 mRNA (Taha et al. 1999; Zelzer et al. 1998). Another translational target of mTORC1 is c-Myc, capable of also inducing expression of various glycolytic genes, thus increasing glycolysis (West et al. 1998; Gordan et al. 2007). Furthermore, AKT-mediated phosphorylation and inhibition of GSK3 prevents GSK3 from phosphorylating and inhibiting its substrate glycogen synthase, promoting glycogen synthesis. In the liver, AKT inhibits gluconeogenesis by blocking FOXO-mediated transcription of gluconeogenic enzymes, such as phosphoenolpyruvate carboxykinase (PEPCK) and glucose-6-phosphatase (G6Pase).

In addition to glucose metabolism, lipid synthesis is also regulated by the PI3K/AKT pathway. The PI3K/AKT pathway has been shown to up-regulate lipogenic gene expression through activating SREBP (sterol regulatory element-binding protein) transcription factors, the master regulators of genes involved in cholesterol, fatty acids, triglycerides, and phospholipids synthesis (Gasic 1994; Sundqvist et al. 2005; Porstmann et al. 2005, 2008; Yecies et al. 2011). AKT has also been reported to directly activate ACL (ATP citrate lyase), which functions in an important step in fatty acid biosynthesis (Berwick et al. 2002).

6.4 PI3K Pathway Alterations in CRC

6.4.1 Known PI3K Pathway Mutations in CRC

CRC formation is a multistep process involving cellular transformation proceeding from normal mucosa to microadenomas, to adenomas with increasing dysplasia, to carcinoma. The stepwise CRC model involves many changes in epithelial morphology and is accompanied by characteristic genetic and epigenetic alterations. Both epidemiological (Yoong et al. 2011) and in vitro (Hanahan and Weinberg 2000) studies have shown that cellular transformation requires the accumulation of at least 6–12 mutational events, possibly more (Wood et al. 2007; Sjoblom et al. 2006), a process facilitated by genetic instability (Lengauer et al. 1997; Hirota et al. 1998).

The PI3K pathway is one of the most frequently mutated pathways and has been implicated in driving the progression of pre-invasive adenoma to CRC. Mutations in one or often more than one member of this pathway are found in the majority of CRCs, providing a challenge, and at the same time a potential target, for the treatment of PI3K-addicted CRC tumors with pan-specific or isoform-specific PI3K inhibitors. Activating mutations in PIK3CA, the gene encoding the p110α catalytic subunit of PI3K, were initially detected in approximately one-third of 234 CRCs, but only in 2 of 76 adenomas (Samuels et al. 2004). More recent data evaluating the mutation frequency of PIK3CA in CRC show that it is almost exclusively mutated in established carcinomas and at a lower rate (~13 %, out of n=9,108 (http://www.sanger.ac.uk/genetics/CGP/cosmic, thereafter COSMIC database) than previously reported, yet still remaining one of the most commonly mutated genes in this tumor type

In CRC, there are three mutation hotspot regions within PIK3CA—the H1047R, the E545K, and the E542K mutations. The H1047R mutation is located at the C-terminal portion of the kinase domain, while E545K and E542K mutations are located within the region encoding the helical domain of the protein (Samuels et al. 2004) (Fig. 6.3a). The three aforementioned PIK3CA mutations account for more than 90 % of all the PIK3CA mutations found in CRC. They are all missense (a type of non-synonymous mutation) and confer constitutive lipid kinase activity that promotes cell growth and invasion of human cancer cells (Samuels et al. 2005). Most of the remaining PIK3CA mutations either code for a different amino acid change or target a region adjacent to the hotspot mutations. A list of complete PIK3CA mutations in CRC is summarized in Fig. 6.1.

Aside from PIK3CA activating mutations, the PI3K pathway can be activated by several other mechanisms in CRC, with the most common being loss or mutation of the PTEN tumor suppressor. PTEN is a haploinsufficient tumor suppressor, hence it is more frequently targeted for hemizygous deletions or inactivating mutations of a single allele, thus retaining a normal wild-type allele. Inactivating mutations of PTEN are detected in 6 % of CRCs (out of n=1,344, COSMIC database), while the overall frequency of PTEN deletion in this tumor type is ~22 % (out of n=161, COSMIC database) with ~5 % of these being focal, affecting in most cases only PTEN. Another PI3K-related gene that is frequently deleted in CRC (~22 % overall, ~7 % focal) is PIK3R1 (Beroukhim et al. 2010), the gene encoding the p85α inhibitory subunit. PIK3R1 is also the target of point mutations in 4 % (out of n=560, COSMIC database) of CRCs (Fig. 6.4).

Given that activation of PI3K signaling begins with the engagement of growth factors by RTKs and recruitment of adaptor proteins, many members of this signaling pathway are altered, although somewhat infrequently, in CRC. Among these, EGFR and the v-erb-b2 erythroblastic leukemia viral oncogene homolog 2 (ERBB2/HER2) are both mutated in 4 % of CRCs (77/2152 for EGFR and 14/365 for ERBB2, COSMIC database), with mutations affecting mostly tumors of the rectum. The ERBB4 gene, on the other hand, has been noted to be mutated in 37 % of CRCs, although the sample size for the mutation detection of this gene is relatively small (out of n=65, COSMIC database).

The RTK KIT is mutated in 4 % of CRC (out of n=367, COSMIC database), although activating mutations in this receptor are one of the main forces driving the progression of gastrointestinal stromal tumors (GIST) and occur in >85 % of GISTs (Lengauer et al. 1998). Mutations in the RTK MET and AKT1 are also detected at very low frequencies (1–2 %), occurring in 7out of 310 and 7 out of 917 CRCs, respectively (COSMIC database).

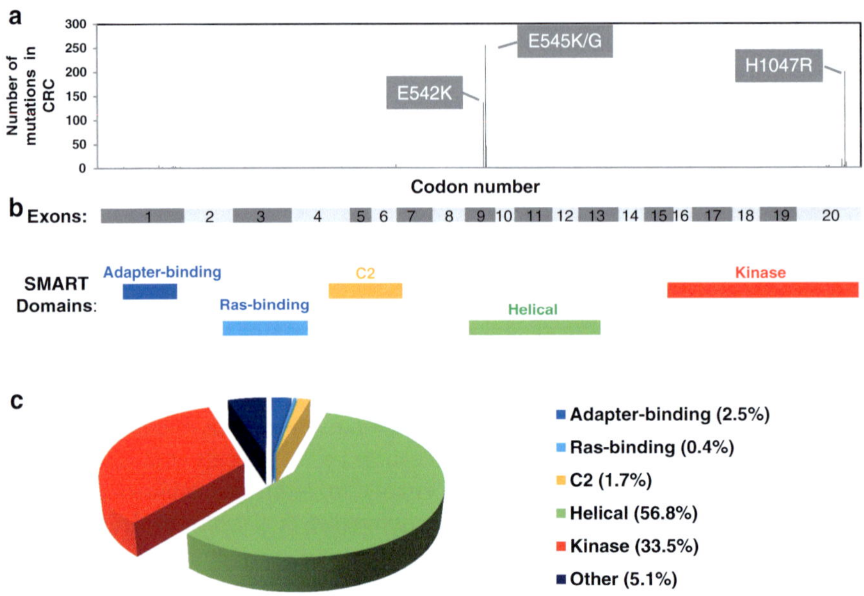

Fig. 6.4 Mutations of PIK3CA identified in colorectal cancer. (**a**) Histogram displaying the position of somatic mutations in the coding sequence of PIK3CA that are identified in CRC (total no of mutations = 787). (**b**) Schematic representation of the PIK3CA exons (1–20) and functional domain of the PIK3CA protein based on the SMART database (http://smart.embl-heidelberg.de/). (**c**) Pie chart showing the proportion of PIK3CA domain mutations in CRC. The frequencies and position of PIK3CA mutations are based on the COSMIC database (http://www.sanger.ac.uk/genetics/CGP/cosmic/)

6.4.1.1 Coexistence and Mutual Exclusivity of PI3K Pathway Mutations in CRC

Cancer progression is considered to be a process of Darwinian evolution, where most genetic alterations are products of selective pressure that drive tumor growth and proliferation. The notion of "the survival of the fittest" would therefore prevent mutations in functionally redundant isoforms of the same gene, and/or genes that lie in the same signaling network. In spite of this, almost every known member of the PI3K/Akt signaling network is frequently altered in most cancers, and certain tumor samples carry two or more mutations in this pathway. Furthermore, the mutation frequency of each of the target genes varies dramatically from one tumor type to another, sometimes even among different tumors of the same origin, suggesting that the genetic profiles of all cancers are determined by distinct somatic evolution that drives the accumulation of mutations. In this section we explore the degree of coexistence and/or mutual exclusivity of PI3K pathway alterations in CRC and speculate on mechanisms which drive the unique characteristics of this tumor type.

There is some degree of mutual exclusivity in PIK3CA and PTEN mutations in breast carcinomas and glioblastomas (Suire et al. 2002; Zhou et al. 2002), where the frequency of coexistent mutations falls lower than that expected by chance. While this is also evident for CRCs (Frattini et al. 2007; Yuan and Cantley 2008), the overall frequency of co-mutation varies significantly from one tumor type to another. Endometrial cancers exhibit the highest incidence of co-mutation between PIK3CA and PTEN (35–36 %). Studies have shown that knocking down PTEN expression in HEC-1B cells that carry mutations in both KRAS and PIK3CA further enhances Akt (S473) phosphorylation (Samuels et al. 2004), suggesting that the functional redundancy in the mutation of PTEN and PIK3CA could be the product of selection for combinatorial and additive effects. Furthermore, the PI3K/AKT pathway is complex and often the target of multiple negative feedback loops. Although mutations in either PTEN or PIK3CA can activate the PI3K pathway, tumors bearing both alterations could circumvent negative feedback loops to sustain tumorigenesis.

One of the best understood forms of genetic instability in CRC is deficiency in the DNA mismatch repair (MMR) system. A possible scenario is that concomitant mutation of PIK3CA and KRAS in CRC is the product of a deficiency in the MMR system. Loss of DNA MMR function can be caused by mutations, deletions, or epigenetic silencing of, both copies of one of the major MMR genes, MLH1, MSH2, MSH6, or PMS2. Failure to repair replication-associated errors due to a defective MMR system allows persistence of mismatch mutations located throughout the genome, especially in regions of repetitive DNA known as microsatellites, giving rise to the phenomenon of microsatellite instability (MSI) (reviewed in (Velho et al. 2005)).

A high frequency of instability at microsatellites (MSI-H) is the hallmark of the most common form of hereditary susceptibility to CRC, known as Lynch syndrome (LS) (previously known as hereditary non-polyposis colorectal cancer syndrome, HNPCC), but is also observed in 15–20 % of sporadic colonic cancers (and rarely in rectal cancers). The mutational targets in MMR-deficient tumors are not random, in that mutations occur preferentially in repetitive sequences; nonetheless there is no a priori favoring of mutation site. However, the mutations in MMR-deficient CRC are positively selected, presumably on the basis that they confer a selective advantage for growth, survival and escape from immune surveillance. Interestingly, there is a strong association between PIK3CA mutations and MSI status ($P=0.0046$) (Sjoblom et al. 2006), despite the fact that there is no repetitive nucleotide sequence within the PIK3CA coding region. On the other hand, the frequency of KRAS mutations is significantly lower in MSI-H colon cancers (Gupta et al. 2007), while PTEN is more frequently mutated in these tumors, particularly in exons 7 and 8 because of the poly-A tracts present in their sequence (Gupta et al. 2007). The co-mutation rate of PI3K/Akt pathway genes is, therefore, likely to be dramatically different between MSI-H and microsatellite stable (MSS) colon cancers and requires further investigation. Uncoupling the coexistence or mutual exclusivity of PI3K alterations will likely be critical in understanding the substantial evolutionary implications that drive the tumorigenic potential of CRC subtypes. A deeper understanding of this phenomenon will guide identification of novel, more effective individualized treatment options necessary to overcome the substantial burden of drug resistance and tumor recurrence.

6.4.2 Relevance of PI3K Pathway Mutations in Human Colorectal Cancer Therapy

The prognostic significance of PI3K-pathway genetic alterations is well established in CRC, yet it appears to be much more complex than initially appreciated. PIK3CA mutations have been associated with poorer survival of CRC patients (Amado et al. 2008), while PIK3CA gene amplification/gain was shown to be independent of PIK3CA point mutations, and was positively correlated with longer survival in patients who received adjuvant chemotherapy and/or radiotherapy (Engelman et al. 2008). However, both the clinical significance of PIK3CA gene amplification and the degree in which this genetic alteration is observed in CRC are unclear. According to recent high-throughput DNA copy number studies, the PIK3CA locus (3q26.32) is the target of frequent amplification in ovarian, breast, and brain tumors (Karapetis et al. 2008), but is very rarely, if ever, amplified in CRCs (Karapetis et al. 2008; Linardou et al. 2008).

The coexistence or mutual exclusivity of PI3K pathway mutations is also likely to have a great impact on the clinical utility of PIK3CA mutations. Interestingly, in a large study of 450 resectable colon cancer biopsies, PIK3CA mutation showed a clear association with higher colon cancer-specific mortality in both univariate and multivariate analyses (Nosho et al. 2008). However, the effect on patient's survival differed significantly among patients with wild-type or mutant KRAS, with the latter group showing no association of PIK3CA mutation with poor prognosis. PTEN loss of expression has also been suggested to serve as a poor prognosis indicator in CRC. One study demonstrated that low PTEN-expressing tumors are significantly associated with shorter median progression-free survival (PFS) (Yuan and Cantley 2008). Likewise, another study showed that loss of PTEN expression was associated with larger tumor size and depth of invasion, lymphatic invasion, lymph node metastasis, higher Dukes' staging, and reduced caspace-3 expression (Abubaker et al. 2009). Several studies have shown that there are multiple effector pathways downstream of PI3K/Akt activation, underlying the therapeutic challenge in treating PI3K-pathway-activated tumors. The degree in which the prognostic significance of PI3K-pathway mutations are influenced by inherent or acquired resistance to the conventional cancer therapeutic approaches is likely to be highly underrated. The standard first-line treatment options for metastatic CRC remain primarily based on traditional cytotoxic chemotherapies consisting of a 5-FU and folinic acid (FA, also known as leucovorin) backbone. In the initial studies, 5-FU/Leucovorin demonstrated a response rate of 20 % and a treatment-induced prolongation survival of 11 months compared with the median 5 month survival of best supportive care (BSC) (Chee and Sinicrope 2010). Subsequent trials demonstrated the survival benefit of adding the nonnephrotoxic platinum analogue oxaliplatin (FOLFOX) or the topoisomerase 1 inhibitor irinotecan (CPT-11)(FOLFIRI) to 5-FU/FA backbone, improving response rates to 35–53 %, PFS of 5–8 months and overall survival of 14–18 months (Laurent-Puig et al. 2009; Li et al. 2009; Loupakis et al. 2009). The first FDA-approved oral chemotherapeutic agent capecitabine is enzymatically converted to 5-FU in vivo, and can be substituted for 5-FU as it has been suggested to have milder side effects (Ogino et al. 2009).

Despite modest survival data, the overall response rate of palliative chemotherapy in CRC is low and often associated with severe side effects, raising the need for the development of targeted cancer therapies. Significant effort has been made in the development of inhibitors that target the VEGF and EGFR pathways in CRC. There are two main classes of EGFR inhibitors currently in clinical use: the anti-EGFR monoclonal antibodies and the small-molecule EGFR tyrosine kinase inhibitors. These inhibitors are not exclusive to the EGFR pathway and can block different RTKs, including VEGF (Chee and Sinicrope 2010). Two clinical trials have evaluated the efficacy of the anti-EGFR monoclonal antibodies panitumumab and cetuximab, and demonstrated that clinical benefit was restricted to wild-type KRAS tumors (De Roock et al. 2010; Li et al. 2010). Evaluation of the predictive and prognostic value of KRAS and PIK3CA mutations in relation to both 5-FU-based first-line chemotherapy treatment and second line cetuximab therapy indicated that KRAS mutations could predict for lack of response ($P=0.002$) and shorter PFS ($P=0.09$) (Souglakos et al. 2009), while PIK3CA mutations were associated with even lower PFS in response to cetuximab treatment ($P=0.01$). Recently, a European consortium studied the effects of KRAS, BRAF, NRAS, and PIK3CA on the efficacy of cetuximab or panitumumab in a large cohort ($n=1{,}022$) of patients with chemotherapy-refractory metastatic CRC (Poulogiannis et al. 2010). This study confirmed that patients with KRAS mutations do not benefit from cetuximab treatment. Importantly, in subset analysis this study showed that only PIK3CA exon 20 mutations are associated with a lack of response to cetuximab in KRAS-WT tumors, with a lower median PFS of 11.5 versus 24 weeks. While there is biologic reason to suspect that exon 20 PIK3CA mutations may predict cetuximab resistance in KRAS-WT CRC, this observation needs to be confirmed in a larger tumor series to exclude the possibility of any confounding factors reflective of the low number of tumors with exon-20 PIK3CA mutation included in the latter study ($n=9$). In contrast, exon 9 PIK3CA mutations were associated with KRAS mutations and did not confer an independent adverse effect on cetuximab response rate. Other PI3K pathway perturbations, including loss of PTEN expression, have also been linked to lack of cetuximab response in metastatic CRC (Sullivan and Kozuch 2011; Naguib et al. 2011; Poulogiannis et al. 2012). However, tumor heterogeneity confounds the immunohistochemical (IHC) assessment of PTEN expression (19–36 %) and necessitates establishment of a more reliable and standardized protocol for PTEN IHC testing.

Despite considerable progress in selecting which CRC patients are suitable for anti-EGFR treatment, and basic understanding of the alternative mechanisms driving resistance to this therapy, most CRC patients who respond to these agents inevitably experience progressive disease after a few months of treatment. The relatively short response durations to second and third line CRC treatments highlight the need for both a better molecular characterization of individual patient tumors and the possibility of combination therapies such as Cetuximab with a pan-PI3K inhibitor which may delay the onset of resistance and translate to clinical benefits in both progression free and hopefully overall survival. Currently, there are ongoing early phase trials of pan-PI3K and PI3K-isoform-specific inhibitors, as well as Akt and mTOR inhibitors, alone or in combination in a wide range of solid tumors and clinical settings, as discussed in the following sections.

6.5 Inhibitors Targeting PI3K Pathways

Development of novel, targeted cancer therapies is rapidly replacing that of traditional, nonspecific cytotoxic drugs, and slowly new targeted therapies are making their way into clinical practice (Yun et al. 2009; Yun 2010). Prior successes in targeted therapy, such as imatinib for chronic myelogenous leukemia (CML), trastuzumab for breast cancer with amplification of HER2 (also known as ERBB2), and erlotinib and gefitinib for lung cancer that expresses mutant EGFR paved the way for future targeted therapy. While, conceptually, targeted therapy is ideal, many roadblocks exist to the development of new targeted therapies. First, the pathway of interest should be both central to maintenance of the malignancy and druggable. Second, the targeted therapy must not be too toxic to surrounding normal tissues. Third, the ideal targeted therapy would have a noninvasive monitoring method and mechanism to study resistance. As discussed previously, genetic and cell line/xenograft studies have shown that deregulated PI3K signaling is vital to the growth and survival of cancer cells, making the PI3K pathway one of the most attractive targets for anticancer therapy. Over the past decade, a number of PI3K pathway inhibitors have been developed and entered into the clinic. Table 6.1 shows a list of drugs in development to exploit the PI3K signaling pathways, and existing clinical trials can be found at http://www.clinicaltrials.gov/. In the following section, we will discuss four different classes of PI3K pathway inhibitors: PI3K inhibitors, AKT inhibitors, mTOR inhibitors, and dual PI3K/mTOR inhibitors.

6.5.1 PI3K Inhibitors

Like the majority of small molecule kinase inhibitors, all existing PI3K inhibitors belong to a class of ATP-competitive inhibitors. The PI3K inhibitors can be further divided into pan-PI3K inhibitors or isoform-specific PI3K inhibitors. The majority of PI3K inhibitors in clinical trials thus far are pan-PI3K inhibitors, inhibiting all of the catalytic subunit isoforms of class I PI3Ks: p110α, p110β, p110γ, and p110δ. Developing isoform-specific inhibitors is challenging because of the highly conserved nature of the ATP-binding pocket. Structural visualization techniques including X-ray crystallography are critical for imaging drug-PI3K complexes and informing rational isoform-specific inhibitor development. Indeed, the X-ray structures of the p110 subunit of PI3Kγ, and more recently of the human p110α/p85α complex, have been crucial in providing a detailed structural analysis of the

Table 6.1 PI3K pathway inhibitors in clinical development

Target(s)/isoforms	Agent	Company	Status
Pan-PI3K	PX-866	Oncothyreon, Inc.	Phase I/II
	XL147	Exelixis/Sanofi-aventis	Phase I
	BKM120	Novartis	Phase I
	GDC0941	Genentech	Phase I
PI3K isoform α	BYL719	Novartis	Phase I
PI3K isoform δ	CAL-101	Calistoga Pharma	Phase I/II
AKT	MK2206	Merck	Phase I/II
	Perifosine	Keryx Biopharmaceuticals	Phase III
	VOD-002 (Triciribine)	VioQuest Pharma	Phase I
mTORC1	RAD 001 (Everolimus)	Novartis	Approved
	CCI-779 (Temsirolimus)	Pfizer	Approved
	Ridaforolimus	Arid/Merck	Phase III
mTOR/catalytic site	AZD8055	AstraZeneca	Phase I/II
	TORKi (CC223)	Celgene	Phase I/II
	OSI-027	Astellas Pharma	Phase I
	INK128	Intellikine	Phase I
Dual PI3K/mTOR	BEZ235	Novartis	Phase I/II
	BGT226	Novartis	Phase I/II
	SF1126	Semafore Pharma	Phase I
	XL765	Exelixis/Sanofi-aventis	Phase I
	PIK-587	Pfizer	Phase I
	GSK2126458	GlaxoSmithKline	Phase I

ATP-binding cleft of class I PI3K, leading to the development of p110γ or p110α isoform-specific PI3K inhibitors (Walker et al. 1999, 2000; Huang et al. 2007).

While there are theoretical benefits to both pan-PI3K and isoform-specific PI3K inhibition, it remains unclear which type of inhibitor, if any, will be more effective clinically. The answer depends on several factors including toxicities resulting from complete inhibition of all PI3K isoforms with pan-PI3K inhibitors, identification of the tumor subtypes in which inhibition of only one or two of the PI3K isoforms will be sufficient, and the time to resistance development. One major potential flaw of PI3K inhibition are the multiple mechanisms by which malignant cells can activate AKT, the major downstream effector of the PI3K signaling cascade.

6.5.1.1 "First Generation" Pan-PI3K Inhibitors: Wortmannin and LY294002

Wortmannin and LY294002 are two well-known, first-generation PI3K inhibitors. Wortmannin was isolated from the fungus *Penicillium wortmannin* in1957 and is an irreversible inhibitor that forms a covalent bond in the ATP-binding pocket of the kinase (Yuan and Cantley 2008). It inhibits PI3K enzymatic activity in the nanomolar range; however, it is not a specific PI3K inhibitor, as it binds to other kinases such as DNA-PK, ATM, ATR, and mTOR. Furthermore, Wortmannin is extremely reactive, with a half-life of only a few minutes in serum, and causes liver dysfunction, lymphocytopenia, and hyperglycemia in animals (Ihle et al. 2004). In 1994, Eli Lilly (Indianapoli, IN) synthesized the reversible PI3K ATP-competitive inhibitor, LY294002. It was developed as a structural analogue of quercetin, a bioflavonoid produced by plants, which can inhibit several protein kinases including PI3K, similar to wortmannin. LY294002 is more stable, but less potent, than Wortmannin. Both Wortmannin and LY294002 substantially inhibit growth of most cancer cell lines when administered as single agents, particularly in cases of excess PI3K activity (Markman et al. 2010) and sensitize tumor cells to other targeted therapeutics such as chemotherapy and radiation (Hu et al. 2002). However, these compounds have not progressed to clinical trials because of unfavorable pharmacokinetic properties, poor selectivity, and high toxicity in animal models (Vlahos et al. 1994).

One approach to bring wortmannin to the clinic involved increasing its stability by PEGylation (Cleary and Shapiro 2010). PWT-458 (Pfizer) is a PEGylated derivative of wortmannin that has a higher therapeutic index in preclinical animal models compared to wortmannin (Yu et al. 2005). Upon intravenous administration, the polyethylene-glycol (PEG) moiety is cleaved, releasing 17-hydroxywortmannin. PWT-458 inhibited AKT kinase and its downstream effectors at nontoxic doses. Inhibition of AKT signaling was accompanied by a slowing of xenograft growth. Moreover, PTW-458 improved the anticancer effects of paclitaxel and PEGylated rapamycin in certain xenograft models (Zhu et al. 2006).

Similar efforts have been applied to improve the pharmacological properties of LY294002. SF1126 (Semafore Phamaceuticals) is a water-soluble prodrug of LY294002 (Garlich et al. 2008; Nutley et al. 2005). The RGD (Arg-Gly-Asp) targeting peptide attached to SF1126 enables the drug to target specific integrins within the tumor compartment. The integrin-targeting RGD peptide moiety causes the drug to preferentially accumulate in endothelial cells and tumor cells. SF1126 inhibits all class I PI3K isoforms and other closely related kinases such as DNA-PK and mTOR. It blocks the phosphorylation and activation of AKT in cell lines with IC_{50} values in the low micromolar range. In preclinical studies, SF1126 has shown antitumor activity in xenograft models of brain, neuroblastoma, non-small cell lung, prostate, myeloma, renal, and colon carcinoma. In addition to its direct activity on cancer cells, it has demonstrated anti-angiogenic activity in xenografted glioma cells by substantially reducing microvessel density. SF1126 is currently in phase I clinical trials. XL147 (under co-development by Exelixis and Sanofi-Aventis) selectively inhibits PI3K without inhibiting mTOR or DNA-PK (Yun et al. 2009). The compound inhibits PI3K signaling in cultured tumor cells and blocks VEGF-induced tubule formation in cell lines. Oral administration slows tumor growth or causes shrinkage of breast, lung, ovarian, prostate, and glioma tumors in xenografts. XL147 is currently in phase I clinical trials and its dose-limiting toxicity is rash, elevated liver function tests, and fatigue. In Phase I clinical trials, increased tumor growth inhibition was achieved by combining XL147 with cytotoxic (carboplatin and paclitaxel) or targeted anticancer agents (erlotinib) without significantly increased toxicity.

6.5.1.2 "Second Generation" Pan-PI3K Inhibitors

The new generation of PI3K compounds was designed to improve upon the pharmaceutical limitations of wortmannin and LY294002. PX-866 (Oncothyreon) is a structural analogue of wortmannin and functions as an irreversible inhibitor of PI3K by making a covalent bond with the PI3K molecule similar to wortmannin (Ihle et al. 2004, 2005; Howes et al. 2007). Compared to Wortmannin, PX-866 exhibits increased stability, reduced toxicity, and enhanced biological activity. In humans and preclinical models, PX-866 is metabolized to produce an active metabolite, 17-OH, that is a more potent PI3K inhibitor than the parent drug and retains the same irreversible mechanism of action. In biochemical assays, PX-866 and the 17-OH metabolite inhibit all four PI3K isoforms and have the greatest potency for PI3K α and β, the two family members that are most strongly associated with solid tumors such as breast, colon, ovarian, and prostate cancers. Preclinical studies demonstrate that PX-866 is efficacious in numerous mouse xenograft models of lung, ovarian, and CRC as a single agent and in combination with chemotherapy, radiation, and targeted cancer drugs, such as EGFR inhibitors. In these studies, PX-866 sustained inhibition of the PI3K pathway, a property that is attributable to its unique, irreversible mechanism of action. Preliminary results from Phase I clinical trials showed several patients with stable disease, with mild side effects including abdominal pain and mild diarrhea (Ihle et al. 2009a). Oncothyreon is currently evaluating PX-866 in Phase I/II and Phase II clinical studies in solid tumors.

BKM120 (Novartis) is an oral pyrimidine-derived pan-PI3K inhibitor. BKM120 inhibits all class I PI3K isoforms at nanomolar concentrations without inhibitory activity against members of the other classes of PI3K or mTOR. In vitro experiments showed a strong anti-proliferative effect of BKM120 on human cancer cell lines exhibiting aberrant PI3K pathway activity. In vivo, BKM120 demonstrated significant antitumor activity in human tumor xenograft models with good correlation between BKM120 treatment and inhibition of the PI3K pathway (Lee et al. 2006; Seki et al. 2004). BKM120 is in phase I clinical trials with colorectal, breast, ovarian, and endometrial cancers patients. Recent preliminary data from phase I trials with 35 patients with advanced solid tumors demonstrated the clinical safety and tolerability of BKM120, as well as its favorable pharmacokinetic profile. The reported side effects were mood alteration, hyperglycemia, and rash (Bendell et al. 2012).

6.5.1.3 Isoform-Specific PI3K Inhibitors

The four isoforms of PI3K, α, β, γ, and δ have distinct biological functions (reviewed in (Liu et al. 2009)). For example, PI3Kα is involved in tumorigenesis and insulin signaling, PIK3β plays a role in platelet aggregation, PI3Kγ is expressed in leukocytes and is a component of the inflammatory response, and PI3Kδ is implicated in allergic responses and hematological cancers. Therefore, a pan-PI3K inhibitor used as an anticancer agent may generate undesirable toxic side effects due to inhibition of all isoforms. p110α-specific inhibitors are of great interest for treating cancers that have PIK3CA mutations. Considering that more than 20 % of CRC have PIK3CA alterations, it will certainly be important to develop p110α-specific inhibitors and evaluate them in CRC lines and animal models with wild-type or mutant PIK3CA. Preclinical models suggest potential advantages of p110α-specific inhibition over inhibition of other PI3K isoforms in certain tumor types. For example, p110α has a critical role in angiogenesis among other Class I PI3K members (Graupera et al. 2008). Thus, specific inhibition of p110α represents a potential method of blocking angiogenesis, a known hallmark of cancer. Since p110α plays a major role in insulin signaling and glucose metabolism, several side effects of p110α can be expected, such as hyperglycemia or glucose intolerance. However, these side effects might be lightened by treatment with peroxisome proliferation-activated receptor gamma (PPARγ) agonists (Ihle et al. 2005). BYL719 (Novartis), a selective inhibitor for p110α, is currently in Phase I clinical trials in patients with advanced solid tumors harboring PIK3CA mutations. While potentially undesirable, a predictable physiological change such as hyperglycemia may offer a minimally invasive clinical surrogate of target inhibition.

The p110β-specific inhibitors are also of interest in treatment of some cancers. Several reports show that p110β is the dominant isoform carrying PI3K activity in PTEN-deficient tumors of brain, breast, prostate, and endometrium both in vitro and in vivo (Jia et al. 2008; Oda et al. 2008; Wee et al. 2008). Since p110β may play a lesser role in insulin response, it is possible that this class of compounds would show fewer side effects compared to p110α-specific inhibitors.

p110δ is mainly expressed in cells of the immune system, where it regulates B-cell maturation and function (Jou et al. 2002; Okkenhaug et al. 2002; Zhang et al. 2011). Therefore, selective inhibitors of p110δ are an attractive therapeutic option in patients with B-cell malignancies. The p110δ-specific inhibitor CAL-101 (Calistoga Phamaceuticals) is being tested in a phase I dose-escalation trial of patients with relapsed or refractory hematologic malignancies (Fruman and Rommel 2011).

Although major advances have been made in the identification of isoform-specific p110 inhibitors, it remains to be seen whether mutant-specific PIK3CA inhibition can be translated into clinical benefit. Ideally, mutant-specific PIK3CA inhibitors would interfere with the oncogenic versions of p110α and leave the important normal functions of wild-type p110α unaffected. Unfortunately, the design of mutant-specific ATP-competitive inhibitors is complicated by the fact that PIK3CA mutations commonly observed in cancers do not alter the ATP-binding site geometry in a manner that can be clearly exploited during drug design. In comparison to wild-type PIK3CA, the crystal structure of PIK3CA containing the hotspot mutation H1047R in the p110α kinase domain (the most commonly observed PIK3CA mutation) revealed no significant structural differences in the ATP-binding site (Mandelker et al. 2009). Structural and biochemical data suggest the H1047R mutation alters the way p110α interacts with lipid membranes, allowing it easier access to the PIP_2 substrate, thereby increasing PI3K pathway activity. Knowledge of the proposed mechanism of action of the H1047R mutation has not yet been exploited to develop novel mutation-specific inhibitors. No crystal structure is currently available of PI3K containing the second most common hotspot mutation

E545K in the p110α helical domain. Nevertheless, similarly to the H1047R mutation, the E545K mutation is not predicted to alter the ATP-binding cleft in a structurally significant manner. Thus, great difficulty exists in designing mutant-specific ATP-competitive PI3K inhibitors. The success of mutant-specific p110α inhibition may depend on alternative inhibitory approaches such as allosteric kinase inhibitors, or antagonists of protein–protein interactions, to provide the desirable activity as well as selectivity profile.

6.5.2 AKT Inhibitors

AKT is another attractive target in inhibiting the PI3K signaling pathway in cancer. AKT is the central node of the PI3K signaling pathway, and both ATP-competitive inhibitors and allosteric inhibitors targeting AKT kinases are under active clinical development. Most ATP-competitive inhibitors are nonselective and target all three isoforms of AKT. GSK690693 (GlaxoSmithKline) is an ATP-competitive AKT kinase inhibitor that targets all three AKT isoforms at low nanomolar range and is active against additional kinases from the cyclic AMP-dependent, cGMP-dependent, and protein kinase C (PKC) family. GSK690693 was recently terminated during clinical phase I trial because of high toxicity. An allosteric dual inhibitor of AKT1 and AKT2 developed by Merck has potent antitumor activity in tumor xenograft models and its analogue MK2206 (Merck) is in Phase I study in patients with locally advanced or metastatic solid tumors (Yap et al. 2011). The most clinically advanced allosteric AKT inhibitor is an alkylphospholipid, perifosine (KRX-0401, Keryx Biopharmaceuticals) (Hilgard et al. 1997; Kondapaka et al. 2003; Van Ummersen et al. 2004). It inhibits AKT activity by disrupting the binding of its PH domain to PIP_3, thereby preventing its membrane translocation and activation by PDK1. In vitro, perifosine inhibits growth of melanoma, colon, lung, prostate, and breast cancer cells in association with inhibition of AKT activity (Kondapaka et al. 2003; Crul et al. 2002). Perifosine has also been found to sensitize cancer cells to apoptosis and cell-cycle arrest induced by radiation in vitro and in vivo (Caron et al. 2005; Vink et al. 2006).

In 2010, perifosine finished clinical phase II and is currently in phase III testing. In a phase II trial of metastatic CRC, perifosine in combination with capecitabine doubled time to progression for metastatic CRCs (Bendell et al. 2011), and this trial led to the FDA assigning Perifosine fast-track status. Another AKT inhibitor, VQD-002 (VioQuest Pharmaceuticals), is a water-soluble tricyclic nucleotide that demonstrated antitumor activity against a wide spectrum of cancers in preclinical and clinical studies. A recent study showed that VQD-002 could play a role in reversing drug resistance in cisplatin treated ovarian cancer (Yang et al. 2008). VQD-002 is currently being tested in phase I/II clinical trials in patients with both solid and hematological malignancies.

The distinct functions of AKT1 and AKT2 in cancers spurred the development of isoform-specific AKT inhibitors, with the promise of effective antitumor activity and fewer toxic side effects compared to compounds that inhibit all three AKT isoforms. In an AKT1 null mouse, glucose homeostasis is unperturbed, but the animals are smaller, consistent with a role for AKT1 in cell growth. In contrast, mice without AKT2 have mild growth defects and show a diabetic phenotype, consistent with data indicating that AKT2 plays an important role in insulin signaling (Cho et al. 2001b; George et al. 2004; Engelman 2009). In this case, it is plausible that AKT1-specific inhibition could shrink tumors with minimal impact on glucose homeostasis. Thus, recent drug-discovery efforts have focused on the development of isoform-specific AKT inhibitors.

Although AKT is the major PI3K downstream effector, PI3K can activate AKT-independent pathways, including the Bruton tyrosine kinase (BTK), the Tec families of non-RTKs, serum-and glucocorticoid-regulated kinase (SGKs), and regulators of small GTPase that are implicated in cell polarity and migration. For example, AKT was a less essential effector of cell survival than SGK3 in a subset of cancers with PIK3CA mutations (Morrow et al. 2005). Further, a recent comprehensive analysis of cancer cells carrying mutant PIK3CA showed that many of these cells exhibit minimal increased activation of AKT and downstream signaling (Vasudevan et al. 2009). Thus, AKT inhibitors alone may not provide adequate inhibition of non-AKT effectors of the PI3K pathway. Additionally, inhibition of AKT may actually increase AKT-independent PI3K signaling via loss of negative feedback loops. The prevalence and importance of AKT-independent effectors of PI3K must be more fully elucidated prior to further clinical testing.

6.5.3 mTOR Inhibitors

mTOR is an important downstream effector of PI3K that regulates protein synthesis, cell proliferation, and angiogenesis. Therefore, mTOR inhibition is another promising approach toward blocking aberrant PI3K signaling in cancer cells, and mTOR inhibitors have been in clinical use for several years. Rapamycin (sirolimus, Wyeth), the prototypical allosteric mTOR inhibitor, is a bacterially derived natural product originally used as antifungal agent. It was later found to have immunosuppressive properties, and was approved for clinical use as an immunosuppressive agent in 1999. Rapamycin binds to its intracellular receptor, FK506-binding protein 12 (FKBP12), which then binds directly to mTORC1, inhibiting mTOR-mediated phosphorylation of its downstream targets, S6K and 4EBP1. Later, derivatives of rapamycin, such as CCI-779 (termsirolimus/Torisel; Wyeth) and RAD001 (Everolimus/Afinitor; Novartis) were developed as anticancer drugs (Granville et al. 2006). These rapamycin analogues (referred to as rapalogues) inhibit mTOR through the same mechanism as rapamycin, but possess more favorable pharmacological properties (Liu et al. 2009). Results from clinical studies with CCI-779 and RAD001 used as single agents showed that these drugs improved survival in patients with advanced renal cell carcinoma (RCC), leading to FDA-approval of in 2007 (CCI-779) and 2009 (RAD001) respectively.

There are several possible mechanisms underlying the limited success of rapalogues outside of RCC and breast cancer. First, the negative feedback loop that is blocked upon mTORC1 inhibition may activate upstream receptor tyrosine signaling through IGF-1R or IRS1, resulting in increased PI3K–Akt signaling (Wan et al. 2007; Baselga 2011). Indeed, tissue samples taken from patients with colon or breast cancer after 4 weeks of treatment with RAD-001 showed higher levels of activated AKT compared to pretreatment samples (O'Reilly et al. 2006). In another study, tumor materials from patients treated with rapalogues also showed increased AKT activity (Tabernero et al. 2008). Second, rapalogues only partially inhibit mTORC1 target phosphorylation. For example, 4E-BP is rephosphorylated and is refractory to long-term rapamycin treatment while S6K phosphorylation remained permanently inhibited under those conditions. Persistent 4E-BP phosphorylation may allow cancer cells to continue proliferating and growing independent of AKT dependency. Last but not least, rapalogues cannot inhibit mTORC2 in acute treatment settings. These data imply that ATP-competitive mTOR inhibitors capable of targeting both mTORC1 and mTORC2 might show broader efficacy than rapalogues, sparkling enthusiasm for mTOR catalytic site inhibitors.

The first reported catalytic mTOR inhibitor was PP242, which potently inhibited both mTORC1 and mTORC2 (Apsel et al. 2008). In a preclinical study, PP242 sustained 4E-BP dephosphorylation and suppressed tumor growth in a mouse model of AKT-driven lymphangiogenesis, whereas rapamycin was ineffective (Hsieh et al. 2010). Interestingly, improved efficacy of PP242 may be the result of more effective mTORC1 inhibition, rather than its additional inhibition of mTORC2 (Feldman et al. 2009). INK128 (Intellikine), a derivative of PP242, is currently in Phase I trials (Hsieh and Ruggero 2010). Three additional mTOR catalytic site inhibitors—TORKi CC223 (Celgene), OS1027 (OSI Pharmaceuticals), and AZD8055 (AstraZeneca)—have been shown to inhibit proliferation of a variety of cancer cell lines and human xenograft models more effectively than rapamycin (Chresta et al. 2010) and each of recently entered Phase I trials. Despite the promising results in preclinical studies involving mTOR catalytic inhibitors, several general concerns exist related to mTOR inhibition. First, these compounds may not inhibit AKT T308 phosphorylation by PDK1. This is concerning because previous studies suggested that loss of AKT S473 phosphorylation was not able to block all downstream effectors of AKT signaling. Indeed, a report demonstrated that the mTOR catalytic site inhibitor PP242 had minimal effects on the phosphorylation state of several AKT substrates despite effectively inhibiting AKT S473 phosphorylation (Feldman et al. 2009). Second, inhibition of mTORC1 may activate AKT-independent PI3K signaling due to loss of feedback inhibition, suggesting a role for dual PI3K/mTOR inhibitors. Third, mTOR is not exclusively regulated by PI3K signaling and is involved in additional cellular functions including protein synthesis, cell growth, survival, and metabolism. These processes can be affected by inhibition of mTOR kinase activity, potentially reducing the therapeutic index. Finally, the genetic factors determining the differential sensitivity of cells to mTOR inhibitors are not clear. Despite mounting cell line and xenograft data the role for mTOR inhibitors in CRC remains to be determined, and response CRC patients enrolled in phase I studies will be important.

6.5.4 Dual PI3K/mTOR Inhibitors

The catalytic domains of the p110 subunits (α, β, δ, and γ) and mTOR are structurally similar, and many PI3K inhibitors under development exhibit concomitant mTOR inhibition (Garcia-Echeverria and Sellers 2008). When compared with other types of PI3K pathway inhibitors, dual PI3K/mTOR inhibitors have the potential advantage of inhibiting all class IA PI3K isoforms (p110α, β, and δ), mTORC1, and mTORC2. The broader spectrum of inhibition offered by dual PI3K/mTOR inhibitors has the added benefit of overcoming feedback inhibition normally observed when either mTOR or PI3K inhibitors are administered alone (O'Reilly et al. 2006; Fan et al. 2007). Numerous dual PI3K/mTOR inhibitors such as NVP-BEZ235 (Novartis), NVP-BGT226 (Novartis), and XL765 (Exelixis) are currently in Phase I/II clinical trials. NVP-BEZ235, an imidazoquinazoline derivative, was generated by structure-based design (Maira et al. 2008). Preclinical data showed that NVP-BEZ235 has effective anti-proliferative activity against tumor xenografts that have aberrant PI3K signaling, especially in the presence of PTEN loss or gain-of-function PIK3CA mutations (Serra et al. 2008). In addition dual PI3K/mTOR inhibitors have been shown to possess anti-angiogenic properties (Schnell et al. 2008). Further, unlike other inhibitors of the PI3K/AKT pathway, in vivo efficacy experiments in mice or rats treated with NVP-BEZ235 demonstrated no statistically significant changes in blood glucose levels. These preclinical data establish the feasibility of effectively blocking the PI3K pathway in vivo without serious effects on glucose regulation (Maira et al. 2008). After promising phase I trials with NVP-BEZ235 the drug is now in phase II study. Early reports on Phase I trials of XL756, another dual mTOR/PI3K inhibitor, demonstrated inhibition of AKT phosphorylation and reduced tumor growth. Five out of 19 patents showed clinical benefit with disease stabilization for at least 3 months, and for longer than 6 months in two cases (Molckovsky and Siu 2008). Due to the fact that the dual PI3K/mTOR inhibitors inhibit multiple kinases, one major concern with this class of inhibitors is off-target toxicities. However, recent clinical data at the 2010 American Society of Clinical Oncology (ASCO) annual meeting indicated there were no significant differences in terms of toxicity profiles among dual PI3K/mTOR, Pan-PI3K, and isoform-specific PI3K inhibitors. The most common side effects reported with dual PI3K/mTOR inhibitors were diarrhea, nausea, vomiting, and fatigue. Interestingly, insulin resistance-hyperinsulinemia or hyperglycemia, originally predicted to be one of the most likely toxicities resulting from on-target effects of PI3K inhibitors, not been widely observed in clinical trials to date.

The ideal combination, sequence, and tumor type for dual PI3K/mTOR, Pan-PI3K, AKT, mTOR and isoform-specific PI3K inhibitors in cancer remains to be seen. Underlying the development of PI3K-modulating drugs is the need for an ongoing paradigm shift in oncology with improved molecular tumor characterization at the individual patient level. The generation of large molecularly annotated tumor registries may increase the identification of patient subsets likely to benefit from PI3K inhibition and streamline pipeline drug development and clinical trial design.

6.5.4.1 Strategies for Targeting PI3K Pathways in CRC Therapy

Single-Agent Therapy

One of the hallmarks of cancer is the accumulation of genomic alterations, and most malignancies accumulate numerous genetic alterations during tumorigenesis and progression. Despite the fact that multiple mutations occur in each cancer during tumor progression, growth, and survival are sometimes highly dependent on one or a few oncogenes, and their growth and survival can often be compromised by the inactivation of a single oncogene. This phenomenon, dubbed as "oncogene addiction" has provided a rationale for targeted cancer therapy (Weinstein and Joe 2008). Recent clinical data suggested that oncogenes that are mutated or amplified represent attractive targets for therapy. This principle is exemplified by the successes of targeted therapies such as imatinib for CML; trastuzumab for breast cancer with amplification of HER2 (also known as ERBB2); and erlotinib and gefitinib for non-small cell lung cancers that express mutant EGFR.

The high frequency of mutations in the PI3K pathway in human cancers strongly supports the critical role of the PI3K pathway during tumorigenesis. Tumors with oncogenic PI3K mutations may be highly susceptible to single agents that target PI3K signaling pathway, and there is mounting preclinical evidence to support this hypothesis. NVP-BEZ 235 (Novartis), a dual PI3K/mTOR inhibitor, inhibited the growth of lung adenocarcinomas in transgenic mice that expressed p110α with a H1047R mutation (Engelman et al. 2008). NVP-BEZ235 or an allosteric AKT inhibitor (AKTi-1/2) suppressed growth of human breast tumor xenografts with PI3KCA mutations (Serra et al. 2008; She et al. 2008). Considering more than 20 % of CRCs have PIK3CA genetic alterations, single agents targeting PI3KCA may be an effective strategy in this subgroup of patients.

Preclinical in vitro and in vivo data have revealed the potential efficacy of single agent PI3K inhibition in CRC treatment. The reversible PI3K inhibitor LY294002 blocked PI3K signaling and specifically inhibited proliferation of CRC cell lines, HCT116 and DLD1with PIK3CA mutations, but not cells with wild-type PIK3CA (Samuels et al. 2005). The GSK3 inhibitors, lithium chloride and SB216763 selectively decreased the proliferation of HCT116 cells with oncogenic PIK3CA mutations (Yoong et al. 2011). Further, oral treatment with lithium preferentially inhibited the growth of xenografts of HCT116 with PIK3CA mutations as compared to isogenic HCT116 containing only wild-type PIK3CA (Yoong et al. 2011). The irreversible PI3K inhibitor Wortmannin reduced anchorage-independent growth of CRC cells in a soft agar assay (Khaleghpour et al. 2004), and small interfering RNA-mediated knockdown of PI3K p85α in CRC cells induced G1-phase arrest (Sun et al. 2009). Ongoing studies will be investigating the effects of p110α isoform-specific inhibitors and p110α mutant-specific inhibitors in CRC cells with PIK3CA mutations.

6.5.5 Combination Therapy

Despite the moderate successes of single agent targeted cancer therapies, preclinical and clinical data suggest a clear role for combination therapy. Combination therapies have the potential advantages of delaying the development of resistance, improving response rate, and improving harder clinical endpoints such as progression-free and overall survival. Most clinically effective targeted therapies are directed against RTKs, such as KIT, EGFR, VEGFR, and HER2, which modulate multiple downstream intracellular pathways. RTK inhibition therefore blocks multiple signaling pathways, not only the PI3K pathway. Despite preclinical evidence of single agent PI3K inhibition in CRC, it remains to be seen whether inhibition of the PI3K pathway alone will offer advantages over upstream RTK inhibition. Other concerns in targeting PI3K pathway components alone is the cross talk between many of the RTK signaling pathways such as the RAS/RAF/MEK/ERK and PI3K/AKT pathways, as well as complex signaling feedback loops.

One of the major pitfalls of targeted therapies is the emergence of resistance. When cancer cells are treated with drugs that block a single molecular target, they are often able to activate alternative pathways as escape mechanisms to overcome the blockade and therefore the effectiveness of these drugs. Recent studies have shown that inhibition of mTORC1 leads to activation of the ERK signaling pathway capable of driving proliferation and growth (Carracedo et al. 2008). Further, mutations in PIK3CA often coexist with other genetic lesions, such as KRAS or BRAF mutations in CRC. KRAS can directly activate the RAF/MEK/ERK and PI3K/AKT signaling pathways by directly binding to RAF proteins and the PI3K subunit p110. Therefore, simultaneous inhibition of MEK/ERK and PI3K/AKT may have potential therapeutic benefit in KRAS-mutant cancers. Engelman et al. showed that KRAS-driven mice lung tumors did not respond when a dual mTOR/PI3K inhibitor (NVP-BEZ235) was administered after tumors were established. However, concomitant inhibition of the PI3K and RAF pathways with NVP-BEZ235 and AZD6244, an inhibitor of MEK, inhibits tumor growth in these mice (Engelman et al. 2008). In human NSCLC cell lines and animal models combination therapy with the MEK inhibitor PD-0329501, and mTOR inhibitor rapamycin, or its derivative AP 23573 (ARIAD Pharmaceuticals/Merck), showed a synergistic effect in the inhibition of cell proliferation and protein translation, indicating that both pathways converge to regulate cell growth and ribosomal biogenesis (Legrier et al. 2007).

Several reports involving CRC cell lines support the advantage of combination therapy. KRAS-mutant CRC cell lines were resistant to MEK inhibition when PTEN deletions or activating PI3KCA mutations exist (Wee et al. 2009). The combination of RTK and MEK inhibitors led to dual inhibition of PI3K and MEK signaling, marked growth suppression and apoptosis of KRAS-mutant CRC in vitro and xenograft models in vivo. The combination of the mTOR inhibitor, rapamycin, and the MEK inhibitor, PD89059, caused cell-cycle arrest and induced apoptosis in KRAS-mutant CRC cell lines (Zhang et al. 2009). Taken together, these studies support the idea that rational combinations of targeted treatments, especially those blocking both the PI3K and RAS/RAF/MEK pathways, to circumvent, reverse, or even stop resistance are necessary for optimal use of molecular targeted therapies in CRC (Engelman 2009; She et al. 2005). Indeed, many pharmaceutical companies are actively looking for promising combination therapies with existing targeted drugs or nonspecific cytotoxic drugs. In a recent phase I trial, the combination of PI3K and MEK inhibitors from Genentech, GDC-0941 and GDC-0973 respectively, was effective against advanced solid tumors at tolerable doses for patients. Additionally, mTOR and MEK inhibitor combinations as well as AKT inhibitor (MK-2206) and MEK inhibitor (AZD6244) combinations are in clinical trials.

6.6 Targeting Angiogenesis via the PI3K Pathway Inhibitors

Angiogenesis, the physiological process involving the growth of new blood vessels from preexisting vessels, plays an essential role in tumor growth and metastasis (Folkman 2007). Tumor growth requires angiogenesis when the size of the tumor reaches 1–2 mm in diameter (Folkman 2007). Endothelial cells express numerous cell-surface RTK receptors such as VEGFR1-3, TIE-1/2, PDGFR-beta, and ERBB1–4 to integrate the VEGFs secreted by tumor and stromal cells (Yuan and Cantley 2008; Hofer and Schweighofer 2007). Molecularly targeted agents against VEGF have been developed, and in clinical trials were shown to augment the efficacy of cytotoxic chemotherapy in patients with advanced CRC (Chee and Sinicrope 2010). Based on improved survival, the anti-VEGF monoclonal antibody bevacizumab (Avastin) was approved by the FDA for the treatment of advanced CRC in 2004. The success of bavacizumab in CRC treatment underscores the idea that angiogenesis inhibition is an effective form of CRC therapy.

Numerous studies using conditional or germline knockout mice suggest the strong connection between the PI3K pathway and angiogenesis. PI3K/AKT signaling can induce expression of VEGF and suppress the expression of anti-angiogenic protein TSP-1 in cancer and endothelial cells (Niu et al. 2004; Wen et al. 2001). FOXO1 germline deletion results in defects in arterial development and early vessel remodeling (Furuyama et al. 2004). Moreover, somatic deletion of all three FOXO genes in mice showed that FOXO regulates endothelial cell homeostasis (Paik et al. 2007). Complete loss of PTEN in endothelial cells resulted in a phenotype indicative of impaired vascular remodeling, resulting in embryonic lethality, whereas heterozygous deletion of PTEN in the endothelium increased tumor growth by enhancing tumor angiogenesis (Hamada et al. 2005). Constitutive AKT1 activation in endothelial cells by transgenic expression of a myristoylated AKT1 caused abnormal vessel patterning and congestion (Sun et al. 2005). A conditional mouse model with homozygous deletion of the class IA PI3K regulatory subunits (p85α, p55α, p50α, and p85β) in the endothelium impaired vessel integrity during development and decreased the rate of tumor growth (Yuan et al. 2008). In another study, embryos with kinase-dead p110α developed gross vascular defects indicating p110α is essential for endothelial cell migration and angiogenesis (Graupera et al. 2008). Overall, the results from genetic mouse models strongly suggest that inhibiting the PI3K pathway may have anti-angiogenic effects in reducing tumor growth. In fact, the traditional PI3K inhibitor LY2904002 inhibited the expression of VEGF in both endothelial cells and ovarian cancer cells (Skinner et al. 2004; Jiang et al. 2000) and blocked EGF-induced expression of VEGF and leptin in the CRC cell line HT-29 (Cascio et al. 2009). Furthermore, LY294001 showed in vivo anti-angiogenic activity by decreasing the microvessel density in tumor tissue of a mouse U87 xenograft model (Su et al. 2003). Newer PI3K inhibitors including SF1126 (Garlich et al. 2008), ZSTK474 (Kong et al. 2009), and PI103 (Raynaud et al. 2007) have also demonstrated anti-angiogenic activity in xenograft models, as we discussed previously. A dual PI3K/mTOR inhibitor, NVP-BEZ235, blocked VEGF-induced angiogenesis in mice (Schnell et al. 2008), and inhibited the growth and proliferation of cancer cells with wild type and mutant p110α (Serra et al. 2008). The mTOR inhibitor rapamycin and its analogues are among the most extensively studied drugs in the clinic as anti-angiogenic agents. Rapamycin markedly reduced production of VEGF (Zhong et al. 2000), and abrogated the response of vascular endothelial cells to stimulation by VEGF in vivo (Guba et al. 2002). In RCC, loss of the von-Hippel-Lindau tumor suppressor (VHL) that normally inhibits HIF1A causes enhanced vascularization, and an mTOR analogue, CCI-779, sensitized the cancer cells to death (Thomas et al. 2006). The efficacy of CCI-779 in RCC likely results from rapamycin inhibition of mTORC1-dependent translation and action of HIF1A, thereby decreasing VEGF production (Hudson et al. 2002; Bernardi et al. 2006). Recently, the catalytic mTOR inhibitors, OSI-027 and OXA-01, have been shown to significantly reduce angiogenesis and tumor growth compared to rapamycin alone (Falcon et al. 2011). Taken together the mounting preclinical data suggest targeting the PI3K pathway in the clinic may have the dual effect of blocking both tumor growth and angiogenesis.

6.7 Future Directions

Since the discovery of PI3K more than 20 years ago, our knowledge of the PI3K pathway in various cellular processes and tumorigenesis has experienced unprecedented advances. The PI3K pathway is unquestionably important in cancers such as CRC, and has demonstrated clear potential as an anticancer target. A number of targeted agents that inhibit the key components of the PI3K pathway are already in clinical trials. The question remains as to how PI3K inhibitors can be strategically applied in patients to maximize patient benefit and minimize toxicity. To understand the optimal role of PI3K inhibitors further work is needed. Rather than testing compounds in mixed patient populations, which occur during the traditional cancer clinical trial, the number of genotype-directed and Bayesian hypothesis-testing trials must be increased. Multiple preclinical studies demonstrated that cancers with PIK3CA mutations might be most sensitive to PI3K inhibition, while cancers with KRAS mutations might be inherently more resistant to single agent PI3K inhibitors (Dan et al. 2010; Sos et al. 2009; Ihle et al. 2009b). Mutation-specific patient clinical trial entry will both address proof of principle and tease out parameters related to treatment failure, ultimately bringing personalized treatment in the clinic. The complexity of PI3K signaling due to numerous downstream targets, feedback loops, and cross talk with multiple pathways leading to drug resistance will likely emerge with single agent therapy. In order to figure out the most effective combination therapies to prevent development of drug resistance, preclinical and clinical studies will need to focus on potential resistance mechanisms of PI3K inhibitors. Furthermore, it is imperative to improve preclinical cancer models and assays capable of predicting drug response or potential drug resistance in clinical trials. For example, development of genetically engineered CRC mouse models with PIK3CA mutations alone, or in combination with other mutations such as APC and/or KRAS, would be an especially valuable tool for understanding drug resistance mechanisms and selection of the most effective method of PI3K pathway inhibition. Finally, although there are growing numbers of isoform-specific and allosteric PI3K pathway inhibitors, the majority of inhibitors in the pathway are ATP-competitive pan inhibitors which may not possess the potency and selectivity necessary for efficacy in patients. The importance of high throughput structural biology techniques, such as X-ray crystallography and kinome wide selectivity profiling will continue aid development of isoform-specific, mutant-specific, and allosteric inhibitors of key components of the PI3K pathway. Direct knowledge of the three dimensional structure of the target affords the best tool for rational inhibitor design and optimization of target selectivity. Crystal structures of p110α mutants, p110β, p110δ, and mTOR will be essential to the advancement of future strategies targeting the PI3K pathway. The success of PI3K pathway inhibitors in CRC and other cancers depends upon the interplay of basic molecular understanding, rational inhibitor design, biomarker advances, and ultimately well-designed clinical trials.

References

Abubaker J, Bavi P, Al-Haqawi W, Jehan Z, Munkarah A, Uddin S, Al-Kuraya KS (2009) PIK3CA alterations in Middle Eastern ovarian cancers. Mol Cancer 8:51

Agoulnik IU, Hodgson MC, Bowden WA, Ittmann MM (2011) INPP4B: the new kid on the PI3K block. Oncotarget 2:321–328

Alessi DR, Caudwell FB, Andjelkovic M, Hemmings BA, Cohen P (1996) Molecular basis for the substrate specificity of protein kinase B; comparison with MAPKAP kinase-1 and p70 S6 kinase. FEBS Lett 399:333–338

Alessi DR, Deak M, Casamayor A, Caudwell FB, Morrice N, Norman DG, Gaffney P, Reese CB, MacDougall CN, Harbison D, Ashworth A, Bownes M (1997) 3-Phosphoinositide-dependent protein kinase-1 (PDK1): structural and functional homology with the Drosophila DSTPK61 kinase. Curr Biol 7:776–789

Amado RG, Wolf M, Peeters M, Van Cutsem E, Siena S, Freeman DJ, Juan T, Sikorski R, Suggs S, Radinsky R, Patterson SD, Chang DD (2008) Wild-type KRAS is required for panitumumab efficacy in patients with metastatic colorectal cancer. J Clin Oncol 26:1626–1634

Apsel B, Blair JA, Gonzalez B, Nazif TM, Feldman ME, Aizenstein B, Hoffman R, Williams RL, Shokat KM, Knight ZA (2008) Targeted polypharmacology: discovery of dual inhibitors of tyrosine and phosphoinositide kinases. Nat Chem Biol 4:691–699

Bader AG, Kang S, Zhao L, Vogt PK (2005) Oncogenic PI3K deregulates transcription and translation. Nat Rev Cancer 5:921–929
Baselga J (2011) Targeting the phosphoinositide-3 (PI3) kinase pathway in breast cancer. Oncologist 16(suppl 1):12–19
Bendell JC, Nemunaitis J, Vukelja SJ, Hagenstad C, Campos LT, Hermann RC, Sportelli P, Gardner L, Richards DA (2011) Randomized placebo-controlled phase II trial of perifosine plus capecitabine as second- or third-line therapy in patients with metastatic colorectal cancer. J Clin Oncol 29:4394–4400
Bendell JC, Rodon J, Burris HA, de Jonge M, Verweij J, Birle D, Demanse D, De Buck SS, Ru QC, Peters M, Goldbrunner M, Baselga J (2012) Phase I, dose-escalation study of BKM120, an oral pan-class I PI3K inhibitor, in patients with advanced solid tumors. J Clin Oncol 30(3):282–290
Bernardi R, Guernah I, Jin D, Grisendi S, Alimonti A, Teruya-Feldstein J, Cordon-Cardo C, Simon MC, Rafii S, Pandolfi PP (2006) PML inhibits HIF-1alpha translation and neoangiogenesis through repression of mTOR. Nature 442:779–785
Beroukhim R, Mermel CH, Porter D, Wei G, Raychaudhuri S, Donovan J, Barretina J, Boehm JS, Dobson J, Urashima M, Mc Henry KT, Pinchback RM, Ligon AH, Cho YJ, Haery L, Greulich H, Reich M, Winckler W, Lawrence MS, Weir BA, Tanaka KE, Chiang DY, Bass AJ, Loo A, Hoffman C, Prensner J, Liefeld T, Gao Q, Yecies D, Signoretti S, Maher E, Kaye FJ, Sasaki H, Tepper JE, Fletcher JA, Tabernero J, Baselga J, Tsao MS, Demichelis F, Rubin MA, Janne PA, Daly MJ, Nucera C, Levine RL, Ebert BL, Gabriel S, Rustgi AK, Antonescu CR, Ladanyi M, Letai A, Garraway LA, Loda M, Beer DG, True LD, Okamoto A, Pomeroy SL, Singer S, Golub TR, Lander ES, Getz G, Sellers WR, Meyerson M (2010) The landscape of somatic copy-number alteration across human cancers. Nature 463:899–905
Berwick DC, Hers I, Heesom KJ, Moule SK, Tavare JM (2002) The identification of ATP-citrate lyase as a protein kinase B (Akt) substrate in primary adipocytes. J Biol Chem 277:33895–33900
Bhaskar PT, Hay N (2007) The two TORCs and Akt. Dev Cell 12:487–502
Bi L, Okabe I, Bernard DJ, Wynshaw-Boris A, Nussbaum RL (1999) Proliferative defect and embryonic lethality in mice homozygous for a deletion in the p110alpha subunit of phosphoinositide 3-kinase. J Biol Chem 274:10963–10968
Brunet A, Bonni A, Zigmond MJ, Lin MZ, Juo P, Hu LS, Anderson MJ, Arden KC, Blenis J, Greenberg ME (1999) Akt promotes cell survival by phosphorylating and inhibiting a Forkhead transcription factor. Cell 96:857–868
Calera MR, Martinez C, Liu H, Jack AK, Birnbaum MJ, Pilch PF (1998) Insulin increases the association of Akt-2 with Glut4-containing vesicles. J Biol Chem 273:7201–7204
Caron RW, Yacoub A, Li M, Zhu X, Mitchell C, Hong Y, Hawkins W, Sasazuki T, Shirasawa S, Kozikowski AP, Dennis PA, Hagan MP, Grant S, Dent P (2005) Activated forms of H-RAS and K-RAS differentially regulate membrane association of PI3K, PDK-1, and AKT and the effect of therapeutic kinase inhibitors on cell survival. Mol Cancer Ther 4:257–270
Carracedo A, Ma L, Teruya-Feldstein J, Rojo F, Salmena L, Alimonti A, Egia A, Sasaki AT, Thomas G, Kozma SC, Papa A, Nardella C, Cantley LC, Baselga J, Pandolfi PP (2008) Inhibition of mTORC1 leads to MAPK pathway activation through a PI3K-dependent feedback loop in human cancer. J Clin Invest 118:3065–3074
Cascio S, Ferla R, D'Andrea A, Gerbino A, Bazan V, Surmacz E, Russo A (2009) Expression of angiogenic regulators, VEGF and leptin, is regulated by the EGF/PI3K/STAT3 pathway in colorectal cancer cells. J Cell Physiol 221:189–194
Castro AF, Rebhun JF, Clark GJ, Quilliam LA (2003) Rheb binds tuberous sclerosis complex 2 (TSC2) and promotes S6 kinase activation in a rapamycin- and farnesylation-dependent manner. J Biol Chem 278:32493–32496
Chee CE, Sinicrope FA (2010) Targeted therapeutic agents for colorectal cancer. Gastroenterol Clin North Am 39:601–613
Chen J, Somanath PR, Razorenova O, Chen WS, Hay N, Bornstein P, Byzova TV (2005) Akt1 regulates pathological angiogenesis, vascular maturation and permeability in vivo. Nat Med 11:1188–1196

Cho H, Thorvaldsen JL, Chu Q, Feng F, Birnbaum MJ (2001a) Akt1/PKBalpha is required for normal growth but dispensable for maintenance of glucose homeostasis in mice. J Biol Chem 276:38349–38352

Cho H, Mu J, Kim JK, Thorvaldsen JL, Chu Q, Crenshaw EB III, Kaestner KH, Bartolomei MS, Shulman GI, Birnbaum MJ (2001b) Insulin resistance and a diabetes mellitus-like syndrome in mice lacking the protein kinase Akt2 (PKB beta). Science 292:1728–1731

Chresta CM, Davies BR, Hickson I, Harding T, Cosulich S, Critchlow SE, Vincent JP, Ellston R, Jones D, Sini P, James D, Howard Z, Dudley P, Hughes G, Smith L, Maguire S, Hummersone M, Malagu K, Menear K, Jenkins R, Jacobsen M, Smith GC, Guichard S, Pass M (2010) AZD8055 is a potent, selective, and orally bioavailable ATP-competitive mammalian target of rapamycin kinase inhibitor with in vitro and in vivo antitumor activity. Cancer Res 70:288–298

Cleary JM, Shapiro GI (2010) Development of phosphoinositide-3 kinase pathway inhibitors for advanced cancer. Curr Oncol Rep 12:87–94

Cook PJ, Doll R, Fellingham SA (1969) A mathematical model for the age distribution of cancer in man. Int J Cancer 4:93–112

Crul M, Rosing H, de Klerk GJ, Dubbelman R, Traiser M, Reichert S, Knebel NG, Schellens JH, Beijnen JH, ten Bokkel Huinink WW (2002) Phase I and pharmacological study of daily oral administration of perifosine (D-21266) in patients with advanced solid tumours. Eur J Cancer 38:1615–1621

Cully M, You H, Levine AJ, Mak TW (2006) Beyond PTEN mutations: the PI3K pathway as an integrator of multiple inputs during tumorigenesis. Nat Rev Cancer 6:184–192

Dan S, Okamura M, Seki M, Yamazaki K, Sugita H, Okui M, Mukai Y, Nishimura H, Asaka R, Nomura K, Ishikawa Y, Yamori T (2010) Correlating phosphatidylinositol 3-kinase inhibitor efficacy with signaling pathway status: in silico and biological evaluations. Cancer Res 70:4982–4994

Datta SR, Dudek H, Tao X, Masters S, Fu H, Gotoh Y, Greenberg ME (1997) Akt phosphorylation of BAD couples survival signals to the cell-intrinsic death machinery. Cell 91:231–241

Datta SR, Katsov A, Hu L, Petros A, Fesik SW, Yaffe MB, Greenberg ME (2000) 14-3-3 proteins and survival kinases cooperate to inactivate BAD by BH3 domain phosphorylation. Mol Cell 6:41–51

De Roock W, Claes B, Bernasconi D, De Schutter J, Biesmans B, Fountzilas G, Kalogeras KT, Kotoula V, Papamichael D, Laurent-Puig P, Penault-Llorca F, Rougier P, Vincenzi B, Santini D, Tonini G, Cappuzzo F, Frattini M, Molinari F, Saletti P, De Dosso S, Martini M, Bardelli A, Siena S, Sartore-Bianchi A, Tabernero J, Macarulla T, Di Fiore F, Gangloff AO, Ciardiello F, Pfeiffer P, Qvortrup C, Hansen TP, Van Cutsem E, Piessevaux H, Lambrechts D, Delorenzi M, Tejpar S (2010) Effects of KRAS, BRAF, NRAS, and PIK3CA mutations on the efficacy of cetuximab plus chemotherapy in chemotherapy-refractory metastatic colorectal cancer: a retrospective consortium analysis. Lancet Oncol 11:753–762

del Peso L, Gonzalez-Garcia M, Page C, Herrera R, Nunez G (1997) Interleukin-3-induced phosphorylation of BAD through the protein kinase Akt. Science 278:687–689

Di Cristofano A, Pesce B, Cordon-Cardo C, Pandolfi PP (1998) Pten is essential for embryonic development and tumour suppression. Nat Genet 19:348–355

Diehl JA, Cheng M, Roussel MF, Sherr CJ (1998) Glycogen synthase kinase-3beta regulates cyclin D1 proteolysis and subcellular localization. Genes Dev 12:3499–3511

Dijkers PF, Birkenkamp KU, Lam EW, Thomas NS, Lammers JW, Koenderman L, Coffer PJ (2002) FKHR-L1 can act as a critical effector of cell death induced by cytokine withdrawal: protein kinase B-enhanced cell survival through maintenance of mitochondrial integrity. J Cell Biol 156:531–542

Dudek H, Datta SR, Franke TF, Birnbaum MJ, Yao R, Cooper GM, Segal RA, Kaplan DR, Greenberg ME (1997) Regulation of neuronal survival by the serine-threonine protein kinase Akt. Science 275:661–665

Easton RM, Cho H, Roovers K, Shineman DW, Mizrahi M, Forman MS, Lee VM, Szabolcs M, de Jong R, Oltersdorf T, Ludwig T, Efstratiadis A, Birnbaum MJ (2005) Role for Akt3/protein kinase Bgamma in attainment of normal brain size. Mol Cell Biol 25:1869–1878

Engelman JA (2009) Targeting PI3K signalling in cancer: opportunities, challenges and limitations. Nat Rev Cancer 9:550–562
Engelman JA, Luo J, Cantley LC (2006) The evolution of phosphatidylinositol 3-kinases as regulators of growth and metabolism. Nat Rev Genet 7:606–619
Engelman JA, Chen L, Tan X, Crosby K, Guimaraes AR, Upadhyay R, Maira M, McNamara K, Perera SA, Song Y, Chirieac LR, Kaur R, Lightbown A, Simendinger J, Li T, Padera RF, Garcia-Echeverria C, Weissleder R, Mahmood U, Cantley LC, Wong KK (2008) Effective use of PI3K and MEK inhibitors to treat mutant Kras G12D and PIK3CA H1047R murine lung cancers. Nat Med 14:1351–1356
Falcon BL, Barr S, Gokhale PC, Chou J, Fogarty J, Depeille P, Miglarese M, Epstein DM, McDonald DM (2011) Reduced VEGF production, angiogenesis, and vascular regrowth contribute to the antitumor properties of dual mTORC1/mTORC2 inhibitors. Cancer Res 71:1573–1583
Fan QW, Cheng CK, Nicolaides TP, Hackett CS, Knight ZA, Shokat KM, Weiss WA (2007) A dual phosphoinositide-3-kinase alpha/mTOR inhibitor cooperates with blockade of epidermal growth factor receptor in PTEN-mutant glioma. Cancer Res 67:7960–7965
Feldman ME, Apsel B, Uotila A, Loewith R, Knight ZA, Ruggero D, Shokat KM (2009) Active-site inhibitors of mTOR target rapamycin-resistant outputs of mTORC1 and mTORC2. PLoS Biol 7:e38
Folkman J (2007) Angiogenesis: an organizing principle for drug discovery? Nat Rev Drug Discov 6:273–286
Foukas LC, Claret M, Pearce W, Okkenhaug K, Meek S, Peskett E, Sancho S, Smith AJ, Withers DJ, Vanhaesebroeck B (2006) Critical role for the p110alpha phosphoinositide-3-OH kinase in growth and metabolic regulation. Nature 441:366–370
Frattini M, Saletti P, Romagnani E, Martin V, Molinari F, Ghisletta M, Camponovo A, Etienne LL, Cavalli F, Mazzucchelli L (2007) PTEN loss of expression predicts cetuximab efficacy in metastatic colorectal cancer patients. Br J Cancer 97:1139–1145
Fruman DA (2004) Phosphoinositide 3-kinase and its targets in B-cell and T-cell signaling. Curr Opin Immunol 16:314–320
Fruman DA, Rommel C (2011) PI3Kdelta inhibitors in cancer: rationale and serendipity merge in the clinic. Cancer Discov 1:562–572
Fruman DA, Meyers RE, Cantley LC (1998) Phosphoinositide kinases. Annu Rev Biochem 67:481–507
Furuyama T, Kitayama K, Shimoda Y, Ogawa M, Sone K, Yoshida-Araki K, Hisatsune H, Nishikawa S, Nakayama K, Ikeda K, Motoyama N, Mori N (2004) Abnormal angiogenesis in Foxo1 (Fkhr)-deficient mice. J Biol Chem 279:34741–34749
Garami A, Zwartkruis FJ, Nobukuni T, Joaquin M, Roccio M, Stocker H, Kozma SC, Hafen E, Bos JL, Thomas G (2003) Insulin activation of Rheb, a mediator of mTOR/S6K/4E-BP signaling, is inhibited by TSC1 and 2. Mol Cell 11:1457–1466
Garcia-Echeverria C, Sellers WR (2008) Drug discovery approaches targeting the PI3K/Akt pathway in cancer. Oncogene 27:5511–5526
Garlich JR, De P, Dey N, Su JD, Peng X, Miller A, Murali R, Lu Y, Mills GB, Kundra V, Shu HK, Peng Q, Durden DL (2008) A vascular targeted pan phosphoinositide 3-kinase inhibitor prodrug, SF1126, with antitumor and antiangiogenic activity. Cancer Res 68:206–215
Garofalo RS, Orena SJ, Rafidi K, Torchia AJ, Stock JL, Hildebrandt AL, Coskran T, Black SC, Brees DJ, Wicks JR, McNeish JD, Coleman KG (2003) Severe diabetes, age-dependent loss of adipose tissue, and mild growth deficiency in mice lacking Akt2/PKB beta. J Clin Invest 112:197–208
Gasic GP (1994) Basic-helix-loop-helix transcription factor and sterol sensor in a single membrane-bound molecule. Cell 77:17–19
George S, Rochford JJ, Wolfrum C, Gray SL, Schinner S, Wilson JC, Soos MA, Murgatroyd PR, Williams RM, Acerini CL, Dunger DB, Barford D, Umpleby AM, Wareham NJ, Davies HA, Schafer AJ, Stoffel M, O'Rahilly S, Barroso I (2004) A family with severe insulin resistance and diabetes due to a mutation in AKT2. Science 304:1325–1328

Gingras AC, Kennedy SG, O'Leary MA, Sonenberg N, Hay N (1998) 4E-BP1, a repressor of mRNA translation, is phosphorylated and inactivated by the Akt(PKB) signaling pathway. Genes Dev 12:502–513
Gordan JD, Thompson CB, Simon MC (2007) HIF and c-Myc: sibling rivals for control of cancer cell metabolism and proliferation. Cancer Cell 12:108–113
Granville CA, Memmott RM, Gills JJ, Dennis PA (2006) Handicapping the race to develop inhibitors of the phosphoinositide 3-kinase/Akt/mammalian target of rapamycin pathway. Clin Cancer Res 12:679–689
Graupera M, Guillermet-Guibert J, Foukas LC, Phng LK, Cain RJ, Salpekar A, Pearce W, Meek S, Millan J, Cutillas PR, Smith AJ, Ridley AJ, Ruhrberg C, Gerhardt H, Vanhaesebroeck B (2008) Angiogenesis selectively requires the p110alpha isoform of PI3K to control endothelial cell migration. Nature 453:662–666
Grothey A (2009) Medical treatment of advanced colorectal cancer in 2009. Ther Adv Med Oncol 1:55–68
Guba M, von Breitenbuch P, Steinbauer M, Koehl G, Flegel S, Hornung M, Bruns CJ, Zuelke C, Farkas S, Anthuber M, Jauch KW, Geissler EK (2002) Rapamycin inhibits primary and metastatic tumor growth by antiangiogenesis: involvement of vascular endothelial growth factor. Nat Med 8:128–135
Gupta S, Ramjaun AR, Haiko P, Wang Y, Warne PH, Nicke B, Nye E, Stamp G, Alitalo K, Downward J (2007) Binding of ras to phosphoinositide 3-kinase p110alpha is required for ras-driven tumorigenesis in mice. Cell 129:957–968
Hamada K, Sasaki T, Koni PA, Natsui M, Kishimoto H, Sasaki J, Yajima N, Horie Y, Hasegawa G, Naito M, Miyazaki J, Suda T, Itoh H, Nakao K, Mak TW, Nakano T, Suzuki A (2005) The PTEN/PI3K pathway governs normal vascular development and tumor angiogenesis. Genes Dev 19:2054–2065
Hanahan D, Weinberg RA (2000) The hallmarks of cancer. Cell 100:57–70
Hilgard P, Klenner T, Stekar J, Nossner G, Kutscher B, Engel J (1997) D-21266, a new heterocyclic alkylphospholipid with antitumour activity. Eur J Cancer 33:442–446
Hirota S, Isozaki K, Moriyama Y, Hashimoto K, Nishida T, Ishiguro S, Kawano K, Hanada M, Kurata A, Takeda M, Muhammad Tunio G, Matsuzawa Y, Kanakura Y, Shinomura Y, Kitamura Y (1998) Gain-of-function mutations of c-kit in human gastrointestinal stromal tumors. Science 279:577–580
Hofer E, Schweighofer B (2007) Signal transduction induced in endothelial cells by growth factor receptors involved in angiogenesis. Thromb Haemost 97:355–363
Howes AL, Chiang GG, Lang ES, Ho CB, Powis G, Vuori K, Abraham RT (2007) The phosphatidylinositol 3-kinase inhibitor, PX-866, is a potent inhibitor of cancer cell motility and growth in three-dimensional cultures. Mol Cancer Ther 6:2505–2514
Hsieh AC, Ruggero D (2010) Targeting eukaryotic translation initiation factor 4E (eIF4E) in cancer. Clin Cancer Res 16:4914–4920
Hsieh AC, Costa M, Zollo O, Davis C, Feldman ME, Testa JR, Meyuhas O, Shokat KM, Ruggero D (2010) Genetic dissection of the oncogenic mTOR pathway reveals druggable addiction to translational control via 4EBP-eIF4E. Cancer Cell 17:249–261
Hu L, Hofmann J, Lu Y, Mills GB, Jaffe RB (2002) Inhibition of phosphatidylinositol 3′-kinase increases efficacy of paclitaxel in in vitro and in vivo ovarian cancer models. Cancer Res 62:1087–1092
Huang XF, Chen JZ (2009) Obesity, the PI3K/Akt signal pathway and colon cancer. Obes Rev 10:610–616
Huang CH, Mandelker D, Schmidt-Kittler O, Samuels Y, Velculescu VE, Kinzler KW, Vogelstein B, Gabelli SB, Amzel LM (2007) The structure of a human p110alpha/p85alpha complex elucidates the effects of oncogenic PI3Kalpha mutations. Science 318:1744–1748
Hudson CC, Liu M, Chiang GG, Otterness DM, Loomis DC, Kaper F, Giaccia AJ, Abraham RT (2002) Regulation of hypoxia-inducible factor 1alpha expression and function by the mammalian target of rapamycin. Mol Cell Biol 22:7004–7014

Ihle NT, Williams R, Chow S, Chew W, Berggren MI, Paine-Murrieta G, Minion DJ, Halter RJ, Wipf P, Abraham R, Kirkpatrick L, Powis G (2004) Molecular pharmacology and antitumor activity of PX-866, a novel inhibitor of phosphoinositide-3-kinase signaling. Mol Cancer Ther 3:763–772

Ihle NT, Paine-Murrieta G, Berggren MI, Baker A, Tate WR, Wipf P, Abraham RT, Kirkpatrick DL, Powis G (2005) The phosphatidylinositol-3-kinase inhibitor PX-866 overcomes resistance to the epidermal growth factor receptor inhibitor gefitinib in A-549 human non-small cell lung cancer xenografts. Mol Cancer Ther 4:1349–1357

Ihle NT, Lemos R, Schwartz D, Oh J, Halter RJ, Wipf P, Kirkpatrick L, Powis G (2009a) Peroxisome proliferator-activated receptor gamma agonist pioglitazone prevents the hyperglycemia caused by phosphatidylinositol 3-kinase pathway inhibition by PX-866 without affecting antitumor activity. Mol Cancer Ther 8:94–100

Ihle NT, Lemos R Jr, Wipf P, Yacoub A, Mitchell C, Siwak D, Mills GB, Dent P, Kirkpatrick DL, Powis G (2009b) Mutations in the phosphatidylinositol-3-kinase pathway predict for antitumor activity of the inhibitor PX-866 whereas oncogenic Ras is a dominant predictor for resistance. Cancer Res 69:143–150

Ikenoue T, Kanai F, Hikiba Y, Obata T, Tanaka Y, Imamura J, Ohta M, Jazag A, Guleng B, Tateishi K, Asaoka Y, Matsumura M, Kawabe T, Omata M (2005) Functional analysis of PIK3CA gene mutations in human colorectal cancer. Cancer Res 65:4562–4567

Inoki K, Zhu T, Guan KL (2003a) TSC2 mediates cellular energy response to control cell growth and survival. Cell 115:577–590

Inoki K, Li Y, Xu T, Guan KL (2003b) Rheb GTPase is a direct target of TSC2 GAP activity and regulates mTOR signaling. Genes Dev 17:1829–1834

Inukai K, Funaki M, Ogihara T, Katagiri H, Kanda A, Anai M, Fukushima Y, Hosaka T, Suzuki M, Shin BC, Takata K, Yazaki Y, Kikuchi M, Oka Y, Asano T (1997) p85alpha gene generates three isoforms of regulatory subunit for phosphatidylinositol 3-kinase (PI 3-Kinase), p50alpha, p55alpha, and p85alpha, with different PI 3-kinase activity elevating responses to insulin. J Biol Chem 272:7873–7882

Jia S, Liu Z, Zhang S, Liu P, Zhang L, Lee SH, Zhang J, Signoretti S, Loda M, Roberts TM, Zhao JJ (2008) Essential roles of PI(3)K-p110beta in cell growth, metabolism and tumorigenesis. Nature 454:776–779

Jiang BH, Zheng JZ, Aoki M, Vogt PK (2000) Phosphatidylinositol 3-kinase signaling mediates angiogenesis and expression of vascular endothelial growth factor in endothelial cells. Proc Natl Acad Sci USA 97:1749–1753

Jou ST, Carpino N, Takahashi Y, Piekorz R, Chao JR, Wang D, Ihle JN (2002) Essential, nonredundant role for the phosphoinositide 3-kinase p110delta in signaling by the B-cell receptor complex. Mol Cell Biol 22:8580–8591

Karapetis CS, Khambata-Ford S, Jonker DJ, O'Callaghan CJ, Tu D, Tebbutt NC, Simes RJ, Chalchal H, Shapiro JD, Robitaille S, Price TJ, Shepherd L, Au HJ, Langer C, Moore MJ, Zalcberg JR (2008) K-ras mutations and benefit from cetuximab in advanced colorectal cancer. N Engl J Med 359:1757–1765

Khaleghpour K, Li Y, Banville D, Yu Z, Shen SH (2004) Involvement of the PI 3-kinase signaling pathway in progression of colon adenocarcinoma. Carcinogenesis 25:241–248

Kishimoto H, Hamada K, Saunders M, Backman S, Sasaki T, Nakano T, Mak TW, Suzuki A (2003) Physiological functions of Pten in mouse tissues. Cell Struct Funct 28:11–21

Kohn AD, Summers SA, Birnbaum MJ, Roth RA (1996) Expression of a constitutively active Akt Ser/Thr kinase in 3T3-L1 adipocytes stimulates glucose uptake and glucose transporter 4 translocation. J Biol Chem 271:31372–31378

Kondapaka SB, Singh SS, Dasmahapatra GP, Sausville EA, Roy KK (2003) Perifosine, a novel alkylphospholipid, inhibits protein kinase B activation. Mol Cancer Ther 2:1093–1103

Kong D, Okamura M, Yoshimi H, Yamori T (2009) Antiangiogenic effect of ZSTK474, a novel phosphatidylinositol 3-kinase inhibitor. Eur J Cancer 45:857–865

Kurosu H, Maehama T, Okada T, Yamamoto T, Hoshino S, Fukui Y, Ui M, Hazeki O, Katada T (1997) Heterodimeric phosphoinositide 3-kinase consisting of p85 and p110beta is synergistically activated by the betagamma subunits of G proteins and phosphotyrosyl peptide. J Biol Chem 272:24252–24256

Laurent-Puig P, Cayre A, Manceau G, Buc E, Bachet JB, Lecomte T, Rougier P, Lievre A, Landi B, Boige V, Ducreux M, Ychou M, Bibeau F, Bouche O, Reid J, Stone S, Penault-Llorca F (2009) Analysis of PTEN, BRAF, and EGFR status in determining benefit from cetuximab therapy in wild-type KRAS metastatic colon cancer. J Clin Oncol 27:5924–5930
Lee YJ, Hong KH, Yun J, Oh SP (2006) Generation of activin receptor type IIB isoform-specific hypomorphic alleles. Genesis 44:487–494
Legrier ME, Yang CP, Yan HG, Lopez-Barcons L, Keller SM, Perez-Soler R, Horwitz SB, McDaid HM (2007) Targeting protein translation in human non small cell lung cancer via combined MEK and mammalian target of rapamycin suppression. Cancer Res 67:11300–11308
Lengauer C, Kinzler KW, Vogelstein B (1997) Genetic instability in colorectal cancers. Nature 386:623–627
Lengauer C, Kinzler KW, Vogelstein B (1998) Genetic instabilities in human cancers. Nature 396:643–649
Li J, Simpson L, Takahashi M, Miliaresis C, Myers MP, Tonks N, Parsons R (1998) The PTEN/MMAC1 tumor suppressor induces cell death that is rescued by the AKT/protein kinase B oncogene. Cancer Res 58:5667–5672
Li XH, Zheng HC, Takahashi H, Masuda S, Yang XH, Takano Y (2009) PTEN expression and mutation in colorectal carcinomas. Oncol Rep 22:757–764
Li FH, Shen L, Li ZH, Luo HY, Qiu MZ, Zhang HZ, Li YH, Xu RH (2010) Impact of KRAS mutation and PTEN expression on cetuximab-treated colorectal cancer. World J Gastroenterol 16:5881–5888
Liang J, Slingerland JM (2003) Multiple roles of the PI3K/PKB (Akt) pathway in cell cycle progression. Cell Cycle 2:339–345
Linardou H, Dahabreh IJ, Kanaloupiti D, Siannis F, Bafaloukos D, Kosmidis P, Papadimitriou CA, Murray S (2008) Assessment of somatic k-RAS mutations as a mechanism associated with resistance to EGFR-targeted agents: a systematic review and meta-analysis of studies in advanced non-small-cell lung cancer and metastatic colorectal cancer. Lancet Oncol 9:962–972
Liu P, Cheng H, Roberts TM, Zhao JJ (2009) Targeting the phosphoinositide 3-kinase pathway in cancer. Nat Rev Drug Discov 8:627–644
Loupakis F, Pollina L, Stasi I, Ruzzo A, Scartozzi M, Santini D, Masi G, Graziano F, Cremolini C, Rulli E, Canestrari E, Funel N, Schiavon G, Petrini I, Magnani M, Tonini G, Campani D, Floriani I, Cascinu S, Falcone A (2009) PTEN expression and KRAS mutations on primary tumors and metastases in the prediction of benefit from cetuximab plus irinotecan for patients with metastatic colorectal cancer. J Clin Oncol 27:2622–2629
Maira SM, Stauffer F, Brueggen J, Furet P, Schnell C, Fritsch C, Brachmann S, Chene P, De Pover A, Schoemaker K, Fabbro D, Gabriel D, Simonen M, Murphy L, Finan P, Sellers W, Garcia-Echeverria C (2008) Identification and characterization of NVP-BEZ235, a new orally available dual phosphatidylinositol 3-kinase/mammalian target of rapamycin inhibitor with potent in vivo antitumor activity. Mol Cancer Ther 7:1851–1863
Mandelker D, Gabelli SB, Schmidt-Kittler O, Zhu J, Cheong I, Huang CH, Kinzler KW, Vogelstein B, Amzel LM (2009) A frequent kinase domain mutation that changes the interaction between PI3Kalpha and the membrane. Proc Natl Acad Sci USA 106:16996–17001
Manning BD, Cantley LC (2007) AKT/PKB signaling: navigating downstream. Cell 129: 1261–1274
Manning BD, Tee AR, Logsdon MN, Blenis J, Cantley LC (2002) Identification of the tuberous sclerosis complex-2 tumor suppressor gene product tuberin as a target of the phosphoinositide 3-kinase/akt pathway. Mol Cell 10:151–162
Markman B, Atzori F, Perez-Garcia J, Tabernero J, Baselga J (2010) Status of PI3K inhibition and biomarker development in cancer therapeutics. Ann Oncol 21:683–691
Mayo LD, Donner DB (2001) A phosphatidylinositol 3-kinase/Akt pathway promotes translocation of Mdm2 from the cytoplasm to the nucleus. Proc Natl Acad Sci USA 98:11598–11603
Mizoguchi M, Nutt CL, Mohapatra G, Louis DN (2004) Genetic alterations of phosphoinositide 3-kinase subunit genes in human glioblastomas. Brain Pathol 14:372–377
Molckovsky A, Siu LL (2008) First-in-class, first-in-human phase I results of targeted agents: highlights of the 2008 American society of clinical oncology meeting. J Hematol Oncol 1:20

Morrow CJ, Gray A, Dive C (2005) Comparison of phosphatidylinositol-3-kinase signalling within a panel of human colorectal cancer cell lines with mutant or wild-type PIK3CA. FEBS Lett 579:5123–5128

Naguib A, Cooke JC, Happerfield L, Kerr L, Gay LJ, Luben RN, Ball RY, Mitrou PN, McTaggart A, Arends MJ (2011) Alterations in PTEN and PIK3CA in colorectal cancers in the EPIC Norfolk study: associations with clinicopathological and dietary factors. BMC Cancer 11:123

Niu Q, Perruzzi C, Voskas D, Lawler J, Dumont DJ, Benjamin LE (2004) Inhibition of Tie-2 signaling induces endothelial cell apoptosis, decreases Akt signaling, and induces endothelial cell expression of the endogenous anti-angiogenic molecule, thrombospondin-1. Cancer Biol Ther 3:402–405

Nosho K, Kawasaki T, Ohnishi M, Suemoto Y, Kirkner GJ, Zepf D, Yan L, Longtine JA, Fuchs CS, Ogino S (2008) PIK3CA mutation in colorectal cancer: relationship with genetic and epigenetic alterations. Neoplasia 10:534–541

Nutley BP, Raynaud FI, Wilson SC, Fischer PM, Hayes A, Goddard PM, McClue SJ, Jarman M, Lane DP, Workman P (2005) Metabolism and pharmacokinetics of the cyclin-dependent kinase inhibitor R-roscovitine in the mouse. Mol Cancer Ther 4:125–139

O'Reilly KE, Rojo F, She QB, Solit D, Mills GB, Smith D, Lane H, Hofmann F, Hicklin DJ, Ludwig DL, Baselga J, Rosen N (2006) mTOR inhibition induces upstream receptor tyrosine kinase signaling and activates Akt. Cancer Res 66:1500–1508

Oda K, Okada J, Timmerman L, Rodriguez-Viciana P, Stokoe D, Shoji K, Taketani Y, Kuramoto H, Knight ZA, Shokat KM, McCormick F (2008) PIK3CA cooperates with other phosphatidylinositol 3′-kinase pathway mutations to effect oncogenic transformation. Cancer Res 68:8127–8136

Ogino S, Nosho K, Kirkner GJ, Shima K, Irahara N, Kure S, Chan AT, Engelman JA, Kraft P, Cantley LC, Giovannucci EL, Fuchs CS (2009) PIK3CA mutation is associated with poor prognosis among patients with curatively resected colon cancer. J Clin Oncol 27:1477–1484

Okkenhaug K, Bilancio A, Farjot G, Priddle H, Sancho S, Peskett E, Pearce W, Meek SE, Salpekar A, Waterfield MD, Smith AJ, Vanhaesebroeck B (2002) Impaired B and T cell antigen receptor signaling in p110delta PI 3-kinase mutant mice. Science 297:1031–1034

Ozes ON, Mayo LD, Gustin JA, Pfeffer SR, Pfeffer LM, Donner DB (1999) NF-kappaB activation by tumour necrosis factor requires the Akt serine-threonine kinase. Nature 401:82–85

Paik JH, Kollipara R, Chu G, Ji H, Xiao Y, Ding Z, Miao L, Tothova Z, Horner JW, Carrasco DR, Jiang S, Gilliland DG, Chin L, Wong WH, Castrillon DH, DePinho RA (2007) FoxOs are lineage-restricted redundant tumor suppressors and regulate endothelial cell homeostasis. Cell 128:309–323

Pause A, Belsham GJ, Gingras AC, Donze O, Lin TA, Lawrence JC Jr, Sonenberg N (1994) Insulin-dependent stimulation of protein synthesis by phosphorylation of a regulator of 5′-cap function. Nature 371:762–767

Peterson RT, Schreiber SL (1998) Translation control: connecting mitogens and the ribosome. Curr Biol 8:R248–R250

Philp AJ, Campbell IG, Leet C, Vincan E, Rockman SP, Whitehead RH, Thomas RJ, Phillips WA (2001) The phosphatidylinositol 3′-kinase p85alpha gene is an oncogene in human ovarian and colon tumors. Cancer Res 61:7426–7429

Porstmann T, Griffiths B, Chung YL, Delpuech O, Griffiths JR, Downward J, Schulze A (2005) PKB/Akt induces transcription of enzymes involved in cholesterol and fatty acid biosynthesis via activation of SREBP. Oncogene 24:6465–6481

Porstmann T, Santos CR, Griffiths B, Cully M, Wu M, Leevers S, Griffiths JR, Chung YL, Schulze A (2008) SREBP activity is regulated by mTORC1 and contributes to Akt-dependent cell growth. Cell Metab 8:224–236

Poulogiannis G, Ichimura K, Hamoudi RA, Luo F, Leung SY, Yuen ST, Harrison DJ, Wyllie AH, Arends MJ (2010) Prognostic relevance of DNA copy number changes in colorectal cancer. J Pathol 220:338–347

Poulogiannis G, Luo F, Arends MJ (2012) RAS signalling in the colorectum in health and disease. Cell Commun Adhes 19(1):1–9

Pullen N, Thomas G (1997) The modular phosphorylation and activation of p70s6k. FEBS Lett 410:78–82

Pullen N, Dennis PB, Andjelkovic M, Dufner A, Kozma SC, Hemmings BA, Thomas G (1998) Phosphorylation and activation of p70s6k by PDK1. Science 279:707–710

Raynaud FI, Eccles S, Clarke PA, Hayes A, Nutley B, Alix S, Henley A, Di-Stefano F, Ahmad Z, Guillard S, Bjerke LM, Kelland L, Valenti M, Patterson L, Gowan S, de Haven Brandon A, Hayakawa M, Kaizawa H, Koizumi T, Ohishi T, Patel S, Saghir N, Parker P, Waterfield M, Workman P (2007) Pharmacologic characterization of a potent inhibitor of class I phosphatidylinositide 3-kinases. Cancer Res 67:5840–5850

Roche S, Downward J, Raynal P, Courtneidge SA (1998) A function for phosphatidylinositol 3-kinase beta (p85alpha-p110beta) in fibroblasts during mitogenesis: requirement for insulin- and lysophosphatidic acid-mediated signal transduction. Mol Cell Biol 18:7119–7129

Samuels Y, Wang Z, Bardelli A, Silliman N, Ptak J, Szabo S, Yan H, Gazdar A, Powell SM, Riggins GJ, Willson JK, Markowitz S, Kinzler KW, Vogelstein B, Velculescu VE (2004) High frequency of mutations of the PIK3CA gene in human cancers. Science 304:554

Samuels Y, Diaz LA Jr, Schmidt-Kittler O, Cummins JM, Delong L, Cheong I, Rago C, Huso DL, Lengauer C, Kinzler KW, Vogelstein B, Velculescu VE (2005) Mutant PIK3CA promotes cell growth and invasion of human cancer cells. Cancer Cell 7:561–573

Sarbassov DD, Guertin DA, Ali SM, Sabatini DM (2005) Phosphorylation and regulation of Akt/PKB by the rictor-mTOR complex. Science 307:1098–1101

Sarbassov DD, Ali SM, Sengupta S, Sheen JH, Hsu PP, Bagley AF, Markhard AL, Sabatini DM (2006) Prolonged rapamycin treatment inhibits mTORC2 assembly and Akt/PKB. Mol Cell 22:159–168

Sasaki T, Irie-Sasaki J, Jones RG, Oliveira-dos-Santos AJ, Stanford WL, Bolon B, Wakeham A, Itie A, Bouchard D, Kozieradzki I, Joza N, Mak TW, Ohashi PS, Suzuki A, Penninger JM (2000) Function of PI3Kgamma in thymocyte development, T cell activation, and neutrophil migration. Science 287:1040–1046

Schnell CR, Stauffer F, Allegrini PR, O'Reilly T, McSheehy PM, Dartois C, Stumm M, Cozens R, Littlewood-Evans A, Garcia-Echeverria C, Maira SM (2008) Effects of the dual phosphatidylinositol 3-kinase/mammalian target of rapamycin inhibitor NVP-BEZ235 on the tumor vasculature: implications for clinical imaging. Cancer Res 68:6598–6607

Seki T, Hong KH, Yun J, Kim SJ, Oh SP (2004) Isolation of a regulatory region of activin receptor-like kinase 1 gene sufficient for arterial endothelium-specific expression. Circ Res 94:e72–e77

Serra V, Markman B, Scaltriti M, Eichhorn PJ, Valero V, Guzman M, Botero ML, Llonch E, Atzori F, Di Cosimo S, Maira M, Garcia-Echeverria C, Parra JL, Arribas J, Baselga J (2008) NVP-BEZ235, a dual PI3K/mTOR inhibitor, prevents PI3K signaling and inhibits the growth of cancer cells with activating PI3K mutations. Cancer Res 68:8022–8030

She QB, Solit DB, Ye Q, O'Reilly KE, Lobo J, Rosen N (2005) The BAD protein integrates survival signaling by EGFR/MAPK and PI3K/Akt kinase pathways in PTEN-deficient tumor cells. Cancer Cell 8:287–297

She QB, Chandarlapaty S, Ye Q, Lobo J, Haskell KM, Leander KR, DeFeo-Jones D, Huber HE, Rosen N (2008) Breast tumor cells with PI3K mutation or HER2 amplification are selectively addicted to Akt signaling. PLoS One 3:e3065

Sjoblom T, Jones S, Wood LD, Parsons DW, Lin J, Barber TD, Mandelker D, Leary RJ, Ptak J, Silliman N, Szabo S, Buckhaults P, Farrell C, Meeh P, Markowitz SD, Willis J, Dawson D, Willson JK, Gazdar AF, Hartigan J, Wu L, Liu C, Parmigiani G, Park BH, Bachman KE, Papadopoulos N, Vogelstein B, Kinzler KW, Velculescu VE (2006) The consensus coding sequences of human breast and colorectal cancers. Science 314:268–274

Skinner HD, Zheng JZ, Fang J, Agani F, Jiang BH (2004) Vascular endothelial growth factor transcriptional activation is mediated by hypoxia-inducible factor 1alpha, HDM2, and p70S6K1 in response to phosphatidylinositol 3-kinase/AKT signaling. J Biol Chem 279:45643–45651

Sos ML, Fischer S, Ullrich R, Peifer M, Heuckmann JM, Koker M, Heynck S, Stuckrath I, Weiss J, Fischer F, Michel K, Goel A, Regales L, Politi KA, Perera S, Getlik M, Heukamp LC, Ansen S, Zander T, Beroukhim R, Kashkar H, Shokat KM, Sellers WR, Rauh D, Orr C, Hoeflich KP,

Friedman L, Wong KK, Pao W, Thomas RK (2009) Identifying genotype-dependent efficacy of single and combined PI3K- and MAPK-pathway inhibition in cancer. Proc Natl Acad Sci USA 106:18351–18356
Souglakos J, Philips J, Wang R, Marwah S, Silver M, Tzardi M, Silver J, Ogino S, Hooshmand S, Kwak E, Freed E, Meyerhardt JA, Saridaki Z, Georgoulias V, Finkelstein D, Fuchs CS, Kulke MH, Shivdasani RA (2009) Prognostic and predictive value of common mutations for treatment response and survival in patients with metastatic colorectal cancer. Br J Cancer 101:465–472
Stambolic V, Woodgett JR (2006) Functional distinctions of protein kinase B/Akt isoforms defined by their influence on cell migration. Trends Cell Biol 16:461–466
Stephens L, Anderson K, Stokoe D, Erdjument-Bromage H, Painter GF, Holmes AB, Gaffney PR, Reese CB, McCormick F, Tempst P, Coadwell J, Hawkins PT (1998) Protein kinase B kinases that mediate phosphatidylinositol 3,4,5-trisphosphate-dependent activation of protein kinase B. Science 279:710–714
Stokoe D (2001) Pten. Curr Biol 11:R502
Su JD, Mayo LD, Donner DB, Durden DL (2003) PTEN and phosphatidylinositol 3′-kinase inhibitors up-regulate p53 and block tumor-induced angiogenesis: evidence for an effect on the tumor and endothelial compartment. Cancer Res 63:3585–3592
Suire S, Hawkins P, Stephens L (2002) Activation of phosphoinositide 3-kinase gamma by Ras. Curr Biol 12:1068–1075
Suire S, Coadwell J, Ferguson GJ, Davidson K, Hawkins P, Stephens L (2005) p84, a new Gbetagamma-activated regulatory subunit of the type IB phosphoinositide 3-kinase p110gamma. Curr Biol 15:566–570
Sullivan KM, Kozuch PS (2011) Impact of KRAS mutations on management of colorectal carcinoma. Patholog Res Int 2011:219309
Sun JF, Phung T, Shiojima I, Felske T, Upalakalin JN, Feng D, Kornaga T, Dor T, Dvorak AM, Walsh K, Benjamin LE (2005) Microvascular patterning is controlled by fine-tuning the Akt signal. Proc Natl Acad Sci USA 102:128–133
Sun Y, Zhao S, Tian H, Xie X, Xiao F, Li K, Song Y (2009) Depletion of PI3K p85alpha induces cell cycle arrest and apoptosis in colorectal cancer cells. Oncol Rep 22:1435–1441
Sundqvist A, Bengoechea-Alonso MT, Ye X, Lukiyanchuk V, Jin J, Harper JW, Ericsson J (2005) Control of lipid metabolism by phosphorylation-dependent degradation of the SREBP family of transcription factors by SCF(Fbw7). Cell Metab 1:379–391
Tabernero J, Rojo F, Calvo E, Burris H, Judson I, Hazell K, Martinelli E, Ramon y Cajal S, Jones S, Vidal L, Shand N, Macarulla T, Ramos FJ, Dimitrijevic S, Zoellner U, Tang P, Stumm M, Lane HA, Lebwohl D, Baselga J (2008) Dose- and schedule-dependent inhibition of the mammalian target of rapamycin pathway with everolimus: a phase I tumor pharmacodynamic study in patients with advanced solid tumors. J Clin Oncol 26:1603–1610
Taha C, Liu Z, Jin J, Al-Hasani H, Sonenberg N, Klip A (1999) Opposite translational control of GLUT1 and GLUT4 glucose transporter mRNAs in response to insulin. Role of mammalian target of rapamycin, protein kinase b, and phosphatidylinositol 3-kinase in GLUT1 mRNA translation. J Biol Chem 274:33085–33091
Thomas GV, Tran C, Mellinghoff IK, Welsbie DS, Chan E, Fueger B, Czernin J, Sawyers CL (2006) Hypoxia-inducible factor determines sensitivity to inhibitors of mTOR in kidney cancer. Nat Med 12:122–127
Tran H, Brunet A, Griffith EC, Greenberg ME (2003) The many forks in FOXO's road. Sci STKE 2003:RE5
Tschopp O, Yang ZZ, Brodbeck D, Dummler BA, Hemmings-Mieszczak M, Watanabe T, Michaelis T, Frahm J, Hemmings BA (2005) Essential role of protein kinase B gamma (PKB gamma/Akt3) in postnatal brain development but not in glucose homeostasis. Development 132:2943–2954
Ueki K, Algenstaedt P, Mauvais-Jarvis F, Kahn CR (2000) Positive and negative regulation of phosphoinositide 3-kinase-dependent signaling pathways by three different gene products of the p85alpha regulatory subunit. Mol Cell Biol 20:8035–8046

Van Ummersen L, Binger K, Volkman J, Marnocha R, Tutsch K, Kolesar J, Arzoomanian R, Alberti D, Wilding G (2004) A phase I trial of perifosine (NSC 639966) on a loading dose/maintenance dose schedule in patients with advanced cancer. Clin Cancer Res 10:7450–7456

Vander Heiden MG, Cantley LC, Thompson CB (2009) Understanding the Warburg effect: the metabolic requirements of cell proliferation. Science 324:1029–1033

Vasudevan KM, Barbie DA, Davies MA, Rabinovsky R, McNear CJ, Kim JJ, Hennessy BT, Tseng H, Pochanard P, Kim SY, Dunn IF, Schinzel AC, Sandy P, Hoersch S, Sheng Q, Gupta PB, Boehm JS, Reiling JH, Silver S, Lu Y, Stemke-Hale K, Dutta B, Joy C, Sahin AA, Gonzalez-Angulo AM, Lluch A, Rameh LE, Jacks T, Root DE, Lander ES, Mills GB, Hahn WC, Sellers WR, Garraway LA (2009) AKT-independent signaling downstream of oncogenic PIK3CA mutations in human cancer. Cancer Cell 16:21–32

Velho S, Oliveira C, Ferreira A, Ferreira AC, Suriano G, Schwartz S Jr, Duval A, Carneiro F, Machado JC, Hamelin R, Seruca R (2005) The prevalence of PIK3CA mutations in gastric and colon cancer. Eur J Cancer 41:1649–1654

Villunger A, Michalak EM, Coultas L, Mullauer F, Bock G, Ausserlechner MJ, Adams JM, Strasser A (2003) p53- and drug-induced apoptotic responses mediated by BH3-only proteins puma and noxa. Science 302:1036–1038

Vink SR, Lagerwerf S, Mesman E, Schellens JH, Begg AC, van Blitterswijk WJ, Verheij M (2006) Radiosensitization of squamous cell carcinoma by the alkylphospholipid perifosine in cell culture and xenografts. Clin Cancer Res 12:1615–1622

Vivanco I, Sawyers CL (2002) The phosphatidylinositol 3-Kinase AKT pathway in human cancer. Nat Rev Cancer 2:489–501

Vlahos CJ, Matter WF, Hui KY, Brown RF (1994) A specific inhibitor of phosphatidylinositol 3-kinase, 2-(4-morpholinyl)-8-phenyl-4H-1-benzopyran-4-one (LY294002). J Biol Chem 269:5241–5248

Voigt P, Dorner MB, Schaefer M (2006) Characterization of p87PIKAP, a novel regulatory subunit of phosphoinositide 3-kinase gamma that is highly expressed in heart and interacts with PDE3B. J Biol Chem 281:9977–9986

Walker KS, Deak M, Paterson A, Hudson K, Cohen P, Alessi DR (1998) Activation of protein kinase B beta and gamma isoforms by insulin in vivo and by 3-phosphoinositide-dependent protein kinase-1 in vitro: comparison with protein kinase B alpha. Biochem J 331(Pt 1):299–308

Walker EH, Perisic O, Ried C, Stephens L, Williams RL (1999) Structural insights into phosphoinositide 3-kinase catalysis and signalling. Nature 402:313–320

Walker EH, Pacold ME, Perisic O, Stephens L, Hawkins PT, Wymann MP, Williams RL (2000) Structural determinants of phosphoinositide 3-kinase inhibition by wortmannin, LY294002, quercetin, myricetin, and staurosporine. Mol Cell 6:909–919

Wan X, Harkavy B, Shen N, Grohar P, Helman LJ (2007) Rapamycin induces feedback activation of Akt signaling through an IGF-1R-dependent mechanism. Oncogene 26:1932–1940

Wang S, Gao J, Lei Q, Rozengurt N, Pritchard C, Jiao J, Thomas GV, Li G, Roy-Burman P, Nelson PS, Liu X, Wu H (2003) Prostate-specific deletion of the murine Pten tumor suppressor gene leads to metastatic prostate cancer. Cancer Cell 4:209–221

Warburg O (1956) On the origin of cancer cells. Science 123:309–314

Wee S, Wiederschain D, Maira SM, Loo A, Miller C, deBeaumont R, Stegmeier F, Yao YM, Lengauer C (2008) PTEN-deficient cancers depend on PIK3CB. Proc Natl Acad Sci USA 105:13057–13062

Wee S, Jagani Z, Xiang KX, Loo A, Dorsch M, Yao YM, Sellers WR, Lengauer C, Stegmeier F (2009) PI3K pathway activation mediates resistance to MEK inhibitors in KRAS mutant cancers. Cancer Res 69:4286–4293

Wei W, Jin J, Schlisio S, Harper JW, Kaelin WG Jr (2005) The v-Jun point mutation allows c-Jun to escape GSK3-dependent recognition and destruction by the Fbw7 ubiquitin ligase. Cancer Cell 8:25–33

Weinstein IB, Joe A (2008) Oncogene addiction. Cancer Res 68:3077–3080; discussion 3080

Welcker M, Singer J, Loeb KR, Grim J, Bloecher A, Gurien-West M, Clurman BE, Roberts JM (2003) Multisite phosphorylation by Cdk2 and GSK3 controls cyclin E degradation. Mol Cell 12:381–392

Wen S, Stolarov J, Myers MP, Su JD, Wigler MH, Tonks NK, Durden DL (2001) PTEN controls tumor-induced angiogenesis. Proc Natl Acad Sci USA 98:4622–4627

West MJ, Stoneley M, Willis AE (1998) Translational induction of the c-myc oncogene via activation of the FRAP/TOR signalling pathway. Oncogene 17:769–780

White MF (1998) The IRS-signalling system: a network of docking proteins that mediate insulin action. Mol Cell Biochem 182:3–11

Wood LD, Parsons DW, Jones S, Lin J, Sjoblom T, Leary RJ, Shen D, Boca SM, Barber T, Ptak J, Silliman N, Szabo S, Dezso Z, Ustyanksky V, Nikolskaya T, Nikolsky Y, Karchin R, Wilson PA, Kaminker JS, Zhang Z, Croshaw R, Willis J, Dawson D, Shipitsin M, Willson JK, Sukumar S, Polyak K, Park BH, Pethiyagoda CL, Pant PV, Ballinger DG, Sparks AB, Hartigan J, Smith DR, Suh E, Papadopoulos N, Buckhaults P, Markowitz SD, Parmigiani G, Kinzler KW, Velculescu VE, Vogelstein B (2007) The genomic landscapes of human breast and colorectal cancers. Science 318:1108–1113

Wullschleger S, Loewith R, Hall MN (2006) TOR signaling in growth and metabolism. Cell 124:471–484

Yang H, Kong W, He L, Zhao JJ, O'Donnell JD, Wang J, Wenham RM, Coppola D, Kruk PA, Nicosia SV, Cheng JQ (2008) MicroRNA expression profiling in human ovarian cancer: miR-214 induces cell survival and cisplatin resistance by targeting PTEN. Cancer Res 68:425–433

Yap TA, Yan L, Patnaik A, Fearen I, Olmos D, Papadopoulos K, Baird RD, Delgado L, Taylor A, Lupinacci L, Riisnaes R, Pope LL, Heaton SP, Thomas G, Garrett MD, Sullivan DM, de Bono JS, Tolcher AW (2011) First-in-man clinical trial of the oral pan-AKT inhibitor MK-2206 in patients with advanced solid tumors. J Clin Oncol 29:4688–4695

Yecies JL, Zhang HH, Menon S, Liu S, Yecies D, Lipovsky AI, Gorgun C, Kwiatkowski DJ, Hotamisligil GS, Lee CH, Manning BD (2011) Akt stimulates hepatic SREBP1c and lipogenesis through parallel mTORC1-dependent and independent pathways. Cell Metab 14:21–32

Yeh E, Cunningham M, Arnold H, Chasse D, Monteith T, Ivaldi G, Hahn WC, Stukenberg PT, Shenolikar S, Uchida T, Counter CM, Nevins JR, Means AR, Sears R (2004) A signalling pathway controlling c-Myc degradation that impacts oncogenic transformation of human cells. Nat Cell Biol 6:308–318

Yoong J, Michael M, Leong T (2011) Targeted therapies for gastric cancer: current status. Drugs 71:1367–1384

Yu J, Zhang Y, McIlroy J, Rordorf-Nikolic T, Orr GA, Backer JM (1998a) Regulation of the p85/p110 phosphatidylinositol 3′-kinase: stabilization and inhibition of the p110alpha catalytic subunit by the p85 regulatory subunit. Mol Cell Biol 18:1379–1387

Yu J, Wjasow C, Backer JM (1998b) Regulation of the p85/p110alpha phosphatidylinositol 3′-kinase. Distinct roles for the n-terminal and c-terminal SH2 domains. J Biol Chem 273:30199–30203

Yu K, Lucas J, Zhu T, Zask A, Gaydos C, Toral-Barza L, Gu J, Li F, Chaudhary I, Cai P, Lotvin J, Petersen R, Ruppen M, Fawzi M, Ayral-Kaloustian S, Skotnicki J, Mansour T, Frost P, Gibbons J (2005) PWT-458, a novel pegylated-17-hydroxywortmannin, inhibits phosphatidylinositol 3-kinase signaling and suppresses growth of solid tumors. Cancer Biol Ther 4:538–545

Yuan TL, Cantley LC (2008) PI3K pathway alterations in cancer: variations on a theme. Oncogene 27:5497–5510

Yuan TL, Choi HS, Matsui A, Benes C, Lifshits E, Luo J, Frangioni JV, Cantley LC (2008) Class 1A PI3K regulates vessel integrity during development and tumorigenesis. Proc Natl Acad Sci USA 105:9739–9744

Yun J (2010) Allosteric AKT inhibitors as a targeted cancer therapy. Cancer Biol Ther 9:504–506

Yun J, Rago C, Cheong I, Pagliarini R, Angenendt P, Rajagopalan H, Schmidt K, Willson JK, Markowitz S, Zhou S, Diaz LA Jr, Velculescu VE, Lengauer C, Kinzler KW, Vogelstein B, Papadopoulos N (2009) Glucose deprivation contributes to the development of KRAS pathway mutations in tumor cells. Science 325:1555–1559

Zelzer E, Levy Y, Kahana C, Shilo BZ, Rubinstein M, Cohen B (1998) Insulin induces transcription of target genes through the hypoxia-inducible factor HIF-1alpha/ARNT. EMBO J 17:5085–5094

Zhang YJ, Tian XQ, Sun DF, Zhao SL, Xiong H, Fang JY (2009) Combined inhibition of MEK and mTOR signaling inhibits initiation and progression of colorectal cancer. Cancer Invest 27:273–285

Zhang J, Roberts TM, Shivdasani RA (2011) Targeting PI3K signaling as a therapeutic approach for colorectal cancer. Gastroenterology 141:50–61

Zhong H, Chiles K, Feldser D, Laughner E, Hanrahan C, Georgescu MM, Simons JW, Semenza GL (2000) Modulation of hypoxia-inducible factor 1alpha expression by the epidermal growth factor/phosphatidylinositol 3-kinase/PTEN/AKT/FRAP pathway in human prostate cancer cells: implications for tumor angiogenesis and therapeutics. Cancer Res 60:1541–1545

Zhou BP, Liao Y, Xia W, Zou Y, Spohn B, Hung MC (2001) HER-2/neu induces p53 ubiquitination via Akt-mediated MDM2 phosphorylation. Nat Cell Biol 3:973–982

Zhou XP, Loukola A, Salovaara R, Nystrom-Lahti M, Peltomaki P, de la Chapelle A, Aaltonen LA, Eng C (2002) PTEN mutational spectra, expression levels, and subcellular localization in microsatellite stable and unstable colorectal cancers. Am J Pathol 161:439–447

Zhu T, Gu J, Yu K, Lucas J, Cai P, Tsao R, Gong Y, Li F, Chaudhary I, Desai P, Ruppen M, Fawzi M, Gibbons J, Ayral-Kaloustian S, Skotnicki J, Mansour T, Zask A (2006) Pegylated wortmannin and 17-hydroxywortmannin conjugates as phosphoinositide 3-kinase inhibitors active in human tumor xenograft models. J Med Chem 49:1373–1378

Chapter 7
TGF-ß Signaling Pathway and Colorectal Cancer

William M. Grady

Abstract The Transforming Growth Factor Beta (TGF-ß) signaling pathway is one of the most commonly disrupted pathways in colorectal cancer. Its deregulation appears to mediate cancer formation through a variety of mechanisms. TGF-ß is the canonical member of a family of secreted proteins that include the TGF-ß isoforms, TGF-ß1, TGF-ß2, and TGF-ß3; activins; Growth and Differentiation Factors (GDFs); bone morphogenetic proteins (BMP); inhibin, nodal, and anti-Mullerian hormone. These ligands all mediate biological activities in cells through binding to cell surface receptor complexes that are composed of type I and type II heteromeric receptors. In the colon, TGF-ß can inhibit cell proliferation, induce apoptosis, and induce terminal differentiation, which suggests this pathway has tumor suppressor activities in colorectal cancers. This role of a tumor suppressor pathway is supported by the identification of inactivating mutations and epigenetic alterations in many TGF-ß pathway genes, including *TGFBR2*, *SMAD4*, *SMAD2*, *BMPR2A*, and *ACVR2*. Interestingly, some studies have suggested that in certain contexts TGF-ß may promote the invasive or metastatic behavior of established cancer cells suggesting TGF-ß has a paradoxical role in primary human cancers that appears to depend on the stage of cancer. This chapter will focus on the tumor suppressor activity of the TGF-ß signaling pathway in colorectal cancer and will highlight mechanisms through which TGF-ß signaling mediates its antitumor effects.

W.M. Grady (✉)
Fred Hutchinson Cancer Research Center, University of Washington Medical School,
1100 Fairview Avenue North, D4-100, Seattle, WA 98109, USA
e-mail: wgrady@fhcrc.org

K.M. Haigis (ed.), *Molecular Pathogenesis of Colorectal Cancer*,
DOI 10.1007/978-1-4614-8412-7_7,

Abbreviations

MMP	Matrix metalloprotease 2
MMP9	Matrix metalloprotease 9
TSP1	Thrombospondin 1

7.1 Introduction

Transforming growth factor ß is the canonical member of a family of secreted proteins that include the TGF-ß isoforms (TGF-ß1, TGF-ß2, and TGF-ß3), activins, Growth and Differentiation Factors (GDFs), Bone Morphogenetic Proteins (BMPs), inhibin, nodal, and anti-Mullerian hormone. These ligands all induce their effects on cells through binding to ligand specific cell surface receptor complexes that are composed of type I and type II heteromeric receptors. The TGF-ß superfamily has been the subject of intense investigation since its discovery in 1982 (Anzano et al. 1982), and studies of this cytokine have revealed a role for the TGF-ß superfamily in development and cancer biology. In epithelial cells, including in the intestines, TGF-ß can inhibit cell proliferation, induce apoptosis, and induce terminal differentiation, which has suggested that this pathway has tumor suppressor activities in epithelial tumors. Indeed, a large body of evidence has established that elements of the TGF-ß signaling pathway have a prominent role as tumor suppressor genes in epithelial organs and are frequently inactivated in gastrointestinal tract cancers. Interestingly, more recent studies have suggested that in certain contexts TGF-ß may promote the invasive or metastatic behavior of established cancer cells, suggesting that TGF-ß has a paradoxical role in primary human cancers. This paradoxical behavior appears to occur more commonly in late-stage cancers and likely depends on the presence of other mutant genes and deregulated signaling pathways. This chapter will focus on the tumor suppressor activity of the TGF-ß signaling pathway since the vast majority of the literature demonstrates tumor suppressor effects of the TGF-ß superfamily in the colon.

7.2 Overview of TGF-ß Signaling Pathway Elements and Role in Tumor Suppression (Fig. 7.1)

TGF-ß is secreted in an inactive state in a protein complex, which includes latent activating peptide and LTBPs (Latent Transforming Growth Factor Binding Proteins), and undergoes activation by extracellular proteases, including MMP2, MMP9, and plasmin, as well as by proteins that induce conformational changes in the secreted LAP:TGF-ß complex, such as TSP1 (Yu and Stamenkovic 2000; Crawford et al. 1998; Munger et al. 1999; Kanzaki et al. 1990). Active TGF-ß mediates its effects on cells through a heteromeric TGF-ß receptor complex that consists of type I (TGFBR1) and type II (TGFBR2) components. TGFBR1 and TGFBR2 are

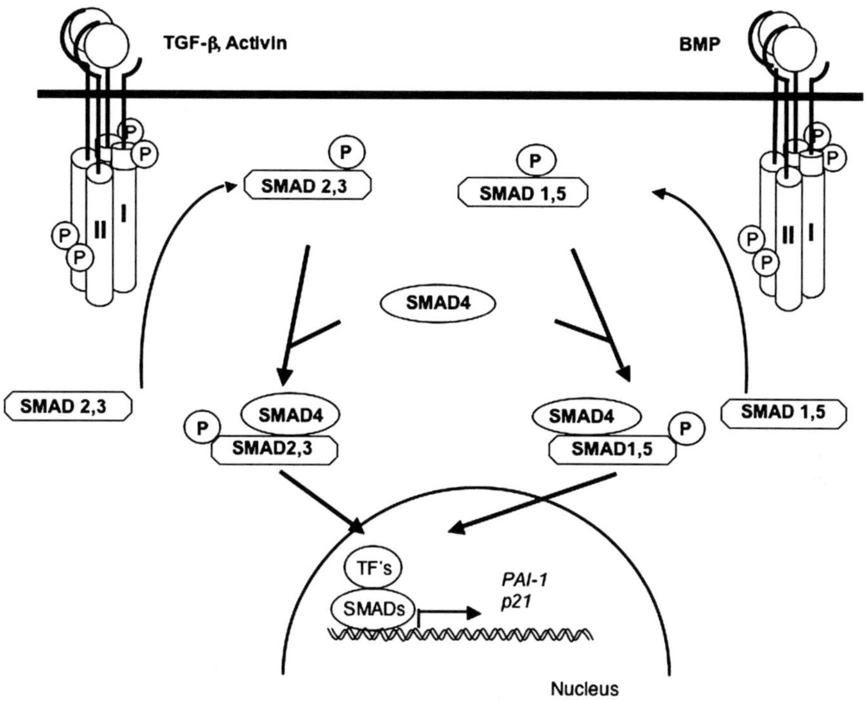

Fig. 7.1 Representation of the Transforming growth factor-β (TGF-β) signaling pathway and BMP signaling pathway, which are frequently deregulated in colorectal cancer. The pathways play a critical role in the regulation of cell growth, differentiation, and development in a wide range of biological systems, largely by affecting gene transcription. In general, signaling is initiated by ligand-induced oligomerization of the TGFBR2 and TGFBR1 serine/threonine receptor kinases followed by phosphorylation of the cytoplasmic signaling molecules Smad2 and Smad3. Carboxy-terminal phosphorylation of Smads by activated receptors results in their association with the signaling transducer Smad4, and translocation to the nucleus. Activated Smads regulate diverse biological effects by binding to transcription factors, resulting in modulation of transcription

cell membrane-associated single-pass transmembrane serine-threonine kinases that phosphorylate downstream signaling proteins upon activation (Massague 1996). After becoming activated by TGF-ß, TGFBR2 activates TGFBR1 through phosphorylation in the GS box region. TGFBR1 then propagates the signal from the receptor to the nucleus through the phosphorylation of downstream proteins, including the Smad proteins (Smad2 and Smad3) and non-Smad proteins (including PI3K, p38MAPK, PKA, and RhoA) (Markowitz and Roberts 1996; Wakefield and Roberts 2002; Chowdhury et al. 2011; Zhang 2009). The Smad pathway is the most extensively characterized post-TGF-ß-receptor pathway. For most of the non-Smad pathways, it is not apparent whether the pathway activation is a direct or an indirect effect of TGF-ß receptor activation.

The downstream transcriptional targets of the TGF-ß signaling pathway are involved in the regulation of a variety of cellular functions, including cell proliferation, extracellular matrix (ECM) production, and immune surveillance. These functions are not only an integral part of tissue homeostasis but also are logical targets whose dysregulation can promote colorectal carcinogenesis. Elements involved in fundamental cell behaviors that have been clearly shown to be influenced, at least in part, by TGF-ß include growth regulation proteins, such as MYC, CYCLIN D1, CDK4, p21, p27, p15, and RB, as well as proteins involved in differentiation, apoptosis, angiogenesis, and ECM remodeling (Grady et al. 2006; Geng and Weinberg 1993; Howe et al. 1991; Ewen et al. 1993; Alexandrow and Moses 1995a; Hannon and Beach 1994; Fava et al. 1990). It appears that certain effector proteins, such as p21 and cdk4, are regulated differentially, depending on the degree of TGF-ß signal pathway activation and/or on the interaction with other signaling pathways, creating another layer of regulatory control on TGF-ßs effects on proliferation that could be disrupted in cancer cells (Wang et al. 2004a; Gong et al. 2003; Seoane et al. 2004; Rojas et al. 2009). A number of micro RNAs (miRNAs) have been identified that modulate members of the TGF-ß pathway, including miR99a, miR99b, and the miR17-92 axis (Turcatel et al. 2012; Dews et al. 2010). In addition, the Smad pathway can also modulate miRNA production, which creates another mechanism through which the TGF-ß signaling pathway can regulate cell behavior (Davis et al. 2010; Blahna and Hata 2012). With regards to the effects of TGF-ß signaling pathway deregulation on other cell behaviors, ECM proteins, regulators of ECM proteins (fibronectin, tenascin, and plasminogen activator inhibitor 1), apoptosis-associated proteins, and senescence related proteins also appear to be altered by TGF-ß pathway mutations in colorectal cancers as well (Keeton et al. 1991; Zhao 1999; Grady and Markowitz 2008). Thus, deregulation of the TGF-ß signaling pathway can occur at multiple levels, and these cancer-related alterations in the pathway ultimately appear to promote the progression of these tumors through a myriad of effects on cell proliferation, apoptosis, senescence, extracellular matrix remodeling, etc.

7.3 TGF-ß Signaling Deregulation and Effects on Tumor Initiation and Progression

Studies of mouse models of intestinal cancer and of somatic mutations in colorectal cancer demonstrate that TGF-ß signaling inactivation in epithelial cancers promotes the progression of neoplasms but does not commonly initiate these tumors (Grady et al. 1998; Biswas et al. 2004; Munoz et al. 2006). This effect on tumor promotion, rather than tumor initiation, is a result of the fact that TGF-ß mediates its effects through cooperation with concurrent mutant genes and deregulated signaling pathways. Consequently, the normal integration of the TGF-ß signaling pathway with other pathways can be deregulated by mutations of tumor suppressor genes and oncogenes in conjunction with mutations in genes in the TGF-ß signaling pathway (Oft et al. 2002; Fujimoto et al. 2001). TGF-ß signaling, either through the Smad pathway or through non-Smad pathways, has been shown to interact with a variety of signaling pathways that are deregulated in cancer, including the Ras-Raf-ERK, PI3K-AKT, Wnt, Rho-like GTPase, and p38MAPK pathways (Wakefield and Roberts 2002; Zhang 2009; Derynck et al. 2001). With regards to the effect of

TGF-ß signaling inactivation on intestinal epithelial cells, studies of intestinal cancer models, *Fabp*$^{4xat-132}$*-Cre*; *Tgfbr2*$^{flx/flx}$ mice, and *Apc*^{1638N}; *Villin-Cre*; *Tgfbr2*$^{flx/flx}$ mice demonstrate that TGF-ß signaling inactivation alone is not sufficient to induce tumors, but that Tgfbr2 inactivation promotes the transformation of initiated tumors (Biswas et al. 2004; Munoz et al. 2006). These findings, and previously demonstrated interactions between oncogenic *HRAS*, *KRAS*, and *TP53* with the TGF-ß signaling pathway, provide evidence that the effects of TGF-ß signaling inactivation in many types of epithelial cancer is dependent on concurrent signal pathway deregulation and mutated oncogenes and tumor suppressor genes (Oft et al. 2002; Dupont et al. 2004; Cordenonsi et al. 2003; Akhurst and Derynck 2001). This context dependence of the TGF-ß signaling pathway with regards to cancer-related behaviors, as well as regarding the effects of TGF-ß on different epithelial cell types, may explain the differences observed in mutation frequencies in genes such as *TGFBR2*, *SMAD4*, *TGFBR1* in different types of cancer. The differences in mutation spectra and frequency in genes that encode for members of the TGF-ß signaling pathway may also reflect differences between primary tumor types regarding the expression of other potentially redundant TGF-ß family members or differences in the complexity of TGF-ß-mediated responses in the different epithelial organs. In a similar vein, the paradoxical role of TGF-ß on advanced cancers appears to be most prominent in certain types cancer, such as breast cancer, and less pronounced in other types of cancer, such as colon cancer (Bierie and Moses 2006). Exceptions to this general principle that TGF-ß signaling deregulation acts predominantly to drive the progression of initiated tumors, rather than to primarily initiate tumors, come from mouse models in which Smad4 has been deregulated. In some of these models *Smad4* deletion can initiate tumors, which may be a result of inactivation of multiple TGF-ß family signaling pathways or because of the effects of Smad4 loss on non-epithelial cells, as these mouse models are in general constitutional knock-out mice (Taketo and Takaku 2000).

It is interesting to note that genome-wide single nuclear polymorphism (SNP) studies demonstrate an association with SNPs in many TGF-ß signal pathway genes and colorectal cancer risk (Slattery et al. 2011; Slattery et al. 2012) (Table 7.1). Slattery and colleagues found both two SNPs in TGFβ1 (rs1800469 and rs4803455) were associated with a decreased risk of colon cancer [odds ratio (OR) = 0.65 and 1.43, 95 % CI = 0.51–0.84 and 1.18–1.73, respectively] but not rectal cancer and that 1/3 tagSNPs for TGFβR1, 2/4 tagSNPs for Smad2, and 4/37 Smad3 tagSNPs were associated with colon cancer (Slattery et al. 2011). They further assessed SNPs in the BMP ligands and receptors and found that genetic variation in BMPR1A, BMPR1B, BMPR2, BMP2, and BMP4 was associated with a risk of developing colon cancer, with 20–30 % increased risk for the most high-risk genotypes. A summary of high-risk genotypes showed over a twofold increase in colon cancer risk in the upper risk category (OR 2.49; 95 % CI 1.95, 3.18). BMPR2, BMPR1B, BMP2, and GDF10 were associated with rectal cancer. The risk associated with the highest category of the summary score for rectal cancer was 2.97 (95 % CI 1.87, 4.72). Of interest, genes in the BMP signaling pathway were consistently associated with the CpG Island Methylator Phenotype (CIMP) status in combination with both mutant KRAS and MSI cancers (Slattery et al. 2012). Furthermore, recent genome-wide

Table 7.1 Mutations and polymorphisms in genes in TGF-ß superfamily members

Gene	Alteration	Cancer	Effect	References
TGFBR2	Somatic mutations: Frameshift mutations; missense mutations; homozygous deletion	Colon cancer-58–92 % of MSI colon cancers; 15 % of MSS cancers Stomach cancer-common in MSI stomach caners, unusual in MSS cancers Pancreatic-4 % Breast: rare Cervical cancer: <10 %		Markowitz et al. (1995a), Grady et al. (1999a), Goggins et al. (1998), Parsons et al. (1995b), Chu et al. (1999), Kim et al. (2000b)
TGFBR1	Polymorphism: 1. TßR-I(6A) 2. Int7G24A	• TGFBR1(6A): Risk of breast, colon, hematological, and, ovarian • Int7G24A: increased risk of renal cell cancer (RCC), bladder cancer, and upper urinary tract transitional cell cancer (TCC), and lung cancer; no association with cervical cancer	• TGFBR1*6A: heterozygotes with cancer risk OR 1.26, 95 % CI 1.07–1.49; homozygous TGFBR1*6A with cancer risk OR 2.53, 95 % CI 1.39–4.61 • Int7G24A: G/A or A/A associated with RCC OR 2.20, 95 % CI 1.22–3.96 and TCC OR 2.45, 95 % CI 1.89–3.16	Pasche et al. (1998), Chen et al. (2004, 2006)
SMAD2/MADH2	Somatic mutations	Colon cancer-6 %, nonsmall lung cancer-2 %, liver cancer: uncommon		Elliott and Blobe (2005), Eppert et al. (1996), Takenoshita et al. (1998)
SMAD4/MADH4	Somatic mutations: Missense mutations; homozygous deletions	Pancreatic cancer-50 %, colon-approx. 20 %, nonsmall cell lung cancer-7 %	MSI tumors carry mutations in exon10 8 bp polyA tract	Hahn et al. (1996), Miyaki et al. (1999)
SMAD4	Germline mutation		Juvenile polyposis syndrome	Howe et al. (1998b, 2002)
BMPR1A	Germline mutation		Juvenile polyposis syndrome	Howe et al. (2001)
ACVR2	Somatic mutation	MSI colon cancer-58–72 %, MSI gastric cancer-44 %, pancreatic cancer		Mori et al. (2001), Deacu et al. (2004), Hempen et al. (2003)

association studies (GWAS) have consistently found SNPs in TGF-ß related genes to be associated with risk for colorectal cancer, supporting the idea that the TGF-ß pathway genes influence susceptibility to colorectal cancer (Tenesa and Dunlop 2009). These results suggest that constitutional alterations in the TGF-ß and BMP signaling pathways may create a predisposition to tumor formation in the intestines either by promoting polyp formation or the progression of polyps to cancer.

7.3.1 TGF-β and Inflammation-Associated Cancer

Inflammation has been generally shown to predispose tissues to cancer formation, presumably through the induction of tumor promoting cytokines and induction of DNA mutations by free radicals and oxidants (Li et al. 2006). Similar to the role TGF-ß can play in sporadic human cancer, studies of mice suggest that TGF-ß signaling can play a paradoxical role in inflammation-associated cancers. In the colon, *Helicobacter hepaticus*-induced inflammation can result in the formation of cancer in the setting of TGF-ß signaling inactivation (Engle et al. 1999a; Maggio-Price et al. 2006). In both the *Tgfb1*$^{-/-}$; *Prkdc*$^{scid/scid}$ mouse and in the *Smad3*$^{-/-}$ mouse, *H. hepaticus* infection results in colon cancer formation (Engle et al. 1999a; Maggio-Price et al. 2006; Engle et al. 2002). However, in the skin, TGF-ß1 overexpression in the suprabasal layer appears to induce a psoriasis-like phenotype through chemotactic effects of TGF-ß1 on inflammatory cells and the recruitment of pro-inflammatory cytokines, like IL-1, IL-2, TNF-alpha, MCP1, etc., which are produced by recruited inflammatory cells, These inflammatory cytokines can override TGF-ß-mediated effects on the epidermal cells and cause epidermal proliferation, angiogenesis, and basement membrane degradation, which should promote tumorigenesis (Li et al. 2006). These studies provide further evidence for the context dependence of TGF-ß's effects with regards to cancer formation and also demonstrate that, in the colon, the TGF-ß signaling pathway acts predominantly as a tumor suppressor.

7.3.2 Effects of the TGF-β Pathway on Fundamental Biological Activities in Cancer

7.3.2.1 TGF-ß and Regulation of Normal Intestinal Epithelial Cell Behavior

TGF-ß has been shown in in vitro and in animal models to regulate a variety of fundamental cell behaviors. These behaviors include cell proliferation, programmed cell death, differentiation, senescence, cell shape, cell:cell and cell:ECM adhesion, and cell motility. TGF-ß can inhibit the proliferation of epithelial cells in developing organs and in adult organs in which the epithelium has been induced to grow, such as liver epithelium after partial hepatectomy and colon epithelium after exposure to DSS (Hahm et al. 2002; Romero-Gallo et al. 2005). TGF-ß's ability to inhibit epithelial cell proliferation has been extensively studied in a variety of types of epithelium including in the intestines, as well as in the mammary gland, liver, etc. (Ten Dijke et al. 2002) Moreover, in the developing epithelium, BMP signaling plays a central role in crypt formation (Haramis et al. 2004; Sancho et al. 2004). Mouse models that lack the Bmpr1a receptor or overexpress Noggin, a soluble antagonist of the BMPs, develop abnormal cystic structures in the intestines and have abnormal villus formation (Haramis et al. 2004).

TGF-ß signaling can also induce apoptosis in many of types of epithelial cells, including thyroid cells, hepatocytes, colon epithelial cells, and mammary epithelium (Nguyen and Pollard 2000; Bravo et al. 2003; Hofmann et al. 2003; Wang et al. 1995a). The ability of TGF-ß to induce apoptosis appears to be restricted to certain types of epithelium and in the intestines it is believed to occur at the luminal surface of the colon or villus tips of the small intestine (Avery et al. 1993; Barnard et al. 1993). In addition to effects on cell growth and apoptosis, TGF-ß also can regulate cell shape, adhesion, epithelial-to-mesenchymal transition (EMT), and motility. Thus, it is clear that TGF-ß has both potentially tumor promoting and tumor suppressing effects on normal epithelial cells, but in the colon, the intact TGF-ß pathway acts predominantly as a tumor suppressor through its effects on proliferation, apoptosis, and differentiation (Mishra et al. 2005).

7.4 Effects of TGF-ß on Specific Cell Behaviors

7.4.1 TGF-ß Regulation of Cell Proliferation

Cell proliferation is regulated by growth stimulatory and growth inhibitory stimuli that converge on the regulators of the cell cycle. The TGF-ß ligands inhibit the proliferation of many cell types, including most epithelial cells (Moses et al. 1990), by blocking progression of cells from G1 into S phase of the cell cycle (Howe et al. 1991). Proposed mechanisms have included: (1) suppression of expression of proteins necessary for G1/S progression, including c-myc, cyclins A and D1, cdk4, and the CDK activator, Cdc25A; (2) the induction of the CDK inhibitors, p15, p21^{Waf1,Cip1}, and p27^{Kip1}; and (3) the inhibition of phosphorylation of Retinoblastoma (Rb) or its cognates p107 and p130 (Markowitz and Roberts 1996; Grady et al. 2006; Hannon and Beach 1994; Romero-Gallo et al. 2005; Polyak et al. 1994; Datto et al. 1995; Moustakas and Kardassis 1998; Alexandrow and Moses 1995b; Iavarone and Massague 1997; Brown et al. 2004; Pietenpol et al. 1990). Importantly, TGF-ß-mediated effects on different regulators of G1 and G1/S progression vary from one cell type to another, demonstrating that there is more than one mechanism for TGF-ß-induced G1 arrest (Alexandrow and Moses 1995b). It appears that certain effector proteins, such as p21^{CIP} and p15^{INK4b}, are regulated differentially depending on the degree of TGF-ß signal pathway activation and/or on the interaction of the TGF-ß signaling pathway with other signaling pathways, creating another layer of regulatory control on TGF-ß's effects on proliferation that is likely disrupted in cancer cells (Gong et al. 2003; Seoane et al. 2004; Wang et al. 2004b; Rojas et al. 2008). It also appears that interactions between multiple TGF-ß regulated proteins can dictate the eventual cellular response. For instance, TGF-ß-induced down regulation of c-myc appears to be required in some systems for TGF-ß-induced $p15^{INK4b}$ expression via Miz-1 (Seoane et al. 2001). Thus, it appears that the "cellular context" is a particularly important factor for governing the growth inhibitory effects of TGF-ß. The growth inhibitory effects are mediated by the Smad pathway as well as by Smad independent pathways, including the MAPK and PI3K pathways and PP2A/p70S6K pathway (Zhang 2009; Hu and Zuckerman 2001; Petritsch et al. 2000). In addition, TGF-ß can inhibit the growth of cancer cell lines that carry inactivating mutations in *MADH4/SMAD4* demonstrating that Smad4 is not always required for TGF-ß-mediated growth inhibition, although it does appear to be required for some transcriptional responses and to play a role in mediating the growth inhibitory effects of TGF-ß and BMP (Fink et al. 2001, 2003; Grau et al. 1997; Freeman et al. 2012). Suppression of Smad3 has also been shown to inhibit TGF-ß-mediated growth inhibition, indicating that impairment of Smad signaling can release cells from the growth inhibitory effects of TGF-ß (Han et al. 2004).

The majority of colon cancers studied to date demonstrate resistance to TGF-ß-mediated growth inhibition, and the mechanisms mediating this resistance varies between tumors. Mutations in *TGFBR2* and *SMAD4* appear to be the most common causes of TGF-ß resistance in colon cancer, whereas in other tumor types, such as prostate, lung, and breast cancer, the mechanism appears to be decreased expression of TGFBR2 (Kim et al. 2000a; Anumanthan et al. 2005; Brattain et al. 1996). Other proposed mechanisms include increased expression of the inhibitory smad, Smad7; repression of TGF-ß signaling by Myc, E1A, Ras, Ski/SnoN, Evi-1; suppression of other pathways that regulate TGF-ß signaling by proteins such as MENIN, DAB2, and RUNX3; and increased expression of miRNAs that silence members of the TGF-ß signaling pathway (Blahna and Hata 2012; Elliott and Blobe 2005; Deng et al. 2013).

7.4.2 TGF-ß Regulation of Apoptosis

A second mechanism through which TGF-ß acts as a tumor suppressor in colon cancer is as an inducer of apoptosis in epithelial cells and lymphocytes. TGF-ß has been shown to mediate apoptosis in a variety of cell types, including intestinal epithelial cells, colon adenoma cells, hepatocytes, prostate cancer cells, and lymphocytes (Elliott and Blobe 2005; Mithani et al. 2004; Brown et al. 1999; Guo and Kyprianou 1999). TGF-ß and TGF-ß receptor localization studies in the intestine demonstrate that the pathway is active at the tips of the villi and the tops of the colon crypts, where apoptosis is occurring (Eskinazi et al. 1998). As with TGF-ß-mediated inhibition of proliferation, the mechanisms through which TGF-ß regulates its effects are cell and context dependent. Pathways implicated in mediating TGF-ß-induced apoptosis include the Smad, JNK, and p38MAPK pathways (Schuster and Krieglstein

2002; Atfi et al. 1997; Yamamura et al. 2000). Potential mechanisms of TGF-ß-mediated apoptosis that have been identified to date include the induction of pro-apoptotic genes such as TIEG-1, TIEG-2, p53 and Bax; suppression of bcl-xL, DAP kinase, and Daxx; release of cytochrome c; and activation of caspase 3 (Freathy et al. 2000; Jang et al. 2002; Perlman et al. 2001). Notably, TGF-ß's effects on apoptosis can be attenuated by the activation of other signaling pathways including the p38MAPK, PI3K-AKT, and RAS-RAF-MAPK pathways (Chen et al. 1998a, 1999; Shih et al. 2000). One potential explanation for the variable apoptotic response of cells to TGF-ß is that the degree of Smad, PI3K-AKT, and JNK/p38MPAK pathway activation may dictate the sensitivity of cells to TGF-ß-mediated apoptosis (Zhang 2009; Conery et al. 2004; Remy et al. 2004). These studies of TGF-ß's effects on apoptosis again highlight the complexity of the responses that are likely a reflection of the cellular context in which the TGF-ß signaling pathway is operating.

7.4.3 TGF-ß and Genomic Instability

Maintenance of DNA fidelity is an integral process in normal somatic cells and appears to be a central mechanism that prevents cancer formation. There are a variety of mechanisms that monitor and maintain the genome including p53, mitotic checkpoint regulators, the Mutation Mismatch Repair system (MMR), Base-Excision Repair system (BER), and DNA double-strand break repair systems (Grady 2004). Cancer cells typically display inactivation of one or more of these systems. TGF-ß has been shown to regulate the expression and/or activity of some these DNA repair proteins, such as Rad51 and p53 (Ewan et al. 2002; Kanamoto et al. 2002). Glick and coworkers showed that malignant transformation of *Tgfb1*$^{-/-}$ keratinocytes, in combination with a *v-Ras* oncogene, is preceded by aneuploidy (Glick et al. 1999). They also showed inactivation of TGF-ß signaling in keratinocytes using a dominant negative *TGFBR2* adenovirus can cause aneuploidy, and that exogenous TGF-ß suppressed the development of aneuploidy and malignant transformation in *Tgfb1*$^{-/-}$ keratinocytes. These studies suggest that the maintenance of genomic stability may be a tumor suppressor activity of TGF-ß, although none of the mouse models of intestinal cancer in which the TGF-ß pathway has been disrupted have been shown to display genomic instability to date (Munoz et al. 2006; Trobridge et al. 2009).

7.4.4 TGF-ß and Senescence

Senescence is a central biological process that limits the ability of somatic cells to replicate and is critical for maintaining homeostasis in organisms. Normally, cells replicate until they reach the Hayflick limit of cell doublings at which time they undergo crisis and cell death. The Hayflick limit is triggered by the progressive shortening of the telomeres that occurs with successive rounds of DNA replication and cell divisions. Cancer cells demonstrate resistance to senescence and appear to become immortalized through mechanisms that maintain the length of telomeres, either by up-regulating telomerase or by an alternative lengthening of telomeres (ALT) mechanism, which maintains telomeres through a recombination-based system.

The most common mechanism through which cancer cells up-regulate telomerase is through increasing the transcription of the catalytic component of telomerase, hTERT (Hahn 2003). Autocrine and exogenous TGF-ß suppress *hTERT* transcription, possibly through MAD1, MENIN, or SIP1/ZEB-2 regulation (Katakura et al. 1999; Lin and Elledge 2003). Autocrine TGF-ß can suppress telomerase activity by repressing *hTERT* mRNA expression through a *SIP1*-mediated process (Lin and Elledge 2003; Yang et al. 2001). TGF-ß's effects on *hTERT* suggest that it acts to suppress tumor formation by regulating senescence in non-neoplastic cells, although the role of TGF-ß-mediated senescence in colorectal cancer remains to be further defined.

7.4.5 TGF-β regulation of mechanisms involved with invasive behavior of cancer

In general, the effects of TGF-ß on mechanisms implicated in invasion, such as EMT, cell adhesion, cell motility, and ECM remodeling have been proposed to mediate tumor promoting effects as opposed to tumor suppressing effects. TGF-ß has been implicated in the regulation of invasive behavior of cancer cells through its effects on EMT and cell migration, although the vast majority of this data is derived from cell line systems or is extrapolated from studies of developmental processes (Nawshad et al. 2005; Bakin et al. 2000, 2002). Indeed, in a skin cancer model system in which keratinocytes overexpress TGF-ß1 and a dominant negative TGFBR2, metastatic cancers form through EMT independent mechanisms (Han et al. 2005). Other potential mechanisms through which TGF-ß may affect tumor progression are on cell adhesion and ECM remodeling, which can affect cell migration, proliferation, and invasion (Wang et al. 1995b). TGF-ß has been shown to modulate the expression of ECM proteins and their integrin receptors in human fibroblasts and the colon cancer cell lines FET and MOSER (Wang et al. 2004a, 1995b). Depending on the cellular context of the cancer cells, TGF-ß signaling inactivation may decrease cancer cell motility and cell-ECM adhesion. The regulation of the response appears to be dependent on whether TGF-ß receptor can activate non-Smad pathways and on the activation state of other signaling pathways (Bakin et al. 2002; O'Mahony et al. 1999; Horowitz et al. 2004). Studies of in vivo mouse models provide strong evidence that TGF-ß signaling acts to suppress the transformation and invasion of benign colon intestinal adenomas to adenocarcinomas (Grady et al. 1998; Biswas et al. 2004; Munoz et al. 2006). In primary tumors, *TGFBR2* mutation has been associated with improved prognosis in stage III colon cancer (Watanabe et al. 2001a) In contrast, decreased SMAD4 expression associates with a worse prognosis in colon cancer and gastric cancer demonstrating the complexity of interpreting the effects of TGF-ß signaling deregulation on the behavior of cancer (Xiangming et al. 2001; Alhopuro et al. 2005). Thus, in the gastrointestinal tract, it appears that TGF-ß signaling suppresses the invasive behavior of neoplastic cells, but in skin and breast cancer TGF-ß signaling promotes the progression of advanced tumors (Wakefield and Roberts 2002; Munoz et al. 2006; Trobridge et al. 2009; Han et al. 2005; Muraoka-Cook et al. 2005).

7.5 Mechanisms for Inhibiting the TGF-ß Signaling Pathway in Cancer

In light of the effects of TGF-ß on epithelial cells in in vitro systems, it appears to have a prominent role as a tumor suppressor in a variety of tumor types. However, the strongest evidence for the tumor suppressor activity of the TGF-ß signaling pathway comes from studies of the molecular genetics of cancers, as well as from animal model studies using mice that are have been engineered to over or under express different elements in the TGF-ß signaling pathway (Tables 7.1 and 7.2). The identification of somatic mutations and deletions of the type II TGF-ß receptor gene (*TGFBR2*) and in the *SMAD* genes (*SMAD2* and *SMAD4*) has provided strong support for the role of TGF-ß receptor as a tumor suppressor (Markowitz et al. 1995a; Grady et al. 1999a; Eppert et al. 1996). In addition, polymorphisms have been identified in *TGFB1* and *TGFBR1* that provide additional evidence that the TGF-ß signaling pathway is a tumor suppressor pathway for breast and colon cancer as well as for a variety of other tumor types (Bellam and Pasche 2010).

7.5.1 *TGF-β1*

Genetic association studies, studies of in vitro tissue culture systems, and mouse model studies indicate that TGF-ß acts as a tumor suppressor in colon cancer (Slattery et al. 2011; Engle et al. 1999a; Grady et al. 1999a). There is evidence that TGF-ß can act as a tumor suppressor in experimental mouse models. Mice that are $Tgfb1^{-/-}$ or $Tgfb1^{-/wt}$ are predisposed to cancer formation. $Tgfb1^{-/-}$; $Prkdc^{scid/scid}$ mice develop colon adenomas and adenocarcinomas in the setting of *Helicobacter hepaticus* colonization in the gut, and $Tgb1^{-/wt}$ mice are more susceptible to carcinogen-induced lung and liver neoplasms (Engle et al. 2002; Tang et al. 1998). However, in contrast to these results in colon mouse models, Alb-*Tgfb1* mice, which constitutively express Tgfb1 in the liver, paradoxically develop fibrosis and hepatocellular adenomas and adenocarcinomas spontaneously, and this process can be enhanced by c-myc and by treatment with the hepatocarcinogen, diethylnitrosamine (DEN) (Factor et al. 1997). Furthermore, in skin tumor models, although Tgfb1 represses the formation of carcinogen-induced papillomas, it induces the transformation of these papillomas to spindle-cell cancers (Cui et al. 1996). These studies provide evidence that the TGF-ß ligand is a tumor suppressor in colon cancer, but that there is likely to be a significant tissue specific effect of TGF-ß on

Table 7.2 Examples of experimental genetic animal models of TGF-ß deregulation and cancer

Organ system	Model	Experimental modification	Results	Biological effect of TGF-signaling deregulation	References
Intestines	*Tgfb1*$^{-/-}$; *Prkdc*$^{scid/scid}$	Constitutional TGF-ß1 null	Colon adenomas and adenocarcinomas if *Helicobacter hepaticus* present	Tumor suppressor effects	Engle et al. (1999a, 2002)
	Fabp$^{4xat-132}$Cre; *Tgfbr2*$^{flx/flx}$	TGFBR2 null colon epithelial cells	Increased adenocarcinoma formation after treatment with azoxymethane	Tumor suppressor effects: Increased proliferation	Biswas et al. (2004)
	Villin-Cre; *Tgfbr2*$^{flx/flx}$; *Apc*^{1638N}	TGFBR2 null intestinal epithelial cells + *Apc* mutation	Increased adenocarcinoma formation in mice with null Tgfbr2 and Apc mutation	Tumor suppressor effects: Increased MMP production	Munoz et al. (2006)
	Villin-Cre; *Tgfbr2*$^{flx/flx}$; *LSL-Kras*^{G12D}	TGFBR2 null intestinal epithelial cells + *Kras* mutation	Increased adenocarcinoma formation in mice with null Tgfbr2 and oncogenic Kras	Tumor suppressor effects: Increased proliferation and EGF signaling	Trobridge et al. (2009)
	ITF-dnRII	Dominant negative-TGFBR2 in intestines; inhibited TGFBR2	Increased ACFs and colon neoplasms after azoxymethane treatment	Tumor suppressor effects	Hahm et al. (2002)
	pS2-dnRII	Dominant negative-TGFBR2 in gastric epithelium; inhibited TGFBR2 activity	Gastric adenocarcinoma after infection with *H. pylori* (ATCC 43504, CagA, VacA +)	Tumor suppressor effects	Hahm et al. (2002)
	Smad4$^{-/+}$/*Apc*$^{\Delta716}$	Smad4 haploinsufficiency in setting of *Apc* mutation	Intestinal adenocarcinoma formation compared to *Apc*$^{\Delta716}$ mice which develop adenomas but not adenocarcinomas	Tumor suppressor effects in initiated intestinal adenomas	Harada et al. (1999)
	Smad4$^{-/+}$	Smad4 haploinsufficiency	Gastric and intestinal juvenile polyps		Takaku et al. (1999)
	Smad4^{E6sad}	Deregulated *Smad4* mRNA processing	Intestinal serrated adenomas	Tumor suppressor effects	Hohenstein et al. (2003)
	Smad3$^{-/-}$	Smad3 null	Spontaneous colon adenocarcinoma in one model, no tumors in two other Smad3 null mouse lines	Tumor suppressor effects	Zhu et al. (1998), Datto et al. (1999), Yang et al. (1999)
	Apc^{580D}; *Smad2*$^{-/wt}$	Smad2 haploinsufficiency in setting of *Apc* mutation	No increase in total number of tumors compared to *Apc*^{580D} mice, increased number of tumors >6 mm in diameter	Tumor suppressor effects	Hamamoto et al. (2002)
	Smad2$^{-/+}$/*Apc*$^{\Delta716}$	Smad2 haploinsufficiency in setting of *Apc* mutation	No difference in number or histology of tumors compared to *Apc*$^{\Delta716}$ mice	No effect on tumor formation possibly because this alle may be partially active	Takaku et al. (2002)

tumorigenesis in humans. In fact, tumor cells often show increased production of TGF-ß despite its tumor suppressor effects suggesting that it can have tumor promoting effects, perhaps through paracrine effects on host stromal cells or by autocrine effects on established cancer cells that carry numerous deregulated pathways and have altered post-receptor responses to TGF-ß.

7.5.2 *TGFBR2*

Evidence of TGF-ß's role as a tumor suppressor for colon cancer formation came first from studies that demonstrated epithelial cell lines were growth inhibited by TGF-ß1 and by studies of cancer cell lines that demonstrated these cell lines were resistant to the normal growth inhibitory effects of TGF-ß (Moses et al. 1987; Hoosein et al. 1989). The role of TGF-ß receptor inactivation in mediating the resistance of epithelial tumors to TGF-ß has been clearly demonstrated in a variety of tumor types and has been most carefully evaluated in colon cancer. Colon cancer cell lines have been extensively studied for TGF-ß sensitivity and are generally resistant to the growth inhibitory effects of TGF-ß1 (Hoosein et al. 1989; Grady et al. 1999b). A major breakthrough in our understanding of the mechanism mediating this resistance occurred when genetic alterations in the gene for the type II receptor gene, *TGFBR2*, in colon cancers that display microsatellite instability (MSI) were identified. In the study that first elucidated the role of *TGFBR2* mutations in colon cancer formation, Markowitz et al. demonstrated that mutational inactivation of *TGFBR2* is an extremely common event in cancers that display microsatellite instability. *TGFBR2* has a microsatellite-like region in exon 3 that consists of a 10 bp polyadenine tract making it particularly susceptible to mutation in the setting of MSI (Markowitz et al. 1995b; Parsons et al. 1995a; Myeroff et al. 1995). The mutations in this region, which has been named *BAT-RII* (*B*ig *A*denine *T*ract in TGF-ß *R*eceptor type *II*), are frameshift mutations that result in the insertion or deletion of one or two adenines between nucleotides 709 and 718 introducing nonsense mutations that encode a truncated TGFBR2 protein. This truncated protein is only 129–161 amino acids in length compared to the wild-type protein (565 amino acids) and lacks the receptor's transmembrane domain and intracellular kinase domain (Markowitz et al. 1995b). In a series of 110 MSI colon cancers, 100 were found to carry *BAT-RII* mutations and in almost all of these cases the mutations were biallelic consistent with the tumor suppressor function of *TGFBR2* (Parsons et al. 1995a). Of interest, the tumors or cell lines that did not possess biallelic *BAT-RII* mutations were found to have missense mutations in the residual *TGFBR2* allele (e.g., codon 452, Pro→Leu; codon 454, Pro→Leu) (Parsons et al. 1995a). Functional evidence of *TGFBR2*'s role as a tumor suppressor gene in colon cancer has been further elucidated by studies showing reconstitution of wild-type *TGFBR2* in HCT116, a MSI colon cancer cell line with a mutant *TGFBR2* gene, and in microsatellite stable colon cancer cell lines, V400 and V410, suppresses the tumor phenotype of these cell lines (Wang et al. 1995c). Functionally significant alterations of *TGFBR2* have now been identified in up to 30 % of colon cancers and are the most common mechanism identified to date for inactivating the TGF-ß signaling pathway in colon cancer (Grady et al. 1999b; Markowitz et al. 1995b). No alterations in the genes for the type I (*TGFBR1*) or the type III TGF-ß receptor (*TGFBR3*) have been observed in studies of TGF-ß resistant colon cancer cell lines suggesting mutational inactivation of *TGFBR2* is a particularly favorable event that leads to tumor formation (personal communication, Markowitz).

In addition to mutational inactivation of *TGFBR2*, transcriptional repression of *TGFBR2* appears to be a second means through which cancers can inhibit the TGF-ß receptor. Although the mechanisms responsible for suppressing the expression of *TGFBR2* have not been identified in primary tumors, potential mechanisms include FL1/EWS-mediated transcriptional repression, inhibition of Ets-mediated transcription, and aberrant DNA methylation (Kim et al. 2000a). Reconstitution of *TGFBR2* in cancer cell lines that have low expression of *TGFBR2* has been shown to restore TGF-ß transcriptional responses or growth inhibition indicating the repression of *TGFBR2* is capable of inactivating the TGF-ß receptor (Anumanthan et al. 2005). Repression of TGFBR2 expression is not a common mechanism for inhibiting TGF-ß signaling in colon cancer.

More recently, several mouse models have been generated that allow the assessment of the effect of TGF-ß signaling inactivation and of Tgbr2 inactivation in vivo. The use of animal models in the study of TGF-ß signaling in cancer was initially limited by embryonic or postnatal mortality (Shull et al. 1992; Oshima et al. 1996). The use of tissue specific promoters to drive the overexpression of *Tgfb1* or of dominant negative *Tgfbr2* and of conditional alleles for *Tgfbr2* has allowed in vivo assessment of the role of TGF-ß signaling deregulation in cancer. Invasive colon tumors develop in $Tgfb1^{-/-}$ mice, and mice that lack TGFBR2 in the colon, $Fabp^{4xat-132}$Cre; $Tgfbr2^{flx/flx}$ mice, are more susceptible to azoxymethane-induced colon neoplasms than are mice with intact TGFBR2 in the colonic epithelium (Biswas et al. 2004; Engle et al. 2002, 1999b). Hahm et al. have demonstrated that mice expressing a dominant negative *Tgfbr2* transgene in the intestinal epithelium (ITF-dnRII) are also more susceptible to azoxymethane-induced colon neoplasms (Hahm et al. 2002). Of perhaps most importance, a variety of models in which intestine-specific deletion of *Tgfbr2* has been carried out have demonstrated that loss of TGF-ß signaling primarily acts to induce the progression of initiated intestinal neoplasms and is not sufficient to independently induce intestinal neoplasm formation (Biswas et al. 2004; Munoz et al. 2006; Trobridge et al. 2009).

7.5.3 TGFBR1

Although *TGFBR1* would appear to be a common target of inactivation given the frequency *TGFBR2* mutations, it is uncommonly a target for mutational inactivation. Somatic mutations in *TGFBR1* have been identified in prostate, biliary, endometrial, ovarian, metastatic breast, cervical, and pancreatic cancer and T-cell lymphomas, although they are uncommon in colon cancers (Elliott and Blobe 2005; Chen et al. 1998b, 2001; Knaus et al. 1996; Knobloch et al. 2001; Nakashima et al. 1999; Goggins et al. 1998). Of note, a *TGFBR1* polymorphism (TßR-I(6A)), which has three alanines deleted from a nine basepair polyadenine tract, has been identified as a modest cancer susceptibility allele for colon, breast, hematologic, and ovarian cancer and to be somatically acquired in metastatic colorectal cancer, but not in breast cancer or head and neck cancer (Pasche et al. 1999, 2005; Kaklamani et al. 2003). The significance of the *TGFBR1*6A* polymorphism in colon cancer is controversial (Guda et al. 2009; Daley et al. 2007).

7.5.4 SMADS

The Smad proteins are a family of proteins that serve as intracellular mediators to regulate TGF-ß superfamily signaling. They comprise an evolutionarily conserved signaling pathway that has been demonstrated in *C. elegans*, *Drosophila melanogaster*, and *Xenopus*, as well as in humans. These proteins are characterized by two regions that are homologous to the Drosophila orthologue, Mad, and that are located at the N- and C- termini of the protein. These regions are termed the Mad-homology domains MH1 and MH2, respectively, and are connected by a less well-conserved, proline-rich linker domain. Numerous studies have identified three major classes of Smad proteins: (1) the receptor-regulated Smads (R-Smads) which are direct targets of the TGF-ß receptor family type I kinases and include Smads1, 2, 3, and 5; (2) the common Smads (Co-Smads: Smad4) which form heteromeric complexes with the R-Smads and propagate the TGF-ß-mediated signal; and (3) the inhibitory Smads (I-Smads: Smad6 and Smad7) which antagonize TGF-ß signaling through the Smad pathway. Ligand binding to the TGF-ß receptor complex results in TGF-ß receptor type I-mediated phosphorylation of Smad2 and Smad3 on two serine residues in a conserved –SS(M/V)S motif located at the C-terminus of the R-Smads (Kretzschmar et al. 1997; Zhang et al. 1996). Phosphorylation of these serine residues is required for downstream signaling pathway activation and for the eventual function of Smads as members of transcription factor complexes (Souchelnytskyi et al. 1997; Abdollah et al. 1997).

In light of the evidence demonstrating that TGF-ß and the TGF-ß receptor commonly mediate tumor suppressor effects, it is not surprising that there is similar evidence for the role of the receptor Smads, SMAD2, and SMAD3, and the co-Smad SMAD4 as tumor suppressors. In addition to the demonstration of inactivating mutations in *SMAD* genes, a variety of in vitro studies have identified other mechanisms that can impair TGF-ß-mediated Smad signaling that are germane to the molecular pathogenesis of cancer, including Ras pathway activation, and the expression of transcriptional repressors including SnoN, Ski, and Evi1 (Deng et al. 2013; Kretzschmar et al. 1999; Kurokawa et al. 1998; Stroschein et al. 1999; Luo et al. 1999).

In addition, the inhibitory Smads, Smad6, and Smad7, have been shown to be overexpressed in a variety of tumors, including the overexpression of Smad7 in colon cancer, as well as in lung cancer, pancreatic cancer, hepatocellular carcinoma, and endometrial cancer (Halder et al. 2005; Park et al. 2004; Dowdy et al. 2005;

Kleeff et al. 1999; Korchynskyi et al. 1999). Smad7 overexpression has been shown to inhibit TGF-ß-mediated growth inhibition and apoptosis in the colorectal cancer cell line FET through interfering with the formation of the Smad2/3/4 complex (Halder et al. 2005). In contrast to these studies, overexpression of Smad7 in melanoma cell lines or the mouse mammary carcinoma cell line, JygMC(A) suppresses the tumorigenicity of these cell lines by inhibiting the metastatic and invasive behavior of the cell lines, demonstrating the complexity of the effects of TGF-ß signaling deregulation in cancer (Javelaud et al. 2005; Azuma et al. 2005).

Genetic studies of primary human cancers have identified inactivating somatic mutations in *MADH2/SMAD2* and *MADH4/SMAD4*. *SMAD2* is mutated in a small proportion of colon cancer (6 %) (Elliott and Blobe 2005). *SMAD4* mutations have been identified in a higher proportion of colon cancers (16–20 %), (Elliott and Blobe 2005). Mutations in the other *SMAD* genes, *SMAD5*, *SMAD6*, *SMAD7*, or *SMAD8*, have not been identified in cancers to date.

The role of the *SMAD* genes as tumor suppressor genes has been best characterized in colon and pancreatic cancer. Loss of heterozygosity (LOH) occurs commonly at 5q, 18q, and 17p in colon cancer and suggests that there are tumor suppressor genes at these loci. *SMAD2* and *SMAD4* are located on chromosome 18q21, which is subject to LOH in approximately 70 % of colon adenocarcinomas. The incidence of 18q LOH is only about 10 % in early stage colon adenomas and is 30 % in later stage, larger adenomas demonstrating that the incidence of LOH involving 18q increases through the adenoma-carcinoma sequence (Vogelstein et al. 1988, 1989). Other genes that are candidate tumor suppressor genes and map at 18q21-qter include *BCL-2*, gastrin-releasing peptide, and the cellular homologue of *YES-1*; however, none of these have been shown to be altered in colorectal cancers (Martinez-Lopez et al. 1998). Thus, based on the results from somatic mutation analysis of these candidate genes, the most likely tumor suppressor genes that are the targets of 18q LOH observed in colon cancer are *SMAD2*, and *SMAD4*.

Experimental animal studies and studies of cancer family syndromes have provided significant insight into the tumor suppressor role of the TGF-ß signaling pathway. The effect of inactivation of Smad signaling on carcinogenesis has been investigated in a number of different animal models, which have provided consistent support for the tumor suppressor role of the *SMAD* genes. One murine model, a compound heterozygote $Smad4^{-/+}/Apc^{\Delta716}$, develops colon cancer, unlike the $Apc^{\Delta716}$ mouse, which only develops small intestinal adenomas (Takaku et al. 1998). This model suggests that *SMAD4* inactivation may play a role in the progression of colon cancers as opposed to their initiation. However, in some contexts *SMAD4* mutations also appear to initiate tumor formation and appear to contribute to tumor initiation while in a state of haploid insufficiency. The $Smad4^{-/+}$ mouse develops gastric and intestinal juvenile polyps and invasive gastric cancer after several months, however, it does not appear to develop colon cancer (Takaku et al. 1999; Xu et al. 2000). Furthermore, germline mutations in *SMAD4/MADH4* have been found in approximately 1/3 of individuals with Juvenile Polyposis syndrome (JPS), an autosomal dominant syndrome characterized by gastrointestinal hamartomatous polyps and an increased risk of gastrointestinal cancer, which is consistent with the concept that haploid insufficiency of *SMAD4* may contribute to tumor initiation (Howe et al.

1998a; Friedl et al. 1999; Roth et al. 1999). Of note, the polyps observed in JPS and the invasive cancers in the *Smad4*$^{-/+}$ mouse have been shown to have allelic loss of *SMAD4*, supporting the idea that biallelic inactivation of *SMAD4* is needed for cancer formation (Xu et al. 2000; Woodford-Richens et al. 2000). A second mouse model of Smad4 deregulation, which carries a mutation in the intron 5/exon 6 splice acceptor site of *Smad4* (*Smad4*^{E6sad}), develops serrated adenomas and intestinal polyposis although it has not been reported to commonly develop cancer (Hohenstein et al. 2003). Mice that are null for *Smad4* in the skin, *K5*-Cre; *Smad4*$^{Co/Co}$, develop spontaneous well-differentiated squamous cell carcinomas, squamous papillomas, and basal cell carcinomas in 70 % of the animals by 12 months of age (Yang et al. 2005). Interestingly and in contrast to the studies of human colon cancer, which do not indicate *SMAD3* is a mutational target, Graff et al. observed in *Smad3*$^{-/-}$ mice a high frequency of invasive colon carcinoma (Zhu et al. 1998). The development of colon neoplasms in the *Smad3*$^{-/-}$ mice appears to require certain environmental factors or genetic modifiers based on the results from two other *Smad3*$^{-/-}$ mouse models that have been generated which do not develop intestinal neoplasms (Derynck et al. 2001). Interestingly, *Smad2*$^{+/-}$ mice do not develop intestinal neoplasms spontaneously, and it does not appear that *Smad2* haploinsufficiency has a significant effect on the formation of intestinal tumors in the *Apc*$^{\Delta 716}$ or *Apc*^{580D} mice (Takaku et al. 2002; Hamamoto et al. 2002). In aggregate, results from studies employing mouse model systems support the role of *Smad3* and *Smad4* as tumor suppressor genes.

In addition to in vitro studies and mouse model studies, studies of the effect of alterations of Smad pathway elements on the clinical behavior of primary cancers have been conducted recently. The effect of 18q LOH and thus presumably inactivation of SMAD2 and/or SMAD4, on the clinical behavior of colon carcinomas has been subjected to intense scrutiny with inconclusive results to date. Several different groups have assayed for LOH of 18q using microsatellite markers in stage II colon cancer and have found either no association with the clinical behavior of the cancer or an association with more aggressive cancer behavior (Martinez-Lopez et al. 1998; Carethers et al. 1998; Jen et al. 1994; Laurent-Puig et al. 1992; Zhou et al. 2002). The reason for the discrepancy is unclear but may be related to different microsatellite loci assessed in each study and thus the specific region of 18q that was assessed by each investigator. Adding to this confusion, *SMAD4* diploidy and *TGFBR2 BAT-RII* mutations have been shown to associate with improved survival after adjuvant chemotherapy (Boulay et al. 2002; Watanabe et al. 2001b).

7.6 TGF-ß Superfamily Signaling Pathways and Cancer

7.6.1 BMP Signaling and Cancer

The TGF-ß superfamily includes TGF-ß1, TGF-ß2, and TGF-ß3, and also the BMPs, activin, nodal, GDF, and inhibin. The BMPs are disulfide-linked dimeric proteins that number at least 15 in total and include BMP-2, BMP-4, and BMP-7

(OP-1). They have a wide range of biological activities including the regulation of morphogenesis of various tissues and organs during development as well as the regulation of growth, differentiation, chemotaxis, and apoptosis in monocytes, epithelial cells, mesenchymal cells, and neuronal cells (Kawabata et al. 1998). The BMPs transduce their signals through a heteromeric receptor that consists of a type I and type II receptor. BMPR1A is one of two different type I BMP receptors (BMPR1A and BMPR1B). It serves to predominantly bind BMP-4 and BMP-2 as well as other BMPs and transduces their signals when partnered with a BMP type II receptor. As with the TGF-ß receptor, the best understood post-BMP receptor pathway is the Smad pathway. The R-Smads, Smads 1 and 5, partner with Smad4 (Co-Smad) to transduce BMP-mediated signals from the BMP receptors (Kawabata et al. 1998).

The identification of germline mutations in signaling elements of the BMP signaling pathway in individuals with JPS, a hereditary colon cancer syndrome, and of somatic mutations in the activin receptor *ACVR2* in colon cancers has globally implicated deregulation of the TGF-ß superfamily in the pathogenesis of cancer. Germline mutations in *BMPR1A*, a type I BMP receptor, and in *MADH4/SMAD4* in families with Juvenile Polyposis have been demonstrated in this subset of hereditary colon cancers. Howe et al. found nonsense and missense germline mutations in *BMPR1A* in four FJP families, 44-47delTGTT, 715C>T, 812G>A, and 961delC affecting exons 1, 7, 7, and 8 respectively (Howe et al. 2001). The identification of both *BMPR1A* and *MADH4/SMAD4* germline mutations in JPS families strongly implicates BMP signaling disruption in the pathogenesis of this syndrome. Furthermore, mice that overexpress Noggin, a soluble antagonist for the BMPs, or a dominant negative *Bmpr1a* in the intestinal epithelium display ectopic crypt formation and a phenotype reminiscent of JPS (Haramis et al. 2004; He et al. 2004). In addition to the genetic studies and studies of animal models, in vitro studies of cell lines have provided data to support a tumor suppressor role for BMP2, BMP4, and BMPR1A, although there are other studies that have suggested that the BMPs may also have tumor promoting functions in the colon (Eckhardt et al. 2006; Beck et al. 2007; Lawrance et al. 2007; Grijelmo et al. 2007). The factors that are modulating the biological effects of the BMPs on colorectal cancer remain to be determined.

7.6.2 Activin Signaling and Cancer

Activin and a related TGF-ß family member, inhibin, were originally identified as substances that could regulate follicle-stimulating hormone release from the anterior pituitary gland (Ying 1988). Activins have been subsequently found to have a variety of activities including embryonic mesoderm induction, left–right patterning in developing organisms, and as a tumor suppressor in certain organs. Activin is a secreted dimeric ligand, composed of either Activin ßA and/or Activin ßB, which activates intracellular signaling pathways that include the SMAD2/3-SMAD4 pathway via a heteromeric receptor that is composed of a type I receptor (ACVR1, ActRIA, or ActRIB) and a type II receptor (ACVR2 or ACVR2B) (de Caestecker 2004).

Inactivating mutations in *ACVR2* have been found to occur in 58–90 % of MSI colon cancers as the result of frameshift mutations in a polyadenine tract in the coding region of the gene (Mori et al. 2001; Deacu et al. 2004). Inactivating somatic mutations in *ACVR1B*, the gene for activin receptor type IB, have been identified in ~1 % of pancreatic cancers and in a subset of pituitary tumors (Su et al. 2001; Alexander et al. 1996). It is also reasonable to consider that the somatic mutations in *MADH4/SMAD4* observed in pancreatic, colon, gallbladder cancer, etc. may promote tumorigenesis through deregulating the activin pathway as well as the TGF-ß signaling pathway.

Functional evidence for a tumor suppressor function of the activin pathway, supporting the data from the molecular genetics of pancreatic and colon cancer, has been derived from in vitro studies as well as from studies of mouse models. Similar to TGF-ß, activin A inhibits cell proliferation through blocking cell cycle progression through the G1-S checkpoint (Chen et al. 2002). Activin A can suppress the proliferation of the breast cancer cell line T47D and the prostate cancer cell line LNCaP (Burdette et al. 2005; Zhang et al. 1997). Activin A has also been demonstrated to affect the growth of neuroblastoma xenografts by suppressing proliferation and angiogenesis (Panopoulou et al. 2005). Furthermore, follistatin, a monomeric glycoprotein that binds activin with high affinity and serves as a soluble antagonist of activins, is expressed by melanoma cell lines and blocks Smad signaling and activin-mediated apoptosis and growth arrest in these cells (Stove et al. 2004). The mouse models targeting the activin/inhibin pathway that have been constructed to date have implicated inhibin-alpha as a tumor suppressor of gonadal tumors (Matzuk et al. 1996). Thus, the identification of mutations that affect activin, TGF-ß, and BMP signaling broadly implicate the TGF-ß family as a tumor suppressor pathway in colon cancer.

7.7 Conclusions

Analysis of TGF-ß and its downstream signaling pathways has provided substantial insights into the molecular pathogenesis of human colorectal cancer. The results of these studies have provided general concepts about TGF-ß's role in cancer formation and have revealed that TGF-ß likely has effects on tumors that are highly tissue-type and context dependent. There is substantial evidence that the TGF-ß signaling pathway is a tumor suppressor pathway that can affect both the initiation and progression of cancers, with a prominent role on suppressing cancer progression in the colon. Furthermore, all the elements in the TGF-ß signaling pathway, the ligands, receptor, and post-receptor signaling pathway elements have been implicated in mediating the tumor suppressor effects of TGF-ß providing strong support for the linear nature of this pathway on key antitumor effects of TGF-ß. Major future challenges related to TGF-ß and cancer lie in the translational of results from in vitro and experimental animal models into diagnostic and/or therapeutic advances in the management of human colorectal cancer.

Acknowledgements Support for these studies was provided by the NIH (RO1CA115513, P30CA15704, UO1CA152756, and 5P01CA077852), and a Burroughs Wellcome Fund Translational Research Award for Clinician Scientist (WMG).

References

Abdollah S, Macias-Silva M, Tsukazaki T, Hayashi H, Attisano L, Wrana JL (1997) TbetaRI phosphorylation of Smad2 on Ser465 and Ser467 is required for Smad2-Smad4 complex formation and signaling. J Biol Chem 272(44):27678–27685

Akhurst RJ, Derynck R (2001) TGF-beta signaling in cancer—a double-edged sword. Trends Cell Biol 11(11):S44–S51

Alexander JM, Bikkal HA, Zervas NT, Laws ER Jr, Klibanski A (1996) Tumor-specific expression and alternate splicing of messenger ribonucleic acid encoding activin/transforming growth factor-beta receptors in human pituitary adenomas. J Clin Endocrinol Metab 81(2):783–790

Alexandrow MG, Moses HL (1995) Transforming growth factor beta and cell cycle regulation. Cancer Res 55(7):1452–1457

Alhopuro P, Alazzouzi H, Sammalkorpi H, Davalos V, Salovaara R, Hemminki A et al (2005) SMAD4 levels and response to 5-fluorouracil in colorectal cancer. Clin Cancer Res 11(17):6311–6316

Anumanthan G, Halder SK, Osada H, Takahashi T, Massion PP, Carbone DP et al (2005) Restoration of TGF-beta signalling reduces tumorigenicity in human lung cancer cells. Br J Cancer 93(10):1157–1167

Anzano MA, Roberts AB, Meyers CA, Komoriya A, Lamb LC, Smith JM et al (1982) Synergistic interaction of two classes of transforming growth factors from murine sarcoma cells. Cancer Res 42(11):4776–4778

Atfi A, Buisine M, Mazars A, Gespach C (1997) Induction of apoptosis by DPC4, a transcriptional factor regulated by transforming growth factor-beta through stress-activated protein kinase/c-Jun N-terminal kinase (SAPK/JNK) signaling pathway. J Biol Chem 272(40):24731–24734

Avery A, Paraskeva C, Hall P, Flanders KC, Sporn M, Moorghen M (1993) TGF-ß expression in the human colon: differential immunostaining along crypt epithelium. Br J Cancer 68:137–139

Azuma H, Ehata S, Miyazaki H, Watabe T, Maruyama O, Imamura T et al (2005) Effect of Smad7 expression on metastasis of mouse mammary carcinoma JygMC(a) cells. J Natl Cancer Inst 97(23):1734–1746

Bakin AV, Tomlinson AK, Bhowmick NA, Moses HL, Arteaga CL (2000) Phosphatidylinositol 3-kinase function is required for transforming growth factor beta -mediated epithelial to mesenchymal transition and cell migration. J Biol Chem 275(47):36803–36810

Bakin AV, Rinehart C, Tomlinson AK, Arteaga CL (2002) p38 mitogen-activated protein kinase is required for TGFbeta-mediated fibroblastic transdifferentiation and cell migration. J Cell Sci 115(Pt 15):3193–3206

Barnard JA, Warwick GJ, Gold LI (1993) Localization of transforming growth factor ß isoforms in the normal murine small intestine and colon. Gastroenterology 105:67–73

Beck SE, Jung BH, Del Rosario E, Gomez J, Carethers JM (2007) BMP-induced growth suppression in colon cancer cells is mediated by p21WAF1 stabilization and modulated by RAS/ERK. Cell Signal 19(7):1465–1472

Bellam N, Pasche B (2010) Tgf-beta signaling alterations and colon cancer. Cancer Treat Res 155:85–103

Bierie B, Moses HL (2006) Tumour microenvironment: TGFbeta: the molecular Jekyll and Hyde of cancer. Nat Rev Cancer 6(7):506–520

Biswas S, Chytil A, Washington K, Romero-Gallo J, Gorska AE, Wirth PS et al (2004) Transforming growth factor beta receptor type II inactivation promotes the establishment and progression of colon cancer. Cancer Res 64(14):4687–4692

Blahna MT, Hata A (2012) Smad-mediated regulation of microRNA biosynthesis. FEBS Lett 586(14):1906–1912

Boulay JL, Mild G, Lowy A, Reuter J, Lagrange M, Terracciano L et al (2002) SMAD4 is a predictive marker for 5-fluorouracil-based chemotherapy in patients with colorectal cancer. Br J Cancer 87(6):630–634

Brattain MG, Ko Y, Banerji SS, Wu G, Willson JK (1996) Defects of TGF-beta receptor signaling in mammary cell tumorigenesis. J Mammary Gland Biol Neoplasia 1(4):365–372

Bravo SB, Pampin S, Cameselle-Teijeiro J, Carneiro C, Dominguez F, Barreiro F et al (2003) TGF-beta-induced apoptosis in human thyrocytes is mediated by p27kip1 reduction and is overridden in neoplastic thyrocytes by NF-kappaB activation. Oncogene 22(49):7819–7830

Brown TL, Patil S, Cianci CD, Morrow JS, Howe PH (1999) Transforming growth factor beta induces caspase 3-independent cleavage of alphaII-spectrin (alpha-fodrin) coincident with apoptosis. J Biol Chem 274(33):23256–23262

Brown KA, Roberts RL, Arteaga CL, Law BK (2004) Transforming growth factor-beta induces Cdk2 relocalization to the cytoplasm coincident with dephosphorylation of retinoblastoma tumor suppressor protein. Breast Cancer Res 6(2):R130–R139

Burdette JE, Jeruss JS, Kurley SJ, Lee EJ, Woodruff TK (2005) Activin a mediates growth inhibition and cell cycle arrest through Smads in human breast cancer cells. Cancer Res 65(17):7968–7975

Carethers JM, Hawn MT, Greenson JK, Hitchcock CL, Boland CR (1998) Prognostic significance of allelic lost at chromosome 18q21 for stage II colorectal cancer. Gastroenterology 114(6):1188–1195

Chen YG, Hata A, Lo RS, Wotton D, Shi Y, Pavletich N et al (1998a) Determinants of specificity in TGF-beta signal transduction. Genes Dev 12(14):2144–2152

Chen T, Carter D, Garrigue-Antar L, Reiss M (1998b) Transforming growth factor ß type I receptor kinase mutant associated with metastatic breast cancer. Cancer Res 58:4805–4810

Chen RH, Chang MC, Su YH, Tsai YT, Kuo ML (1999) Interleukin-6 inhibits transforming growth factor-beta-induced apoptosis through the phosphatidylinositol 3-kinase/Akt and signal transducers and activators of transcription 3 pathways. J Biol Chem 274(33):23013–23019

Chen T, Yan W, Wells RG, Rimm DL, McNiff J, Leffell D et al (2001) Novel inactivating mutations of transforming growth factor-beta type I receptor gene in head-and-neck cancer metastases. Int J Cancer 93(5):653–661

Chen YG, Lui HM, Lin SL, Lee JM, Ying SY (2002) Regulation of cell proliferation, apoptosis, and carcinogenesis by activin. Exp Biol Med (Maywood) 227(2):75–87

Chen T, Jackson C, Costello B, Singer N, Colligan B, Douglass L et al (2004) An intronic variant of the TGFBR1 gene is associated with carcinomas of the kidney and bladder. Int J Cancer 112(3):420–425

Chen T, Jackson CR, Link A, Markey MP, Colligan BM, Douglass LE et al (2006) Int7G24A variant of transforming growth factor-beta receptor type I is associated with invasive breast cancer. Clin Cancer Res 12(2):392–397

Chowdhury S, Howell GM, Rajput A, Teggart CA, Brattain LE, Weber HR et al (2011) Identification of a novel TGFbeta/PKA signaling transduceome in mediating control of cell survival and metastasis in colon cancer. PLoS One 6(5):e19335

Chu TY, Lai JS, Shen CY, Liu HS, Chao CF (1999) Frequent aberration of the transforming growth factor-beta receptor II gene in cell lines but no apparent mutation in pre-invasive and invasive carcinomas of the uterine cervix. Int J Cancer 80(4):506–510

Conery AR, Cao Y, Thompson EA, Townsend CM Jr, Ko TC, Luo K (2004) Akt interacts directly with Smad3 to regulate the sensitivity to TGF-beta induced apoptosis. Nat Cell Biol 6(4):366–372

Cordenonsi M, Dupont S, Maretto S, Insinga A, Imbriano C, Piccolo S (2003) Links between tumor suppressors: p53 is required for TGF-beta gene responses by cooperating with Smads. Cell 113(3):301–314

Crawford SE, Stellmach V, Murphy-Ullrich JE, Ribeiro SM, Lawler J, Hynes RO et al (1998) Thrombospondin-1 is a major activator of TGF-beta1 in vivo. Cell 93(7):1159–1170

Cui W, Fowlis DJ, Bryson S, Duffie E, Ireland H, Balmain A et al (1996) TGFbeta1 inhibits the formation of benign skin tumors, but enhances progression to invasive spindle carcinomas in transgenic mice. Cell 86(4):531–542

Daley D, Morgan W, Lewis S, Willis J, Elston RC, Markowitz SD et al (2007) Is TGFBR1*6A a susceptibility allele for nonsyndromic familial colorectal neoplasia? Cancer Epidemiol Biomarkers Prev 16(5):892–894

Datto M, Li Y, Panus J, Howe D, Xiong Y, Wang X-F (1995) Transforming growth factor ß induces the cyclin-dependent kinase inhibitor p21 through a p53-independent mechanism. Proc Natl Acad Sci USA 92:5545–5549

Datto MB, Frederick JP, Pan L, Borton AJ, Zhuang Y, Wang XF (1999) Targeted disruption of Smad3 reveals an essential role in transforming growth factor beta-mediated signal transduction. Mol Cell Biol 19(4):2495–2504

Davis BN, Hilyard AC, Nguyen PH, Lagna G, Hata A (2010) Smad proteins bind a conserved RNA sequence to promote microRNA maturation by Drosha. Mol Cell 39(3):373–384

de Caestecker M (2004) The transforming growth factor-beta superfamily of receptors. Cytokine Growth Factor Rev 15(1):1–11

Deacu E, Mori Y, Sato F, Yin J, Olaru A, Sterian A et al (2004) Activin type II receptor restoration in ACVR2-deficient colon cancer cells induces transforming growth factor-beta response pathway genes. Cancer Res 64(21):7690–7696

Deng X, Cao Y, Liu Y, Li F, Sambandam K, Rajaraman S et al (2013) Overexpression of Evi-1 oncoprotein represses TGF-beta signaling in colorectal cancer. Mol Carcinog 52(4):255–264

Derynck R, Akhurst RJ, Balmain A (2001) TGF-beta signaling in tumor suppression and cancer progression. Nat Genet 29(2):117–129

Dews M, Fox JL, Hultine S, Sundaram P, Wang W, Liu YY et al (2010) The myc-miR-17 92 axis blunts TGF{beta} signaling and production of multiple TGF{beta}-dependent antiangiogenic factors. Cancer Res 70(20):8233–8246

Dowdy SC, Mariani A, Reinholz MM, Keeney GL, Spelsberg TC, Podratz KC et al (2005) Overexpression of the TGF-beta antagonist Smad7 in endometrial cancer. Gynecol Oncol 96(2):368–373

Dupont S, Zacchigna L, Adorno M, Soligo S, Volpin D, Piccolo S et al (2004) Convergence of p53 and TGF-beta signaling networks. Cancer Lett 213(2):129–138

Eckhardt F, Lewin J, Cortese R, Rakyan VK, Attwood J, Burger M et al (2006) DNA methylation profiling of human chromosomes 6, 20 and 22. Nat Genet 38(12):1378–1385

Elliott RL, Blobe GC (2005) Role of transforming growth factor beta in human cancer. J Clin Oncol 23(9):2078–2093

Engle SJ, Hoying JB, Boivin GP, Ormsby I, Gartside PS, Doetschman T (1999) Transforming growth factor beta1 suppresses nonmetastatic colon cancer at an early stage of tumorigenesis. Cancer Res 59(14):3379–3386

Engle SJ, Ormsby I, Pawlowski S, Boivin GP, Croft J, Balish E et al (2002) Elimination of colon cancer in germ-free transforming growth factor beta 1-deficient mice. Cancer Res 62(22):6362–6366

Eppert K, Scherer S, Ozcelik H, Pirone R, Hoodless P, Kim H et al (1996) Madr2 maps to 18q21 and encodes a TGF-ß regulated Mad-related protein that is functionally mutated in colorectal carcinoma. Cell 86:543–552

Eskinazi R, Resibois A, Svoboda M, Peny MO, Adler M, Robberecht P et al (1998) Expression of transforming growth factor beta receptors in normal human colon and sporadic adenocarcinomas. Gastroenterology 114(6):1211–1220

Ewan KB, Henshall-Powell RL, Ravani SA, Pajares MJ, Arteaga C, Warters R et al (2002) Transforming growth factor-beta1 mediates cellular response to DNA damage in situ. Cancer Res 62(20):5627–5631

Ewen ME, Sluss HK, Whitehouse LL, Livingston DM (1993) TGF beta inhibition of Cdk4 synthesis is linked to cell cycle arrest. Cell 74(6):1009–1020

Factor VM, Kao CY, Santoni-Rugiu E, Woitach JT, Jensen MR, Thorgeirsson SS (1997) Constitutive expression of mature transforming growth factor beta1 in the liver accelerates hepatocarcinogenesis in transgenic mice. Cancer Res 57(11):2089–2095

Fava RA, Casey TT, Wilcox J, Pelton RW, Moses HL, Nanney LB (1990) Synthesis of transforming growth factor-beta 1 by megakaryocytes and its localization to megakaryocyte and platelet alpha-granules. Blood 76(10):1946–1955

Fink S, Swinler S, Lutterbaugh J, Massague J, Thiagalingam S, Kinzler K et al (2001) Transforming growth factor-beta induced growth inhibition in a Smad4 mutant colon adenoma cell line. Cancer Res 61:256–260

Fink SP, Mikkola D, Willson JK, Markowitz S (2003) TGF-beta-induced nuclear localization of Smad2 and Smad3 in Smad4 null cancer cell lines. Oncogene 22(9):1317–1323

Freathy C, Brown DG, Roberts RA, Cain K (2000) Transforming growth factor-beta(1) induces apoptosis in rat FaO hepatoma cells via cytochrome c release and oligomerization of Apaf-1 to form a approximately 700-kd apoptosome caspase-processing complex. Hepatology 32(4 Pt 1):750–760

Freeman TJ, Smith JJ, Chen X, Washington MK, Roland JT, Means AL, et al (2012) Smad4-mediated signaling inhibits intestinal neoplasia by inhibiting expression of beta-catenin. Gastroenterology 142(3):562–571. e2

Friedl W, Kruse R, Uhlhaas S, Stolte M, Schartmann B, Keller KM et al (1999) Frequent 4-bp deletion in exon 9 of the SMAD4/MADH4 gene in familial juvenile polyposis patients. Genes Chromosomes Cancer 25(4):403–406

Fujimoto K, Sheng H, Shao J, Beauchamp RD (2001) Transforming growth factor-beta1 promotes invasiveness after cellular transformation with activated Ras in intestinal epithelial cells. Exp Cell Res 266(2):239–249

Geng Y, Weinberg RA (1993) Transforming growth factor beta effects on expression of G1 cyclins and cyclin-dependent protein kinases. Proc Natl Acad Sci USA 90(21):10315–10319

Glick A, Popescu N, Alexander V, Ueno H, Bottinger E, Yuspa SH (1999) Defects in transforming growth factor-beta signaling cooperate with a Ras oncogene to cause rapid aneuploidy and malignant transformation of mouse keratinocytes. Proc Natl Acad Sci USA 96(26):14949–14954

Goggins M, Shekher M, Turnacioglu K, Yeo CJ, Hruban RH, Kern SE (1998) Genetic alterations of the transforming growth factor beta receptor genes in pancreatic and biliary adenocarcinomas. Cancer Res 58(23):5329–5332

Gong J, Ammanamanchi S, Ko TC, Brattain MG (2003) Transforming growth factor beta1 increases the stability of p21/WAF1/CIP1 protein and inhibits CDK2 kinase activity in human colon carcinoma FET cells. Cancer Res 63(12):3340–3346

Grady WM (2004) Genomic instability and colon cancer. Cancer Metastasis Rev 23(1–2):11–27

Grady WM, Markowitz SD (2008) TGF-ß signaling pathway and tumor suppression. In: Derynck R, Miyazono K (eds) The TGF-ß family, 1st edn. Cold Spring Harbor Laboratory, Cold Spring Harbor, NY, pp 889–938

Grady W, Rajput A, Myeroff L, Liu D, Willis J, Kwon K et al (1998) Mutation of the type II TGF-ß receptor is coincident with the transformation of human colon adenomas to malignant carcinomas. Cancer Res 58:3101–3104

Grady W, Myeroff L, Swinler S, Rajput A, Thiagalingam S, Lutterbaugh J et al (1999) Mutational inactivation of transforming growth factor ß receptor type II in microsatellite stable colon cancers. Cancer Res 59:320–324

Grady WM, Willis JE, Trobridge P, Romero-Gallo J, Munoz N, Olechnowicz J et al (2006) Proliferation and Cdk4 expression in microsatellite unstable colon cancers with TGFBR2 mutations. Int J Cancer 118(3):600–608

Grau AM, Zhang L, Wang W, Ruan S, Evans DB, Abbruzzese JL et al (1997) Induction of p21waf1 expression and growth inhibition by transforming growth factor beta involve the tumor suppressor gene DPC4 in human pancreatic adenocarcinoma cells. Cancer Res 57(18):3929–3934

Grijelmo C, Rodrigue C, Svrcek M, Bruyneel E, Hendrix A, de Wever O et al (2007) Proinvasive activity of BMP-7 through SMAD4/src-independent and ERK/Rac/JNK-dependent signaling pathways in colon cancer cells. Cell Signal 19(8):1722–1732

Guda K, Natale L, Lutterbaugh J, Wiesner GL, Lewis S, Tanner SM et al (2009) Infrequent detection of germline allele-specific expression of TGFBR1 in lymphoblasts and tissues of colon cancer patients. Cancer Res 69(12):4959–4961

Guo Y, Kyprianou N (1999) Restoration of transforming growth factor beta signaling pathway in human prostate cancer cells suppresses tumorigenicity via induction of caspase-1-mediated apoptosis. Cancer Res 59(6):1366–1371
Hahm KB, Lee KM, Kim YB, Hong WS, Lee WH, Han SU et al (2002) Conditional loss of TGF-beta signalling leads to increased susceptibility to gastrointestinal carcinogenesis in mice. Aliment Pharmacol Ther 16(Suppl 2):115–127
Hahn WC (2003) Role of telomeres and telomerase in the pathogenesis of human cancer. J Clin Oncol 21(10):2034–2043
Hahn SA, Schutte M, Hoque ATM, Moskaluk C, da Costa L, Rozenblum E et al (1996) DPC4, a candidate tumor suppressor gene at human chromosome 18q21.1. Science 271(5247):350–353
Halder SK, Beauchamp RD, Datta PK (2005) Smad7 induces tumorigenicity by blocking TGF-beta-induced growth inhibition and apoptosis. Exp Cell Res 307(1):231–246
Hamamoto T, Beppu H, Okada H, Kawabata M, Kitamura T, Miyazono K et al (2002) Compound disruption of smad2 accelerates malignant progression of intestinal tumors in apc knockout mice. Cancer Res 62(20):5955–5961
Han SU, Kim HT, Seong do H, Kim YS, Park YS, Bang YJ et al (2004) Loss of the Smad3 expression increases susceptibility to tumorigenicity in human gastric cancer. Oncogene 23(7):1333–1341
Han G, Lu SL, Li AG, He W, Corless CL, Kulesz-Martin M et al (2005) Distinct mechanisms of TGF-beta1-mediated epithelial-to-mesenchymal transition and metastasis during skin carcinogenesis. J Clin Invest 115(7):1714–1723
Hannon G, Beach D (1994) $p15^{INK4B}$ is a potential effector of TGF-ß-induced cell cycle arrest. Nature 371:257–261
Harada N, Tamai Y, Ishikawa T, Sauer B, Takaku K, Oshima M et al (1999) Intestinal polyposis in mice with a dominant stable mutation of the beta-catenin gene. EMBO J 18(21):5931–5942
Haramis AP, Begthel H, van den Born M, van Es J, Jonkheer S, Offerhaus GJ et al (2004) De novo crypt formation and juvenile polyposis on BMP inhibition in mouse intestine. Science 303(5664):1684–1686
He XC, Zhang J, Tong WG, Tawfik O, Ross J, Scoville DH et al (2004) BMP signaling inhibits intestinal stem cell self-renewal through suppression of Wnt-beta-catenin signaling. Nat Genet 36(10):1117–1121
Hempen PM, Zhang L, Bansal RK, Iacobuzio-Donahue CA, Murphy KM, Maitra A et al (2003) Evidence of selection for clones having genetic inactivation of the activin A type II receptor (ACVR2) gene in gastrointestinal cancers. Cancer Res 63:994–999
Hofmann TG, Stollberg N, Schmitz ML, Will H (2003) HIPK2 regulates transforming growth factor-beta-induced c-Jun NH(2)-terminal kinase activation and apoptosis in human hepatoma cells. Cancer Res 63(23):8271–8277
Hohenstein P, Molenaar L, Elsinga J, Morreau H, van der Klift H, Struijk A et al (2003) Serrated adenomas and mixed polyposis caused by a splice acceptor deletion in the mouse Smad4 gene. Genes Chromosomes Cancer 36(3):273–282
Hoosein N, McKnight M, Levine A, Mulder K, Childress K, Brattain D et al (1989) Differential sensitivity of subclasses of human colon carcinoma cell lines to the growth inhibitory effects of transforming growth factor-ß1. Exp Cell Res 181:442–453
Horowitz JC, Lee DY, Waghray M, Keshamouni VG, Thomas PE, Zhang H et al (2004) Activation of the pro-survival phosphatidylinositol 3-kinase/AKT pathway by transforming growth factor-beta1 in mesenchymal cells is mediated by p38 MAPK-dependent induction of an autocrine growth factor. J Biol Chem 279(2):1359–1367
Howe PH, Draetta G, Leof EB (1991) Transforming growth factor beta 1 inhibition of p34cdc2 phosphorylation and histone H1 kinase activity is associated with G1/S-phase growth arrest. Mol Cell Biol 11(3):1185–1194
Howe JR, Roth S, Ringold JC, Summers RW, Jarvinen HJ, Sistonen P et al (1998) Mutations in the SMAD4/DPC4 gene in juvenile polyposis. Science 280(5366):1086–1088
Howe JR, Bair JL, Sayed MG, Anderson ME, Mitros FA, Petersen GM et al (2001) Germline mutations of the gene encoding bone morphogenetic protein receptor 1A in juvenile polyposis. Nat Genet 28(2):184–187

Howe JR, Shellnut J, Wagner B, Ringold JC, Sayed MG, Ahmed AF et al (2002) Common deletion of SMAD4 in juvenile polyposis is a mutational hotspot. Am J Hum Genet 70(5):1357–1362
Hu X, Zuckerman KS (2001) Transforming growth factor: signal transduction pathways, cell cycle mediation, and effects on hematopoiesis. J Hematother Stem Cell Res 10(1):67–74
Iavarone A, Massague J (1997) Repression of the CDK activator Cdc25A and cell-cycle arrest by cytokine TGF-beta in cells lacking the CDK inhibitor p15. Nature 387(6631):417–422
Jang CW, Chen CH, Chen CC, Chen JY, Su YH, Chen RH (2002) TGF-beta induces apoptosis through Smad-mediated expression of DAP- kinase. Nat Cell Biol 4(1):51–58
Javelaud D, Delmas V, Moller M, Sextius P, Andre J, Menashi S et al (2005) Stable overexpression of Smad7 in human melanoma cells inhibits their tumorigenicity in vitro and in vivo. Oncogene 24(51):7624–7629
Jen J, Kim H, Piantadosi S, Liu ZF, Levitt RC, Sistonen P et al (1994) Allelic loss of chromosome 18q and prognosis in colorectal cancer. N Engl J Med 331(4):213–221
Kaklamani VG, Hou N, Bian Y, Reich J, Offit K, Michel LS et al (2003) TGFBR1*6A and cancer risk: a meta-analysis of seven case-control studies. J Clin Oncol 21(17):3236–3243
Kanamoto T, Hellman U, Heldin CH, Souchelnytskyi S (2002) Functional proteomics of transforming growth factor-beta1-stimulated Mv1Lu epithelial cells: Rad51 as a target of TGFbeta1-dependent regulation of DNA repair. EMBO J 21(5):1219–1230
Kanzaki T, Olofsson A, Moren A, Wernstedt C, Hellman U, Miyazono K et al (1990) TGF-beta 1 binding protein: a component of the large latent complex of TGF-beta 1 with multiple repeat sequences. Cell 61(6):1051–1061
Katakura Y, Nakata E, Miura T, Shirahata S (1999) Transforming growth factor beta triggers two independent-senescence programs in cancer cells. Biochem Biophys Res Commun 255(1):110–115
Kawabata M, Imamura T, Miyazono K (1998) Signal transduction by bone morphogenetic proteins. Cytokine Growth Factor Rev 9(1):49–61
Keeton MR, Curriden SA, van Zonneveld AJ, Loskutoff DJ (1991) Identification of regulatory sequences in the type 1 plasminogen activator inhibitor gene responsive to transforming growth factor beta. J Biol Chem 266(34):23048–23052
Kim SJ, Im YH, Markowitz SD, Bang YJ (2000) Molecular mechanisms of inactivation of TGF-beta receptors during carcinogenesis. Cytokine Growth Factor Rev 11(1–2):159–168
Kleeff J, Ishiwata T, Maruyama H, Friess H, Truong P, Buchler MW et al (1999) The TGF-beta signaling inhibitor Smad7 enhances tumorigenicity in pancreatic cancer. Oncogene 18(39):5363–5372
Knaus P, Lindemann D, DeCoteau J, Perlman R, Yankelev H, Hille M et al (1996) A dominant inhibitory mutant of the type II transforming growth factor ß receptor in the malignant progression of a cutaneous T-cell lymphoma. Mol Cell Biol 16:3480–3489
Knobloch TJ, Lynch MA, Song H, DeGroff VL, Casto BC, Adams EM et al (2001) Analysis of TGF-beta type I receptor for mutations and polymorphisms in head and neck cancers. Mutat Res 479(1–2):131–139
Korchynskyi O, Landstrom M, Stoika R, Funa K, Heldin CH, ten Dijke P et al (1999) Expression of Smad proteins in human colorectal cancer. Int J Cancer 82(2):197–202
Kretzschmar M, Liu F, Hata A, Doody J, Massague J (1997) The TGF-beta family mediator Smad1 is phosphorylated directly and activated functionally by the BMP receptor kinase. Genes Dev 11(8):984–995
Kretzschmar M, Doody J, Timokhina I, Massague J (1999) A mechanism of repression of TGFß/Smad signaling in oncogenic Ras. Genes Dev 13:804–816
Kurokawa M, Mitani K, Irie K, Matsuyama T, Takahashi T, Chiba S et al (1998) The oncoprotein Evi-1 represses TGF-beta signalling by inhibiting Smad3. Nature 394(6688):92–96
Laurent-Puig P, Olschwang S, Delattre O, Remvikos Y, Asselain B, Melot T et al (1992) Survival and acquired genetic alterations in colorectal cancer. Gastroenterology 102(4 Pt 1):1136–1141
Lawrance AK, Deng L, Brody LC, Finnell RH, Shane B, Rozen R (2007) Genetic and nutritional deficiencies in folate metabolism influence tumorigenicity in Apcmin/+ mice. J Nutr Biochem 18(5):305–312

Li AG, Lu SL, Han G, Hoot KE, Wang XJ (2006) Role of TGFbeta in skin inflammation and carcinogenesis. Mol Carcinog 45(6):389–396
Lin SY, Elledge SJ (2003) Multiple tumor suppressor pathways negatively regulate telomerase. Cell 113(7):881–889
Luo K, Stroschein SL, Wang W, Chen D, Martens E, Zhou S et al (1999) The Ski oncoprotein interacts with the Smad proteins to repress TGFbeta signaling. Genes Dev 13(17):2196–2206
Maggio-Price L, Treuting P, Zeng W, Tsang M, Bielefeldt-Ohmann H, Iritani BM (2006) Helicobacter infection is required for inflammation and colon cancer in Smad3-deficient mice. Cancer Res 66(2):828–838
Markowitz S, Roberts A (1996) Tumor supressor activity of the TGF-ß pathway in human cancers. Cytokine Growth Factor Rev 7:93–102
Markowitz S, Wang J, Myeroff L, Parsons R, Sun L, Lutterbaugh J et al (1995) Inactivation of the type II TGF-ß receptor in colon cancer cells with microsatellite instability. Science 268:1336–1338
Martinez-Lopez E, Abad A, Font A, Monzo M, Ojanguren I, Pifarre A et al (1998) Allelic loss on chromosome 18q as a prognostic marker in stage II colorectal cancer. Gastroenterology 114(6):1180–1187
Massague J (1996) TGF-ß signaling: receptors, transducers, and mad proteins. Cell 85:947–950
Matzuk MM, Kumar TR, Shou W, Coerver KA, Lau AL, Behringer RR et al (1996) Transgenic models to study the roles of inhibins and activins in reproduction, oncogenesis, and development. Recent Prog Horm Res 51:123–154, discussion 55–7
Mishra L, Shetty K, Tang Y, Stuart A, Byers SW (2005) The role of TGF-beta and Wnt signaling in gastrointestinal stem cells and cancer. Oncogene 24(37):5775–5789
Mithani SK, Balch GC, Shiou SR, Whitehead RH, Datta PK, Beauchamp RD (2004) Smad3 has a critical role in TGF-beta-mediated growth inhibition and apoptosis in colonic epithelial cells. J Surg Res 117(2):296–305
Miyaki M, Iijima T, Konishi M, Sakai K, Ishii A, Yasuno M et al (1999) Higher frequency of Smad4 gene mutation in human colorectal cancer with distant metastasis. Oncogene 18(20):3098–3103
Mori Y, Yin J, Rashid A, Leggett BA, Young J, Simms L et al (2001) Instabilotyping: comprehensive identification of frameshift mutations caused by coding region microsatellite instability. Cancer Res 61(16):6046–6049
Moses HL, Coffey RJ, Jr., Leof EB, Lyons RM, Keski-Oja J (1987) Transforming growth factor beta regulation of cell proliferation. J Cell Physiol Suppl (Suppl 5):1–7
Moses H, Yang E, Pietonpol J (1990) TGF-ß stimulation and inhibition of cell proliferation: new mechanistic insights. Cell 63:245–247
Moustakas A, Kardassis D (1998) Regulation of the human p21/WAF1/Cip1 promoter in hepatic cells by functional interactions between Sp1 and Smad family members. Proc Natl Acad Sci USA 95(12):6733–6738
Munger JS, Huang X, Kawakatsu H, Griffiths MJ, Dalton SL, Wu J et al (1999) The integrin alpha v beta 6 binds and activates latent TGF beta 1: a mechanism for regulating pulmonary inflammation and fibrosis. Cell 96(3):319–328
Munoz NM, Upton M, Rojas A, Washington MK, Lin L, Chytil A et al (2006) Transforming growth factor beta receptor type II inactivation induces the malignant transformation of intestinal neoplasms initiated by Apc mutation. Cancer Res 66(20):9837–9844
Muraoka-Cook RS, Dumont N, Arteaga CL (2005) Dual role of transforming growth factor beta in mammary tumorigenesis and metastatic progression. Clin Cancer Res 11(2 Pt 2):937s–943s
Myeroff L, Parsons R, Kim S-J, Hedrick L, Cho K, Orth K et al (1995) A transforming growth factor ß receptor type II gene mutation common in colon and gastric but rare in endometrial cancers with microsatellite instability. Cancer Res 55:5545–5547
Nakashima R, Song H, Enomoto T, Murata Y, McClaid MR, Casto BC et al (1999) Genetic alterations in the transforming growth factor receptor complex in sporadic endometrial carcinoma. Gene Expr 8(5–6):341–352
Nawshad A, Lagamba D, Polad A, Hay ED (2005) Transforming growth factor-beta signaling during epithelial-mesenchymal transformation: implications for embryogenesis and tumor metastasis. Cells Tissues Organs 179(1–2):11–23

Nguyen AV, Pollard JW (2000) Transforming growth factor beta3 induces cell death during the first stage of mammary gland involution. Development 127(14):3107–3118
O'Mahony CA, Beauchamp RD, Albo D, Tsujii M, Sheng HM, Shao J et al (1999) Cyclooxygenase-2 alters transforming growth factor-beta 1 response during intestinal tumorigenesis. Surgery 126(2):364–370
Oft M, Akhurst RJ, Balmain A (2002) Metastasis is driven by sequential elevation of H-ras and Smad2 levels. Nat Cell Biol 4(7):487–494
Oshima M, Oshima H, Taketo M (1996) TGF-ß receptor type II deficiency results in defects of yolk sac hematopoiesis and vasulogenesis. Dev Biol 179:297–302
Panopoulou E, Murphy C, Rasmussen H, Bagli E, Rofstad EK, Fotsis T (2005) Activin A suppresses neuroblastoma xenograft tumor growth via antimitotic and antiangiogenic mechanisms. Cancer Res 65(5):1877–1886
Park YN, Chae KJ, Oh BK, Choi J, Choi KS, Park C (2004) Expression of Smad7 in hepatocellular carcinoma and dysplastic nodules: resistance mechanism to transforming growth factor-beta. Hepatogastroenterology 51(56):396–400
Parsons R, Myeroff L, Liu B, Willson J, Markowitz S, Kinzler K et al (1995) Microsatellite instability and mutations of the transforming growth factor ß type II receptor gene in colorectal cancer. Cancer Res 55:5548–5550
Pasche B, Luo Y, Rao P, Nimer S, Dmitrovsky E, Caron P et al (1998) Type I transforming growth factor ß receptor maps to 9q22 and exhibits a polymorphism and a rare variant within a polyadenine tract. Cancer Res 58:2727–2732
Pasche B, Kolachana P, Nafa K, Satagopan J, Chen YG, Lo RS et al (1999) TbetaR-I(6A) is a candidate tumor susceptibility allele. Cancer Res 59(22):5678–5682
Pasche B, Knobloch TJ, Bian Y, Liu J, Phukan S, Rosman D et al (2005) Somatic acquisition and signaling of TGFBR1*6A in cancer. JAMA 294(13):1634–1646
Perlman R, Schiemann WP, Brooks MW, Lodish HF, Weinberg RA (2001) TGF-beta-induced apoptosis is mediated by the adapter protein Daxx that facilitates JNK activation. Nat Cell Biol 3(8):708–714
Petritsch C, Beug H, Balmain A, Oft M (2000) TGF-beta inhibits p70 S6 kinase via protein phosphatase 2A to induce G(1) arrest. Genes Dev 14(24):3093–3101
Pietenpol JA, Stein RW, Moran E, Yaciuk P, Schlegel R, Lyons RM et al (1990) TGF-beta 1 inhibition of c-myc transcription and growth in keratinocytes is abrogated by viral transforming proteins with pRB binding domains. Cell 61(5):777–785
Polyak K, Kato JY, Solomon MJ, Sherr CJ, Massague J, Roberts JM et al (1994) p27Kip1, a cyclin-Cdk inhibitor, links transforming growth factor-beta and contact inhibition to cell cycle arrest. Genes Dev 8(1):9–22
Remy I, Montmarquette A, Michnick SW (2004) PKB/Akt modulates TGF-beta signalling through a direct interaction with Smad3. Nat Cell Biol 6(4):358–365
Rojas A, Meherem S, Kim YH, Washington MK, Willis JE, Markowitz SD et al (2008) The aberrant methylation of TSP1 suppresses TGF-beta1 activation in colorectal cancer. Int J Cancer 123(1):14–21
Rojas A, Padidam M, Cress D, Grady WM (2009) TGF-ß receptor levels regulated the specificity of signaling pathway activation and biological effects of TGF-ß. Biochim Biophys Acta 1793(7):1165–1173
Romero-Gallo J, Sozmen EG, Chytil A, Russell WE, Whitehead R, Parks TW et al (2005) Inactivation of TGF-ß signaling in hepatocytes results in an increased proliferative response after partial hepatectomy. Oncogene 24(18):3028–3041
Roth S, Sistonen P, Salovaara R, Hemminki A, Loukola A, Johansson M et al (1999) SMAD genes in juvenile polyposis. Genes Chromosomes Cancer 26(1):54–61
Sancho E, Batlle E, Clevers H (2004) Signaling pathways in intestinal development and cancer. Annu Rev Cell Dev Biol 20:695–723
Schuster N, Krieglstein K (2002) Mechanisms of TGF-beta-mediated apoptosis. Cell Tissue Res 307(1):1–14
Seoane J, Pouponnot C, Staller P, Schader M, Eilers M, Massague J (2001) TGFbeta influences Myc, Miz-1 and Smad to control the CDK inhibitor p15INK4b. Nat Cell Biol 3(4):400–408

Seoane J, Le HV, Shen L, Anderson SA, Massague J (2004) Integration of Smad and forkhead pathways in the control of neuroepithelial and glioblastoma cell proliferation. Cell 117(2):211–223
Shih WL, Kuo ML, Chuang SE, Cheng AL, Doong SL (2000) Hepatitis B virus X protein inhibits transforming growth factor-beta -induced apoptosis through the activation of phosphatidylinositol 3-kinase pathway. J Biol Chem 275(33):25858–25864
Shull MM, Ormsby I, Kier AB, Pawlowski S, Diebold RJ, Yin M et al (1992) Targeted disruption of the mouse transforming growth factor-beta 1 gene results in multifocal inflammatory disease. Nature 359(6397):693–699
Slattery ML, Herrick JS, Lundgreen A, Wolff RK (2011) Genetic variation in the TGF-beta signaling pathway and colon and rectal cancer risk. Cancer Epidemiol Biomarkers Prev 20(1):57–69
Slattery ML, Lundgreen A, Herrick JS, Kadlubar S, Caan BJ, Potter JD et al (2012) Genetic variation in bone morphogenetic protein and colon and rectal cancer. Int J Cancer 130(3):653–664
Souchelnytskyi S, Tamaki K, Engstrom U, Wernstedt C, ten Dijke P, Heldin CH (1997) Phosphorylation of Ser465 and Ser467 in the C terminus of Smad2 mediates interaction with Smad4 and is required for transforming growth factor-beta signaling. J Biol Chem 272(44):28107–28115
Stove C, Vanrobaeys F, Devreese B, Van Beeumen J, Mareel M, Bracke M (2004) Melanoma cells secrete follistatin, an antagonist of activin-mediated growth inhibition. Oncogene 23(31): 5330–5339
Stroschein SL, Wang W, Zhou S, Zhou Q, Luo K (1999) Negative feedback regulation of TGF-beta signaling by the SnoN oncoprotein. Science 286(5440):771–774
Su GH, Bansal R, Murphy KM, Montgomery E, Yeo CJ, Hruban RH et al (2001) ACVR1B (ALK4, activin receptor type 1B) gene mutations in pancreatic carcinoma. Proc Natl Acad Sci USA 98(6):3254–3257
Takaku K, Oshima M, Miyoshi H, Matsui M, Seldin M, Taketo M (1998) Intestinal tumorigenesis in compound mutant mice of both *Dpc4* (*Smad4*) and *Apc* genes. Cell 92:645–656
Takaku K, Miyoshi H, Matsunaga A, Oshima M, Sasaki N, Taketo MM (1999) Gastric and duodenal polyps in Smad4 (Dpc4) knockout mice. Cancer Res 59(24):6113–6117
Takaku K, Wrana JL, Robertson EJ, Taketo MM (2002) No effects of Smad2 (madh2) null mutation on malignant progression of intestinal polyps in Apc(delta716) knockout mice. Cancer Res 62(16):4558–4561
Takenoshita S, Tani M, Mogi A, Nagashima M, Nagamachi Y, Bennett WP et al (1998) Mutation analysis of the Smad2 gene in human colon cancers using genomic DNA and intron primers. Carcinogenesis 19(5):803–807
Taketo MM, Takaku K (2000) Gastro-intestinal tumorigenesis in Smad4 mutant mice. Cytokine Growth Factor Rev 11(1–2):147–157
Tang B, Bottinger EP, Jakowlew SB, Bagnall KM, Mariano J, Anver MR et al (1998) Transforming growth factor-beta1 is a new form of tumor suppressor with true haploid insufficiency. Nat Med 4(7):802–807
Ten Dijke P, Goumans M-J, Itoh F, Itoh S (2002) Regulation of proliferation by Smad proteins. J Cell Physiol 191:1–16
Tenesa A, Dunlop MG (2009) New insights into the aetiology of colorectal cancer from genome-wide association studies. Nat Rev Genet 10(6):353–358
Trobridge P, Knoblaugh S, Washington MK, Munoz NM, Tsuchiya KD, Rojas A, et al (2009) TGF-beta receptor inactivation and mutant Kras induce intestinal neoplasms in mice via a beta-catenin independent pathway. Gastroenterology 136(5):1680–1688. e7
Turcatel G, Rubin N, El-Hashash A, Warburton D (2012) MIR-99a and MIR-99b modulate TGF-beta induced epithelial to mesenchymal plasticity in normal murine mammary gland cells. PLoS One 7(1):e31032
Vogelstein B, Fearon ER, Hamilton SR, Kern SE, Preisinger AC, Leppert M et al (1988) Genetic alterations during colorectal-tumor development. N Engl J Med 319(9):525–532
Vogelstein B, Fearon ER, Kern SE, Hamilton SR, Preisinger AC, Nakamura Y et al (1989) Allelotype of colorectal carcinomas. Science 244(4901):207–211
Wakefield LM, Roberts AB (2002) TGF-beta signaling: positive and negative effects on tumorigenesis. Curr Opin Genet Dev 12(1):22–29

Wang J, Sun L, Myeroff L, Wang X, Gentry LE, Yang J et al (1995a) Demonstration that mutation of the type II TGF-ß receptor inactivates its tumor suppressor activity in replication error-positive colon carcinoma cells. J Biol Chem 270:22044–22049
Wang D, Zhou GH, Birkenmeier TM, Gong J, Sun L, Brattain MG (1995b) Autocrine transforming growth factor beta 1 modulates the expression of integrin alpha 5 beta 1 in human colon carcinoma FET cells. J Biol Chem 270(23):14154–14159
Wang H, Radjendirane V, Wary KK, Chakrabarty S (2004a) Transforming growth factor beta regulates cell-cell adhesion through extracellular matrix remodeling and activation of focal adhesion kinase in human colon carcinoma Moser cells. Oncogene 23(32):5558–5561
Wang J, Sergina N, Ko TC, Gong J, Brattain MG (2004b) Autocrine and exogenous transforming growth factor beta control cell cycle inhibition through pathways with different sensitivity. J Biol Chem 279(38):40237–40244
Watanabe T, Wu T-T, Catalano PJ, Ueki T, Satriano R, Haller DG et al (2001) Molecular predictors of survival after adjuvant chemotherapy for colon cancer. N Engl J Med 344(16):1196–1206
Woodford-Richens K, Williamson J, Bevan S, Young J, Leggett B, Frayling I et al (2000) Allelic loss at SMAD4 in polyps from juvenile polyposis patients and use of fluorescence in situ hybridization to demonstrate clonal origin of the epithelium. Cancer Res 60(9):2477–2482
Xiangming C, Natsugoe S, Takao S, Hokita S, Ishigami S, Tanabe G et al (2001) Preserved Smad4 expression in the transforming growth factor beta signaling pathway is a favorable prognostic factor in patients with advanced gastric cancer. Clin Cancer Res 7(2):277–282
Xu X, Brodie SG, Yang X, Im YH, Parks WT, Chen L et al (2000) Haploid loss of the tumor suppressor Smad4/Dpc4 initiates gastric polyposis and cancer in mice. Oncogene 19(15):1868–1874
Yamamura Y, Hua X, Bergelson S, Lodish HF (2000) Critical role of smads and AP-1 complex in TGF-{beta}-dependent apoptosis. J Biol Chem 275(46):36295–36302
Yang X, Letterio JJ, Lechleider RJ, Chen L, Hayman R, Gu H et al (1999) Targeted disruption of SMAD3 results in impaired mucosal immunity and diminished T cell responsiveness to TGF-beta. EMBO J 18(5):1280–1291
Yang H, Kyo S, Takatura M, Sun L (2001) Autocrine transforming growth factor beta suppresses telomerase activity and transcription of human telomerase reverse transcriptase in human cancer cells. Cell Growth Differ 12(2):119–127
Yang L, Mao C, Teng Y, Li W, Zhang J, Cheng X et al (2005) Targeted disruption of Smad4 in mouse epidermis results in failure of hair follicle cycling and formation of skin tumors. Cancer Res 65(19):8671–8678
Ying SY (1988) Inhibins, activins, and follistatins: gonadal proteins modulating the secretion of follicle-stimulating hormone. Endocr Rev 9(2):267–293
Yu Q, Stamenkovic I (2000) Cell surface-localized matrix metalloproteinase-9 proteolytically activates TGF-beta and promotes tumor invasion and angiogenesis. Genes Dev 14(2):163–176
Zhang YE (2009) Non-Smad pathways in TGF-beta signaling. Cell Res 19(1):128–139
Zhang Y, Feng X-H, Wu R-Y, Derynck R (1996) Receptor-associated Mad homologues synergize as effectors of the TGF-ß response. Nature 383:168–172
Zhang Z, Zhao Y, Batres Y, Lin MF, Ying SY (1997) Regulation of growth and prostatic marker expression by activin A in an androgen-sensitive prostate cancer cell line LNCAP. Biochem Biophys Res Commun 234(2):362–365
Zhao Y (1999) Transforming growth factor-beta (TGF-beta) type I and type II receptors are both required for TGF-beta-mediated extracellular matrix production in lung fibroblasts. Mol Cell Endocrinol 150(1–2):91–97
Zhou W, Goodman SN, Galizia G, Lieto E, Ferraraccio F, Pignatelli C et al (2002) Counting alleles to predict recurrence of early-stage colorectal cancers. Lancet 359(9302):219–225
Zhu Y, Richardson JA, Parada LF, Graff JM (1998) Smad3 mutant mice develop metastatic colorectal cancer. Cell 94(6):703–714

Chapter 8
The Clinical Significance of Mutations in Colorectal Cancer

Franklin W. Huang, Laura B. Kleiman, and Theodore S. Hong

Abstract Genetic mutations are emerging as critical factors in treatment decisions for colorectal cancer. In this chapter, we review the clinical data for predictive and prognostic genetic markers. While a prognostic role for *KRAS* mutations remains uncertain, these activating mutations are clearly associated with nonresponsiveness to anti-EGFR therapy. We additionally examine the burgeoning clinical implications of *NRAS*, *BRAF*, and *PIK3CA* mutations. Finally, we provide an outlook on the potential impact of personalized cancer medicine to future clinical practice.

More effective treatment options for colorectal cancer (CRC) are urgently needed. While low-grade CRCs can be cured with surgery alone, later stage cancers are treated with some combination of chemotherapy, radiotherapy, and targeted therapies, depending on the anatomic site and staging of the tumor. Different combinations of chemotherapies such as 5-fluorouracil (5-FU), irinotecan, and oxaliplatin are used, including FOLFOX, which combines 5-FU, leucovorin, and oxaliplatin, and FOLFIRI, which combines 5-FU, leucovorin, and irinotecan. Because of their anatomic location, surgical resections are more challenging for rectal cancer compared to colon cancer, and therefore there is a greater risk of local recurrence (Adam et al. 1994; Manfredi et al. 2006). Preoperative radiotherapy decreases the local

F.W. Huang, M.D., Ph.D.
Division of Hematology/Oncology, Massachusetts General Hospital
and Department of Medical Oncology, Dana-Farber Cancer Institute, Boston, MA, USA

L.B. Kleiman, Ph.D.
Molecular Pathology Unit and Center for Cancer Research, Massachusetts General Hospital, Charlestown, MA, USA

T.S. Hong, M.D. (✉)
Department of Radiation Oncology, Massachusetts General Hospital, 55 Fruit Street, Boston, MA 02114, USA
e-mail: tshong1@partners.org

K.M. Haigis (ed.), *Molecular Pathogenesis of Colorectal Cancer*,
DOI 10.1007/978-1-4614-8412-7_8, © Springer Science+Business Media New York 2013

recurrence rate and is used in combination with chemotherapy to treat locally advanced rectal cancer (Folkesson et al. 2005; Hong et al. 2012).

Targeted therapies (small molecule inhibitors and monoclonal antibodies (mAbs)) affect signaling pathways aberrantly activated in cancer cells and are slowly making their way into the clinic for the treatment of various cancers. Three such drugs (all mAbs) are approved to treat CRC: cetuximab (Erbitux) and panitumumab (Vectibix), which inhibit the epidermal growth factor receptor (EGFR/ErbB1), and bevacizumab (Avastin), which inhibits the angiogenesis-promoting vascular endothelial growth factor (VEGF) (Eng 2010). These mAbs have not demonstrated benefit for locally advanced disease (Aklilu and Eng 2011; Debucquoy et al. 2010) and are only used in the setting of metastatic CRC (mCRC). The chimeric IgG_1 cetuximab is used in combination with FOLFIRI for first-line treatment, in combination with irinotecan in patients who are refractory to irinotecan-based chemotherapy, or as a single agent in patients who have failed oxaliplatin- and irinotecan-based chemotherapy or who are intolerant to irinotecan. The fully human IgG_2 panitumumab is approved as a single agent for mCRCs with disease progression on, or following, chemotherapy. The recombinant humanized IgG_1 bevacizumab is used in combination with 5-FU-based chemotherapy for first- or second-line treatment.

EGFR is a prototypical receptor tyrosine kinase that activates multi-kinase phosphorylation cascades to regulate diverse cellular processes including proliferation, survival, and migration (Citri and Yarden 2006). Ligand binding promotes EGFR homodimerization and hetero-oligomerization with three close homologues: ErbB2 (Her2), ErbB3, and ErbB4 (collectively referred to as the ErbB receptors). Receptor conformational changes position the cytoplasmic C-terminal tail of one receptor near the activation loop of the other, thereby facilitating receptor phosphorylation *in trans* (Zhang et al. 2006). Receptor phosphorylation and binding of adaptor proteins leads to transmission of signals through RAS to various downstream targets, including the mitogen-activated protein kinase (MAPK) (RAF-MEK-ERK) and phosphoinositide 3-kinase (PI3K) cascades (Citri and Yarden 2006).

Deficiencies in ErbB signaling are associated with the development of neurodegenerative diseases in humans, and the importance of these receptors during development and in normal adult physiology has become apparent from analyses of genetically modified mice (Bublil and Yarden 2007; Hynes and Lane 2005). Excessive ErbB signaling, on the other hand, is associated with many types of cancer. ErbB1-3 receptors are frequently mutated, overexpressed, or activated by autocrine or paracrine ligand production in solid tumors (Hynes and Lane 2005; Holbro and Hynes 2004; Sharma and Settleman 2009) and have been the target of extensive drug discovery efforts (Sebastian et al. 2006). Multiple tyrosine kinase inhibitors and therapeutic antibodies against ErbB receptors are in clinical use or in development.

EGFR is often overexpressed in CRC (Spano et al. 2005), and increased EGFR gene copy number and protein expression may correlate with a higher stage, aggressiveness, presence of metastases, and poorer prognosis (Bronte et al. 2011). Cetuximab and panitumumab bind to the extracellular domain of EGFR and competitively block the binding of EGFR ligands, thereby inhibiting receptor activation as well as VEGF production (Wheeler et al. 2010). The growth and survival of tumor cells expressing EGFR are inhibited by cetuximab and panitumumab in both in vitro assays and in vivo animal studies. While these mAbs have shown benefit for patients with mCRC, they are effective in only 10–20 % of patients, and response rates do not correlate with EGFR expression (Bronte et al. 2011; Bardelli and Siena 2010).

Genetic mutations found in colorectal tumors have the potential to impact prognosis and guide therapeutic interventions as well as inform our understanding of the biology of CRC. The importance of the genomic era in cancer has been demonstrated by the discoveries of predictive and prognostic genetic markers and the identification of novel drug targets. The clinical significance and implications of genetic alterations can be valued by their predictive and prognostic roles. In CRC, currently the most important of these genetic alterations have been mutations in the proto-oncogene *KRAS*. Other mutations for which prognostic or predictive roles have been proposed and for which supportive evidence exists include *PIK3CA* in up to 20 % of CRCs, *NRAS* in 5 %, and *BRAF* in 10 % of CRCs. As next-generation sequencing efforts are completed, the clinical implications of genetic mutations in CRC will likely continue to increase in importance. Many large clinical trials have retrospectively characterized the frequency of *KRAS* mutations. As further clinical data is obtained regarding other genetic mutations, this will likely impact the design of clinical trials and stratification of patients into different treatment regimens, toward the possibility of personalized cancer medicine.

The discovery and establishment of predictive and prognostic genetic alterations in CRC represent opportunities for guiding prognosis, treatment regimens, the design of clinical trials, and drug development. In this chapter, we will discuss the significance of *KRAS* mutations and its impact on current clinical practice. Additionally, we will discuss the evolving data regarding other potential genes of significance.

8.1 KRAS

Within the *RAS* family of proteins, *KRAS* mutations represent 85 % of *RAS* mutations, while *NRAS* mutations represent 15 %, and *HRAS* mutations <1 %. The *KRAS* proto-oncogene is one of the most frequently mutated genes in human cancers. The vast majority of *KRAS* mutations are located in codons 12 and 13. These mutations compromise GTP hydrolysis stimulated by GTPase-activating proteins (GAPs), resulting in hyperactive RAS protein and uncontrolled proliferation. The high prevalence of *KRAS* mutations in colon cancer has resulted in attempts to understand the clinical significance of these mutations.

Several studies have established that activating mutations in *KRAS* predict for non-responsiveness to anti-EGFR therapy. As a result of these discoveries after the advent of anti-EGFR therapy, genetic testing of colorectal tumors for *KRAS* mutations is now standard of care prior to consideration of anti-EGFR therapy. The determination of those patients who will not benefit from anti-EGFR therapy has avoided the unnecessary utilization of resources and treatment of patients and highlights the role that a genetic biomarker can play in clinical care. Although a predictive role for *KRAS* mutation has been established, a prognostic role of *KRAS* mutation remains unclear.

8.2 A Predictive Role for *KRAS* Mutations and EGFR Therapy

Cetuximab was approved by the FDA in 2004 and its use was further expanded in 2007. The first studies that demonstrated efficacy of anti-EGFR therapy for use in mCRC with cetuximab did not evaluate tumor genotype and therefore did not differentiate between patients based on *KRAS* mutation status.

Beginning in 2006, evidence began to accumulate that *KRAS* mutation status was associated with responsiveness to cetuximab (Lievre et al. 2006). Soon, multiple randomized trials were reevaluated to explore responsiveness to anti-EGFR therapy based on *KRAS* mutation status, establishing that *KRAS* mutation was a negative predictor for responsiveness to EGFR therapy (Bardelli and Siena 2010). In a retrospective analysis of the CO.17 randomized trial, which showed improved overall survival (OS) and progression-free survival (PFS) in refractory CRC with cetuximab monotherapy, Karapetis et al. evaluated whether *KRAS* status modified the effect of cetuximab treatment. *KRAS* mutation status was available for 68.9 % of the total study population. *KRAS* mutation was detected in 40.9 % of the cetuximab treated group and in 42 % of the supportive care group. *KRAS* WT tumors treated with cetuximab plus best supportive care (BSC) vs. BSC alone had improved PFS (3.7 vs. 1.9 month) and OS (9.5 vs. 4.8 month). However, in mutant *KRAS* patients there was no difference in OS or PFS between the groups treated with cetuximab plus BSC vs. BSC alone (Karapetis et al. 2008).

As the clinical benefit from combination chemotherapy with regimens such as FOLFOX and FOLFIRI in mCRC patients became realized, several randomized studies examined the association of *KRAS* mutation status in mCRC patients treated with anti-EGFR therapy and chemotherapy. In the OPUS trial, Bokemeyer et al. randomized patients to FOLFOX-4 + cetuximab or FOLFOX-4 alone (Bokemeyer et al. 2009). In an updated analysis, they were able to evaluate up to 93 % of patients and confirmed results of *KRAS* mutation status as a predictive biomarker (Bokemeyer et al. 2011). The CRYSTAL trial evaluated the efficacy of cetuximab with FOLFIRI in patients with mCRC in the first-line setting. Tumor tissue was available for approximately half of the patients, and a *KRAS* mutation was found in 36 % of these patients. *KRAS* status appeared to predict response to cetuximab. Among patients with *KRAS* wild-type (WT) tumors, the response rate in the cetuximab–FOLFIRI group was 59.3 % and in the FOLFIRI group 43.2 %, whereas in patients with mutant *KRAS* tumors, there was no significant difference in response rate (36.2 % vs. 40.2 %). The hazard ratio for PFS was 0.68 in the *KRAS* WT group treated with cetuximab plus FOLFIRI. Still there was no interaction between treatment group and *KRAS* status in PFS or OS. The authors concluded that cetuximab plus FOLFIRI reduced the risk of PFS and that this effect was primarily seen in patients that had *KRAS* WT tumors (Van Cutsem et al. 2009). *KRAS* status was a predictive factor for response to cetuximab when treated with FOLFIRI. An updated analysis by Van Cutsem et al. found that in patients with WT *KRAS*, FOLFIRI + cetuximab resulted in improvement of OS from 23.5 vs. 20.0 months compared with FOLFIRI alone (Van Cutsem et al. 2011).

In the COIN trial, a large prospective randomized trial evaluating the addition of cetuximab to oxaliplatin-based combination chemotherapy in the first-line metastatic setting, cetuximab showed no benefit in PFS or OS in *KRAS* WT patients. There was only an improvement in response rate in *KRAS* WT patients (Maughan et al. 2011). However, in the first-line setting when cetuximab was added to capecitabine and bevacizumab (CBC), a shorter PFS was seen compared to the capecitabine and bevacizumab alone group (CB) (Tol et al. 2009). Of note, patients with *KRAS* mutant tumors who were treated with cetuximab did worse compared to patients with WT *KRAS* tumors.

Cetuximab has clinical benefit in the setting of chemotherapy-refractory metastatic colon cancer. CA225025 was a randomized trial of 572 patients previously treated EGFR-positive mCRC who were treated with single-agent cetuximab plus BSC vs. BSC alone (Jonker et al. 2007). The group treated with single-agent cetuximab had an improved OS. In patients with WT *KRAS* status, single-agent cetuximab improved OS from 5 to 8.6 months. There was no improvement in OS or PFS in patients with mutant *KRAS*. Based on the retrospective analyses of the OPUS, CRYSTAL and CA225025 studies, the FDA granted approval in July 2012 for the use of cetuximab in combination with FOLFIRI for first-line treatment of patients with *KRAS* mutant-negative mCRC.

Panitumumab is a human anti-EGFR monoclonal antibody that was approved by the FDA in 2006 on the basis of improvement in PFS in patients with mCRC. It has also been assessed in combination with cytotoxic chemotherapy. In the PRIME study, 1,183 patients with mCRC not previously treated were randomly assigned to receive panitumumab-FOLFOX4 vs. FOLFOX4 alone. *KRAS* status was available for 93 % of patients. In the WT *KRAS* patients, panitumumab-FOLFOX4 significantly improved PFS compared with FOLFOX4 (median PFS, 9.6 vs. 8.0 months, respectively; hazard ratio [HR], 0.80; 95 % CI, 0.66–0.97; $P=0.02$). There was a nonsignificant increase in OS observed for panitumumab-FOLFOX4 vs. FOLFOX4 (median OS, 23.9 vs. 19.7 months, respectively; HR, 0.83; 95 % CI, 0.67–1.02; $P=0.072$). In the mutant *KRAS* patients, PFS and OS (15.5 vs. 19.3 month) were significantly reduced in the panitumumab-FOLFOX4 arm vs. the FOLFOX4 arm (Douillard et al. 2010).

In a separate trial examining the use of panitumumab in combination with FOLFIRI, 1,186 patients with mCRC who failed initial treatment were randomly assigned to receive panitumumab plus FOLFIRI vs. FOLFIRI alone. *KRAS* status was examined: 55 % were found to be WT tumors and 45 % mutant *KRAS* tumors. In patients with WT *KRAS* tumors, there was a significant improvement in PFS (5.9 vs. 3.9 months for FOLFIRI alone). There was no significant difference in OS but response rate was improved to 35 % vs. 10 % with addition of panitumumab. In patients with mutant *KRAS* tumors, there was no difference in efficacy with the addition of panitumumab to FOLFIRI (Peeters et al. 2010a).

In the chemotherapy refractory setting, panitumumab was compared with BSC alone in a phase III study of 231 patients (Van Cutsem et al. 2007). Panitumumab led to an improvement in PFS (median PFS of 8 vs. 7.3 weeks in BSC alone) and a lower rate of disease progression after a median follow-up of 35 weeks. On the basis

Table 8.1 Lack of benefit from EGFR therapy in *KRAS* mutant CRC

Study	Cancer stage/ treatment setting	EGFR therapy	Outcome in patients with *KRAS* mutant tumors	References
CRYSTAL	Metastatic/ chemotherapy naive	FOLFIRI +/ –Cetuximab	No benefit	Van Cutsem et al. (2011)
PACCE	Metastatic/ chemotherapy naive	Panitumumab	Worse	Hecht et al. (2009)
COIN	Metastatic/ chemotherapy naive	Cetuximab + combination chemotherapy	No benefit	Maughan et al. (2011)
CAIRO2	Metastatic/ chemotherapy naive	Cetuximab + bevacizumab, capecitabine, oxaliplatin	Worse	Tol et al. (2009)
PRIME	Metastatic/ chemotherapy naive	Panitumumab + FOLFOX	No benefit	Douillard et al. (2010)
OPUS	Metastatic/ chemotherapy naive	Cetuximab	No benefit, worse PFS	Bokemeyer et al. (2011)
Amado et al. (2008)	Metastatic/ chemotherapy refractory	Panitumumab monotherapy	No benefit	Amado et al. (2008)
NCIC-017 (CO.17)	Metastatic/ chemotherapy refractory	Cetuximab monotherapy	No benefit	Karapetis et al. (2008)

of this study, panitumumab was approved for use as a single-agent in EGFR-positive chemotherapy refractory mCRC. Subsequent analysis of *KRAS* mutation status was consistent with other studies finding that WT *KRAS* status was associated with efficacy of panitumumab (Amado et al. 2008).

Among patients treated with anti-EGFR antibodies, Dahabreh performed a meta-analysis of 13 nonoverlapping studies with 1,695 patients in total, and found a summary HR of 1.79, suggesting that *KRAS* mutation status in patients treated with anti-EGFR therapy was associated with a shorter OS (Dahabreh et al. 2011). Among *KRAS* WT patients, while 30–40 % respond to anti-EGFR therapy, the majority of patients (up to 60 %) do not respond to anti-EGFR therapy. It is not yet known what accounts for this lack of response although downstream effectors of the EGFR signaling pathway have been implicated.

Taken together, the preponderance of evidence indicates that *KRAS* WT status is necessary, but not sufficient, to predict responsiveness to anti-EGFR therapy (Table 8.1). Current guidelines recommend *KRAS* mutation testing in all mCRC patients in whom anti-EGFR therapy is being considered. The use of the anti-EGFR agents cetuximab and panitumumab is not recommended in the setting of *KRAS* mutant CRC. Reflecting this consensus view, the FDA has updated labeling of panitumumab and cetuximab to include information about *KRAS* mutation status.

8.3 Is a *KRAS* Mutation a Prognostic Factor?

Although *KRAS* mutation status has a predictive role for non-responsiveness to anti-EGFR therapy, whether *KRAS* mutations in CRC have a significant prognostic role has been an area of controversy. In this area, the data appear to be mixed (Table 8.2). Data from the CRYSTAL and CO.17 trials indicated that *KRAS* status alone does not have a prognostic role (Karapetis et al. 2008). In the CO.17 trial, in the BSC treatment arm, there was no significant difference in median OS between the groups with *KRAS* mutant vs. *KRAS* WT (4.8 vs. 4.6 month). The CO.17 trial enrolled patients who had failed two lines of previous chemotherapy, but seemed to indicate that *KRAS* mutation status did not have a prognostic role. In an analysis of the phase III mCRC trial comparing panitumumab monotherapy to BSC, Amado et al. found that *KRAS* WT status predicted improved OS (Amado et al. 2008). They reported that OS was longer in the WT group than in the mutant group adjusting for stratification factors and randomized treatment with a hazard ratio of 0.67. However, the majority of patients in the BSC arm crossed over to panitumumab and some *KRAS* WT patients benefited from treatment. In patients who received BSC only, there was no difference in OS between *KRAS* WT and mutant patients. The data did not clearly demonstrate whether *KRAS* mutant status was an adverse prognostic factor.

In data from the NCCTG Intergroup Phase III trial (N0147), in which patients with resected stage III colon cancer were treated with adjuvant modified FOLFOX6 with or without cetuximab, *KRAS* mutant patients did worse in both arms (Goldberg et al. 2010). In a prospective analysis of a large cohort of patients from PETACC-3, an adjuvant trial with resected stage II–III colon cancer, *KRAS* mutation status did not appear to have any prognostic value in patients with resected colon cancer treated with adjuvant chemotherapy. The *KRAS* mutation rate was 37 % and *KRAS* mutations did not have any significant prognostic role in determining relapse-free survival or OS (Roth et al. 2010).

Although the COIN trial overall was a negative trial, there was a prognostic effect of mutated oncogenes as *BRAF* and *KRAS/NRAS* mutant patients had worse OS irrespective of treatment (Maughan et al. 2011). When Ogino et al. analyzed the

Table 8.2 *KRAS* mutation status as a prognostic marker in CRC (non-EGFR treatment arms/studies)

Study	Stage of CRC	Outcome in comparison to *KRAS* WT patients	References
PETACC III	Stages II–III	No significant survival difference	Roth et al. (2010)
CALGB 89803	Stage III	No significant survival difference	Ogino et al. (2009a)
N0147	Stage III	No significant survival difference	Goldberg et al. (2010)
QUASAR 1	Stage II	No significant survival difference	Hutchins et al. (2011)
CO.17	Metastatic	No significant survival difference (in BSC arms)	Amado et al. (2008)
FOCUS	Metastatic	Worse overall survival	Richman et al. (2009)
COIN	Metastatic	Worse overall survival in all arms	Maughan et al. (2011)

effect of *KRAS* mutation in stage III colon cancer patients who were treated with 5-FU and leucovorin with or without irinotecan, they did not see any significant difference in the 5-year disease-free or OS between *KRAS* mutant or *KRAS* WT groups (Ogino et al. 2009a). Several other studies have also not found a prognostic role for *KRAS* mutation in the metastatic setting on OS (Price et al. 2011). However, a more recent analysis of *KRAS* mutation status in a large cohort of *BRAF* WT CRC patients suggested that *KRAS* mutations in codon 12, but not in codon 13, appeared to be associated with inferior survival (Imamura et al. 2012). At this time, the data do not appear to support a role for use of *KRAS* mutation status on guiding the need for adjuvant chemotherapy in resected stage II or III colon cancer. Furthermore, *KRAS* mutation status does not appear to offer a strong prognostic role given the mixed data to date. Further studies are needed to clarify the prognostic role of specific *KRAS* mutations.

8.4 NRAS

NRAS, another member of the RAS proto-oncogene family, is also mutated in CRCs, but at a much lower frequency than *KRAS* (2.2–9 %) (Irahara et al. 2010; TCGA 2012). In a study of 225 patients with CRC, Irahara et al. found only 5 patients (2.2 %) with *NRAS* mutations, leading them to conclude that *NRAS* mutations are rare in CRC. All of these tumors were located in the distal colon, sigmoid, or rectum. There appeared to be an association with female sex, although the overall number of patients was small (Irahara et al. 2010).

The prognostic significance of *NRAS* mutations remains to be clearly defined, but current evidence suggests an association with worse overall survival (Wang et al. 2013) as well as resistance to anti-EGFR therapy. Several studies have now presented evidence that patients with *NRAS* mutations do not respond to cetuximab (De Roock et al. 2009, 2010) or panitumumab (Peeters et al. 2010b).

8.5 BRAF

BRAF is a serine-threonine kinase and a member of the RAF family of kinases that acts downstream of RAS. Activating mutations in *BRAF* (e.g., V600E) result in constitutive activation of the MAPK signaling pathway. The *BRAF* (V600E) mutation has been found in 8–12 % of colon cancers. *BRAF* and *KRAS* mutations are mutually exclusive. Several studies have reported an association with *BRAF* mutation status and a worse prognosis. In a study of patients with stage II colon cancer treated with surgery alone and in stage III patients who received 5-fluorouracil in combination with leucovorin chemotherapy, Farina-Sarasqueta et al. found in a multivariate analysis that *BRAF* mutation was a prognostic factor for lower OS (Farina-Sarasqueta et al. 2010). In the PETACC-3 study in stage II and III patients,

BRAF mutations occurred in 7.9 % of tumors and were prognostic for OS but not prognostic of relapse-free survival (Roth et al. 2010). Selective inhibitors of mutant *BRAF* (V600E) have demonstrated remarkable responses in melanoma patients and the first agent (PLX4032, vemurafenib) for clinical use was approved for use in metastatic melanoma in 2011.

8.6 *BRAF* and Metastatic CRC

In a retrospective analysis of patients with mCRC treated with capecitabine, oxaliplatin, and bevacizumab (CB), with and without cetuximab, the authors found an 8.7 % frequency of the *BRAF* V600E mutation. Patients with *BRAF* mutations in both treatment groups did worse compared to *BRAF* WT patients and had lower PFS and lower OS. There was no difference in response rate (Tol et al. 2010). In another retrospective analysis of patients with mCRC, the mutation status in tumors from cetuximab- or panitumumab-treated patients were evaluated for the presence of *BRAF* mutations (Di Nicolantonio et al. 2008). The *BRAF* V600E mutation was detected in 14 % of *KRAS* WT patients (11/79) that were also non-responders to anti-EGFR therapy. No *BRAF* mutations were detected in *KRAS* WT patients that responded to anti-EGFR therapy and none of the responders to treatment carried the *BRAF* mutation. The authors concluded that WT *BRAF* was required for responsiveness to panitumumab or cetuximab. Patients with mCRC with *BRAF* mutations were associated with a lower PFS and OS than *BRAF* WT/*KRAS* WT patients. These results have led to the conclusion that *BRAF* mutation is a poor prognostic biomarker in mCRC. These studies have led investigators to consider that *BRAF* mutation status be considered in determining utility of anti-EGFR therapy, in addition to *KRAS* mutation status.

BRAF mutation appears to be most commonly associated with right-sided colon cancers (Farina-Sarasqueta et al. 2010). In an analysis of 54 *BRAF* mutant stage II and III colon cancers, 92.5 % (50) were classified as right-sided colon cancers (Farina-Sarasqueta et al. 2010). In an analysis of mCRCs, among *BRAF* mutant tumors, 68 % (39/57) were right-sided primary tumors (Tran et al. 2011). MSI has also been reported to be associated with *BRAF* mutant tumors (Tran et al. 2011).

The approval of vemurafanib for treatment of *BRAF* V600E associated metastatic melanoma raised the possibility of treatment of *BRAF* V600E colon cancers by targeting mutant *BRAF*. A Phase I study was conducted using PLX4032 in 23 mCRC patients confirmed to have *BRAF* V600E. The data presented to date suggest modest activity against these tumors with this agent with only one partial response and a median PFS of 3.7 months (Kopetz et al. 2010). At this time it appears that targeting the *BRAF* V600E alone is not sufficient for meaningful clinical efficacy. Targeted inhibition of several signaling pathways may be necessary to achieve clinical efficacy in *BRAF* mutant colon cancer. Recent in vitro studies suggest that resistance to *BRAF* inhibition in colon cancer cells may be mediated by EGFR signaling and suggest that inhibition of EGFR signaling by tyrosine kinase inhibitors may be a possible target in *BRAF* mutant colon cancer (Prahallad et al. 2012).

8.7 PIK3CA

PIK3CA, the gene encoding the p110α PI3K catalytic subunit, is estimated to be mutated in up to ~20 % of CRCs. Mutations in *PIK3CA* occur largely in exon 9 and exon 20 and result in constitutive activation of the PI3K pathway. Current limited evidence does not support a prognostic role for *PIK3CA* mutation in CRC. However, *PIK3CA* likely plays a predictive role to certain therapies and could be predictive for local failure and may be associated with recurrence. In a large analysis of mCRC tumors, De Roock et al. found that *PIK3CA* mutations in exon 20 may provide clinically meaningful information with respect to lack of response to anti-EGFR therapy. In their study, patients with exon 20 *PIK3CA* mutations had lower OS compared to *PIK3CA* WT (De Roock et al. 2010).

Additionally, a secondary analysis of the Dutch TME trial demonstrated that *PIK3CA* mutation correlated with increased risk of local recurrence in patients treated with surgery alone (He et al. 2009). Furthermore, these mutations are potentially targetable with targeted therapies. Thus, understanding whether these mutations impact response to therapy, local control, and survival is important for establishing a baseline for future genotype directed studies (He et al. 2009).

In an analysis of 110 patients with mCRC treated with anti-EGFR therapy, Sartore-Bianchi et al. identified 13.9 % bearing *PIK3CA* mutations. Patients with *PIK3CA* mutations did not respond to anti-EGFR therapy. Unlike the *BRAF* V600E mutation, there was no association between *PIK3CA* mutation and location of primary tumor (Sartore-Bianchi et al. 2009). In this study, neither *PIK3CA* nor *KRAS* mutations were associated with OS. Taking into account these data, the authors suggest that up to 70 % of patients who do not respond to anti-EGFR therapy could be accounted for by mutation status.

In an analysis of resected stage I–III colon cancer patients, Ogino et al. detected *PIK3CA* mutations in 18 % of the tumors. Although there was no difference in OS, *PIK3CA* mutated tumors appeared to be associated with worse cancer-specific survival. However, there appeared to be a difference between patients with *KRAS* WT and *KRAS* mutated tumors. Among those with *PIK3CA* mutations, there was a significant increase in cancer-specific mortality in *KRAS* WT patients but *PIK3CA* mutations did not appear to be associated with any difference in cancer-specific mortality in *KRAS* mutant colon cancer patients (Ogino et al. 2009b).

In a large analysis of CRC patients, Liao et al. suggested that concomitant *PIK3CA* mutations in exons 9 and 20 may be associated with inferior survival while a mutation in either exon alone was not associated with survival (Liao et al. 2012a). More recent work has shown that patients with mutated-*PIK3CA* CRCs who received regular aspirin use after diagnosis were associated with improved overall survival and cancer-specific survival. Patients with CRCs with WT *PIK3CA* who had regular aspirin use did not show any association with improved cancer-specific or overall survival (Liao et al. 2012b). This intriguing interaction between aspirin use and *PIK3CA* mutations requires further investigation.

Targeted therapy for *PIK3CA* mutations is currently undergoing clinical trials. The promise of small molecules targeting *PIK3CA* in CRC remains to be determined.

8.8 The Interaction of *KRAS* and Radiation Therapy

As with any therapy, the response to radiation in the clinical setting can be variable. As one of the first oncogenes identified in human tumors, *KRAS* was an obvious candidate for evaluation as a putative cause for radiation resistance. Multiple studies have implicated *ras* in radioresistance by showing that transfection of an activated *ras* allele into rodent cells made them radioresistant (Bernhard et al. 1998, 2000; Brunner et al. 2003, 2005; Cengel et al. 2007; Gupta et al. 2001; McKenna et al. 2003). In human colon cancer, knocking out the activated *ras* allele restored radiosensitization (Bernhard et al. 2000).

For rectal cancer, only limited small studies have evaluated the impact of *KRAS* status on tumor regression with chemoradiation. Bengala and colleagues conducted a study of 40 patients treated with cetuximab-based chemoradiation and analyzed the study by *KRAS* status (Bengala et al. 2009). Of the 39 assessable patients, 30 had tumors that were *KRAS* WT and 9 tumors were *KRAS* mutant. Eleven of 30 *KRAS* WT tumors had a robust response to chemoradiation, with a tumor regression grade (TRG) 3–4, while only 1 of 9 *KRAS* mutant tumors had a TRG 3–4. Another cetuximab-based chemoradiation study (Debucquoy et al. 2009) evaluated 41 patient tumors (~30 % *KRAS* mutant) and showed a trend toward better response using Wheeler score ($P=0.09$). A study evaluating mutational status of genes implicated in CRC development also evaluated the impact of *KRAS* status on response to conventional chemoradiation (Zauber et al. 2009). Fifty-three patients were evaluated, of whom 34 % had *KRAS* mutant tumors. Stage I (yp T0-2, N0) regression was seen in 49 % of the *KRAS* WT tumors, compared with 33 % of the *KRAS* mutant tumors. This result was not statistically significant, owing to the small numbers of patients evaluated. However, in all studies to date, response is lower in *KRAS* mutant tumors.

The largest evaluation of *KRAS* status and non-cetuximab containing chemoradiation for rectal cancer evaluated 132 patients with rectal cancer homogeneously treated with standard radiation with 5-FU (Garcia-Aguilar et al. 2011). Twenty-three candidate genes were evaluated by Sanger sequencing. Of the 23 candidate genes, mutations in three genes were associated with non-pCR: *KRAS*, cyclin D1, and methylenetetrahydrofolate reductase (MTHFR). MTHFR is a gene associated with 5FU metabolism. Cyclin D1 mutations occurred at a low frequency (14 %). *KRAS* mutations were more frequently found in the non-pCR group (49 %) than in the pCR group (24 %) ($P=0.0145$).

8.9 Future Directions

In CRC, the potential for targeted therapy has not yet been fully realized. Evidence to date has not supported *BRAF* inhibitors to be effective as monotherapies in *BRAF* V600E CRCs, in contrast to their clear efficacy in *BRAF* V600E metastatic melanomas. It remains a distinct possibility that other genetic or epigenetic alterations in concert with *BRAF* V600E mutation will need to be inhibited in order for critical pathway disruption and clinical benefit. Therapeutic drugs targeting *PIK3CA* mutations that have been characterized to date are undergoing clinical trials in multiple tumor types including breast and CRC.

As multiple signaling pathways are dissected, the promise of targeted therapy remains in the goal of finding "actionable" mutations in colon cancer has become increasingly important. Several large-scale sequencing projects including The Cancer Genome Atlas will likely identify genetic variants for which existing drugs can target or that can be druggable. Whether tumor sequencing will become a part of routine clinical care will likely depend on the utility of the genetic alterations found and the validation that alterations found can lead to meaningful clinical benefit for patients. Whether targeting a single pathway with a single drug or multiple pathways with multiple drugs will be more effective remains to be determined. The demonstration of drug resistance in *BRAF* V600E melanomas has led to drug trials aimed at co-targeting *BRAF* and MEK.

In mutant *KRAS* CRC, it is thought that mutant *KRAS* can activate both ERK and PI3K signaling, thereby rendering resistance to targeting of receptor tyrosine kinases. As a result, it is hypothesized that targeting both the MEK and PI3K pathways will be necessary for an effective cancer treatment, at least in mutant *KRAS* tumors (Ebi et al. 2011). Whether combination targeted therapy of activated pathways in CRC will result in improvement in clinical outcomes remains to be seen. The discovery of actionable mutations, that is, mutations that lead to choosing one treatment regimen over another or enrollment on a clinical trial for a targeted agent will continue to remain an active area of research in CRC.

References

Adam IJ et al (1994) Role of circumferential margin involvement in the local recurrence of rectal cancer. Lancet 344(8924):707–711

Aklilu M, Eng C (2011) The current landscape of locally advanced rectal cancer. Nat Rev Clin Oncol 8(11):649–659

Amado RG et al (2008) Wild-type KRAS is required for panitumumab efficacy in patients with metastatic colorectal cancer. J Clin Oncol 26(10):1626–1634

Bardelli A, Siena S (2010) Molecular mechanisms of resistance to cetuximab and panitumumab in colorectal cancer. J Clin Oncol 28(7):1254–1261

Bengala C et al (2009) Epidermal growth factor receptor gene copy number, K-ras mutation and pathological response to preoperative cetuximab, 5-FU and radiation therapy in locally advanced rectal cancer. Ann Oncol 20(3):469–474

Bernhard EJ et al (1998) Inhibiting Ras prenylation increases the radiosensitivity of human tumor cell lines with activating mutations of ras oncogenes. Cancer Res 58(8):1754–1761

Bernhard EJ et al (2000) Direct evidence for the contribution of activated N-ras and K-ras oncogenes to increased intrinsic radiation resistance in human tumor cell lines. Cancer Res 60(23):6597–6600

Bokemeyer C et al (2009) Fluorouracil, leucovorin, and oxaliplatin with and without cetuximab in the first-line treatment of metastatic colorectal cancer. J Clin Oncol 27(5):663–671

Bokemeyer C et al (2011) Efficacy according to biomarker status of cetuximab plus FOLFOX-4 as first-line treatment for metastatic colorectal cancer: the OPUS study. Ann Oncol 22(7):1535–1546

Bronte G et al (2011) EGFR genomic alterations in cancer: prognostic and predictive values. Front Biosci (Elite Ed) 3:879–887

Brunner TB et al (2003) Farnesyltransferase inhibitors: an overview of the results of preclinical and clinical investigations. Cancer Res 63(18):5656–5668

Brunner TB et al (2005) Pancreatic cancer cell radiation survival and prenyltransferase inhibition: the role of K-Ras. Cancer Res 65(18):8433–8441

Bublil EM, Yarden Y (2007) The EGF receptor family: spearheading a merger of signaling and therapeutics. Curr Opin Cell Biol 19(2):124–134

Cengel KA et al (2007) Oncogenic K-Ras signals through epidermal growth factor receptor and wild-type H-Ras to promote radiation survival in pancreatic and colorectal carcinoma cells. Neoplasia 9(4):341–348

Citri A, Yarden Y (2006) EGF-ERBB signalling: towards the systems level. Nat Rev Mol Cell Biol 7(7):505–516

Dahabreh IJ et al (2011) Systematic review: anti-epidermal growth factor receptor treatment effect modification by KRAS mutations in advanced colorectal cancer. Ann Intern Med 154(1): 37–49

De Roock W, Lambrechts D, Tejpar S (2009) K-ras mutations and cetuximab in colorectal cancer. N Engl J Med 360(8):834; author reply 835–836

De Roock W et al (2010) Effects of KRAS, BRAF, NRAS, and PIK3CA mutations on the efficacy of cetuximab plus chemotherapy in chemotherapy-refractory metastatic colorectal cancer: a retrospective consortium analysis. Lancet Oncol 11(8):753–762

Debucquoy A et al (2009) Molecular response to cetuximab and efficacy of preoperative cetuximab-based chemoradiation in rectal cancer. J Clin Oncol 27(17):2751–2757

Debucquoy A et al (2010) Integration of epidermal growth factor receptor inhibitors with preoperative chemoradiation. Clin Cancer Res 16(10):2709–2714

Di Nicolantonio F et al (2008) Wild-type BRAF is required for response to panitumumab or cetuximab in metastatic colorectal cancer. J Clin Oncol 26(35):5705–5712

Douillard JY et al (2010) Randomized, phase III trial of panitumumab with infusional fluorouracil, leucovorin, and oxaliplatin (FOLFOX4) versus FOLFOX4 alone as first-line treatment in patients with previously untreated metastatic colorectal cancer: the PRIME study. J Clin Oncol 28(31):4697–4705

Ebi H et al (2011) Receptor tyrosine kinases exert dominant control over PI3K signaling in human KRAS mutant colorectal cancers. J Clin Invest 121(11):4311–4321

Eng C (2010) The evolving role of monoclonal antibodies in colorectal cancer: early presumptions and impact on clinical trial development. Oncologist 15(1):73–84

Farina-Sarasqueta A et al (2010) The BRAF V600E mutation is an independent prognostic factor for survival in stage II and stage III colon cancer patients. Ann Oncol 21(12):2396–2402

Folkesson J et al (2005) Swedish Rectal Cancer Trial: long lasting benefits from radiotherapy on survival and local recurrence rate. J Clin Oncol 23(24):5644–5650

Garcia-Aguilar J et al (2011) Identification of a biomarker profile associated with resistance to neoadjuvant chemoradiation therapy in rectal cancer. Ann Surg 254(3):486–492; discussion 492–3

Goldberg RM, Sargent DJ, Thibodeau SN, Mahoney MR, Shields AF, Chan E, Gill S, Kahlenberg MS, Nair S, Alberts SR (2010) Adjuvant mFOLFOX6 plus or minus cetuximab (Cmab) in patients (pts) with KRAS mutant (m) resected stage III colon cancer (CC): NCCTG Intergroup Phase III Trial N0147. J Clin Oncol 28(suppl):15s; abstr 3508

Gupta AK et al (2001) The Ras radiation resistance pathway. Cancer Res 61(10):4278–4282

He Y et al (2009) PIK3CA mutations predict local recurrences in rectal cancer patients. Clin Cancer Res 15(22):6956–6962

Hecht JR et al (2009) A randomized phase IIIB trial of chemotherapy, bevacizumab, and panitumumab compared with chemotherapy and bevacizumab alone for metastatic colorectal cancer. J Clin Oncol 27(5):672–680

Holbro T, Hynes NE (2004) ErbB receptors: directing key signaling networks throughout life. Annu Rev Pharmacol Toxicol 44:195–217

Hong TS, Clark JW, Haigis KM (2012) Cancers of the colon and rectum: identical or fraternal twins? Cancer Discov 2(2):117–121

Hutchins G et al (2011) Value of mismatch repair, KRAS, and BRAF mutations in predicting recurrence and benefits from chemotherapy in colorectal cancer. J Clin Oncol 29(10):1261–1270

Hynes NE, Lane HA (2005) ERBB receptors and cancer: the complexity of targeted inhibitors. Nat Rev Cancer 5(5):341–354

Imamura Y et al (2012) Specific mutations in KRAS codons 12 and 13, and patient prognosis in 1075 BRAF wild-type colorectal cancers. Clin Cancer Res 18(17):4753–4763

Irahara N et al (2010) NRAS mutations are rare in colorectal cancer. Diagn Mol Pathol 19(3):157–163

Jonker DJ et al (2007) Cetuximab for the treatment of colorectal cancer. N Engl J Med 357(20):2040–2048

Karapetis CS et al (2008) K-ras mutations and benefit from cetuximab in advanced colorectal cancer. N Engl J Med 359(17):1757–1765

Kopetz S, Desai J, Chan E, Hecht JR, O'Dwyer PJ, Lee RJ, Nolop KB, Saltz L (2010) PLX4032 in metastatic colorectal cancer patients with mutant BRAF tumors. J Clin Oncol 28(suppl):15s; abstr 3534)

Liao X et al (2012a) Prognostic role of PIK3CA mutation in colorectal cancer: cohort study and literature review. Clin Cancer Res 18(8):2257–2268

Liao X et al (2012b) Aspirin use, tumor PIK3CA mutation, and colorectal-cancer survival. N Engl J Med 367(17):1596–1606

Lievre A et al (2006) KRAS mutation status is predictive of response to cetuximab therapy in colorectal cancer. Cancer Res 66(8):3992–3995

Manfredi S et al (2006) Incidence and patterns of recurrence after resection for cure of colonic cancer in a well defined population. Br J Surg 93(9):1115–1122

Maughan TS et al (2011) Addition of cetuximab to oxaliplatin-based first-line combination chemotherapy for treatment of advanced colorectal cancer: results of the randomised phase 3 MRC COIN trial. Lancet 377(9783):2103–2114

McKenna WG et al (2003) The RAS signal transduction pathway and its role in radiation sensitivity. Oncogene 22(37):5866–5875

Ogino S et al (2009a) KRAS mutation in stage III colon cancer and clinical outcome following intergroup trial CALGB 89803. Clin Cancer Res 15(23):7322–7329

Ogino S et al (2009b) PIK3CA mutation is associated with poor prognosis among patients with curatively resected colon cancer. J Clin Oncol 27(9):1477–1484

Peeters M et al (2010a) Randomized phase III study of panitumumab with fluorouracil, leucovorin, and irinotecan (FOLFIRI) compared with FOLFIRI alone as second-line treatment in patients with metastatic colorectal cancer. J Clin Oncol 28(31):4706–4713

Peeters M, et al (2010) Use of massively parallel, next-generation sequencing to identify gene mutations beyond KRAS that predict response to panitumumab in a randomized, phase 3, monotherapy study of metastatic colorectal cancer (mCRC) [Abstract]. In: Proceedings of the 101st annual meeting of the American Association for Cancer Research, Washington, DC

Prahallad A et al (2012) Unresponsiveness of colon cancer to BRAF(V600E) inhibition through feedback activation of EGFR. Nature 483(7387):100–103

Price TJ et al (2011) Impact of KRAS and BRAF gene mutation status on outcomes from the phase III AGITG MAX trial of capecitabine alone or in combination with bevacizumab and mitomycin in advanced colorectal cancer. J Clin Oncol 29(19):2675–2682

Richman SD et al (2009) KRAS and BRAF mutations in advanced colorectal cancer are associated with poor prognosis but do not preclude benefit from oxaliplatin or irinotecan: results from the MRC FOCUS trial. J Clin Oncol 27(35):5931–5937

Roth AD et al (2010) Prognostic role of KRAS and BRAF in stage II and III resected colon cancer: results of the translational study on the PETACC-3, EORTC 40993, SAKK 60–00 trial. J Clin Oncol 28(3):466–474

Sartore-Bianchi A et al (2009) PIK3CA mutations in colorectal cancer are associated with clinical resistance to EGFR-targeted monoclonal antibodies. Cancer Res 69(5):1851–1857
Sebastian S et al (2006) The complexity of targeting EGFR signalling in cancer: from expression to turnover. Biochim Biophys Acta 1766(1):120–139
Sharma SV, Settleman J (2009) ErbBs in lung cancer. Exp Cell Res 315(4):557–571
Spano JP et al (2005) Impact of EGFR expression on colorectal cancer patient prognosis and survival. Ann Oncol 16(1):102–108
TCGA (2012) Comprehensive molecular characterization of human colon and rectal cancer. Nature 487(7407):330–337
Tol J et al (2009) Chemotherapy, bevacizumab, and cetuximab in metastatic colorectal cancer. N Engl J Med 360(6):563–572
Tol J et al (2010) Markers for EGFR pathway activation as predictor of outcome in metastatic colorectal cancer patients treated with or without cetuximab. Eur J Cancer 46(11):1997–2009
Tran B et al (2011) Impact of BRAF mutation and microsatellite instability on the pattern of metastatic spread and prognosis in metastatic colorectal cancer. Cancer 117(20):4623–4632
Van Cutsem E et al (2007) Open-label phase III trial of panitumumab plus best supportive care compared with best supportive care alone in patients with chemotherapy-refractory metastatic colorectal cancer. J Clin Oncol 25(13):1658–1664
Van Cutsem E et al (2009) Cetuximab and chemotherapy as initial treatment for metastatic colorectal cancer. N Engl J Med 360(14):1408–1417
Van Cutsem E et al (2011) Cetuximab plus irinotecan, fluorouracil, and leucovorin as first-line treatment for metastatic colorectal cancer: updated analysis of overall survival according to tumor KRAS and BRAF mutation status. J Clin Oncol 29(15):2011–2019
Wang Y et al (2013) Mutant N-RAS protects colorectal cancer cells from stress-induced apoptosis and contributes to cancer development and progression. Cancer Discov 3(3):294–307
Wheeler DL, Dunn EF, Harari PM (2010) Understanding resistance to EGFR inhibitors-impact on future treatment strategies. Nat Rev Clin Oncol 7(9):493–507
Zauber NP et al (2009) Molecular genetic changes associated with colorectal carcinogenesis are not prognostic for tumor regression following preoperative chemoradiation of rectal carcinoma. Int J Radiat Oncol Biol Phys 74(2):472–476
Zhang X et al (2006) An allosteric mechanism for activation of the kinase domain of epidermal growth factor receptor. Cell 125(6):1137–1149

Chapter 9
Colorectal Cancer Genome and Its Implications

Nickolas Papadopoulos

Abstract The iconic genetic model for colorectal cancer progression associated the genetic and epigenetic changes known at the time with well-defined pathological and histological stages of colorectal cancer from early benign lesions to metastatic cancer. The basic features of the model transcended colorectal cancer and were emulated in other cancer types. Almost 20 years later, colorectal cancers were the first cancers in which all of the protein-coding genes were sequenced, heralding the era of global identification of genetic landscapes of cancers. Recently, the whole genome in a series of colorectal cancers was sequenced, adding more information to the landscape of these common cancers. An important goal of determining the genetic landscape of cancers is to provide information that can be translated into clinical applications. This chapter will examine the genetic complexity of colorectal cancers, and the implications of this information on current and future targeted therapies, as well as in the development of diagnostics and early detection.

Keywords Sequencing • Genomics • Cancer genetics • Genetic landscape • Targeted therapy • Diagnostics

9.1 Introduction

Colorectal cancer (CRC) is a major contributor to cancer mortality and morbidity. CRC has the third highest incidence, excluding basal and squamous skin cancer. The estimated new cases in 2012 are 143,460 in the USA and close to a million

N. Papadopoulos, Ph.D. (✉)
Department of Oncology, Ludwig Center for Cancer Genetics & Therapeutics,
Sidney Kimmel Comprehensive Cancer Center, The Johns Hopkins Institutions,
CRB1, Room 585, 1650 Orleans Street, Baltimore, MD 21231, USA
e-mail: npapado1@jhmi.edu

K.M. Haigis (ed.), *Molecular Pathogenesis of Colorectal Cancer*,
DOI 10.1007/978-1-4614-8412-7_9,

worldwide, while the estimated deaths in the USA are 51,690 (Siegel et al. 2012). In the USA there are approximately 1.1 million alive people who had a history of CRC. The etiological factors underlying CRC development appear to be complex and include environmental, behavioral, and genetic causes. Approximately 15 % of CRCs are associated with inherited predispositions or familial clustering (Kinzler and Vogelstein 1996; Lynch and de la Chapelle 2003). Irrespective of the etiology, what we have learned the last 30 years is that CRC and cancer overall is, in essence, a genetic disease caused by the sequential accumulation of genetic changes in tumor suppressors and oncogenes. Genetic and epigenetic alterations in these genes drive the process by providing selective advantage to the cancer cells over neighboring cells. The molecular genetics of CRC have been elegantly described elsewhere (Markowitz and Bertagnolli 2009; Fearon 2011). This chapter will focus on the studies that helped build the progression model, the genomic studies that provided the genetic landscape of CRC, its interpretation, and its clinical implications.

9.2 Progression Model of Colorectal Tumorigenesis

Colorectal tumors provide a unique system for studying the genetics of human carcinogenesis. Histophathological data suggested that the vast majority of the carcinomas (malignant tumors) arise from adenomas (benign tumors) (Sugarbaker et al. 1985) allowing the comparison of the genetic status of the two states. Furthermore, tumors of various stages of development—from small polyps, intermediate adenomas, to large metastatic cancers—can be readily obtained. The clonal nature of tumors was another critical feature of the somatic mutation-based clonal evolution of carcinogenesis (Nowell 1976). According to this model, a single cell acquires a mutation that provides a growth advantage, allowing its progeny to outgrow the neighboring cells. Later, a single cell derived from these progeny acquires another mutation, in a different gene, allowing further clonal expansion. This process continues for additional rounds of clonal expansion. That colorectal tumors were clonal was first demonstrated by techniques looking at X chromosome inactivation. Only a single copy of chromosome X is activated in a somatic cell and, although the inactivation is established in random during embryogenesis, it is transmitted in a stable manner in the progeny. Using such analysis in a female population, colorectal tumors exhibited a monoclonal pattern of X inactivation, while normal colonic mucosa did not (Vogelstein et al. 1985, 1987). Chromosomal losses in cancer cells were also shown to be clonal (Fearon et al. 1987), consistent with cytogenetic analyses that had demonstrated clonal chromosomal abnormalities (Mitelman et al. 1974; Reichmann et al. 1981). Finally, the identification of mutations in oncogenes and tumor suppressor genes provided conclusive proof. The identification of *KRAS* mutations in 50 % of CRCs was the first molecular genetics breakthrough in colorectal tumors (Bos et al. 1987; Forrester et al. 1987). Losses of chromosomal regions were believed to harbor tumor suppressor genes. Chromosome arms 5q, 8p, 17p, and 18q were the most common losses associated with CRC (Monpezat et al. 1988;

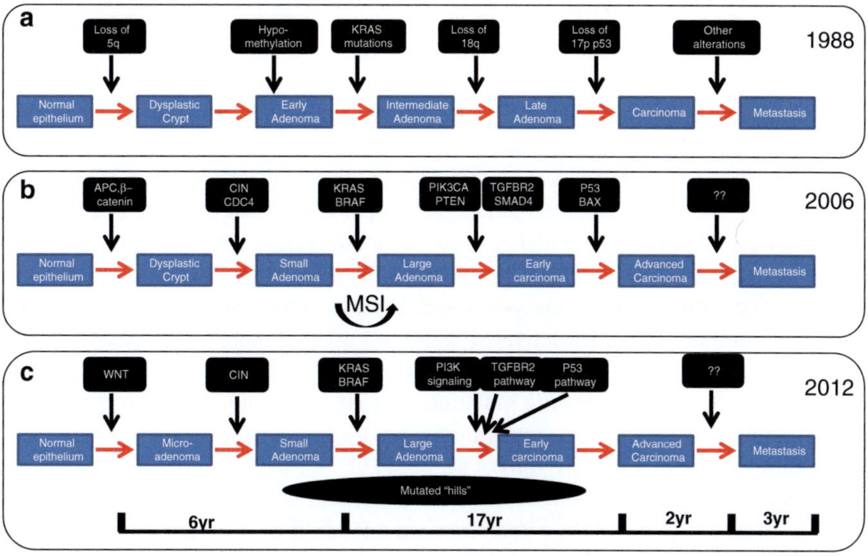

Fig. 9.1 The progression of colorectal cancer. More information has been added through the years; however the fundamentals of the model have not changed. Panels (**a**), (**b**) and (**c**) show the evolution of the model over time

Solomon et al. 1987; Vogelstein et al. 1989). These known genetic changes were evaluated in a series of adenomas of different stages and carcinomas. While *KRAS* mutations were present in 50 % of adenomas greater than 1 cm, they were present in less than 10 % in smaller adenomas. Similarly, losses of 17p were present in 80 % of carcinomas, but they were largely absent in early adenomas, while chromosome 5q losses were equivalent in early adenomas, late adenomas, and carcinomas. This analysis provided the grounds to build the progression model of colon cancer and the association of genetic changes with specific histopathological stages of CRC development (Vogelstein et al. 1988). The transition from normal epithelium to early adenomas was associated with chromosome 5q losses, while *KRAS* mutations and 18q losses were associated with later stages of adenoma development, and finally *TP53* mutations and loss were associated with the transition from the late adenoma to carcinoma (Fig. 9.1a). Early events affect gate-keeping mechanisms that allow the process to start, while subsequent events provide selective advantage suited to microenvironmental pressures allowing the cancer cells to survive while normal cells do not.

Later studies associated the chromosomal losses mentioned above with mutations in specific genes. Some of the breakthroughs were provided from the determination of the genetic basis of hereditary forms of CRC. Genes whose mutations were inherited in the predispositions to colon cancer were also mutated in the sporadic tumors. Mutated *APC*, localized on chromosome 5q, was identified as the culprit of familial adenomatous polyposis (FAP) (Kinzler et al. 1991; Joslyn et al.

1991; Nishisho et al. 1991; Groden et al. 1991). Adenomas from FAP patients acquire a second inactivating hit on the remaining wild-type allele of *APC*, following the classic definition of a tumor suppressor gene. In sporadic adenomas and cancers, inactivating mutations of *APC* were also coupled with losses of chromosome 5q. Biallelic inactivating *APC* mutations have been identified in very early neoplastic lesions, microadenomas, and aberrant crypts in greater than 85 % of the cases tested. Thus, it is considered the gatekeeper of CRC development. Losses of chromosome 18q provided evidence for a tumor suppressor gene, the same way that losses of chromosome 17p and chromosome 5q provided evidence for *TP53* and *APC*, respectively. Detailed analysis of deletions in chromosome 18q in pancreatic cancers identified *SMAD4* inactivated in at least 30 % of pancreatic cancers (Hahn et al. 1996). Mutational analysis of *SMAD4* and *SMAD2*, also located on chromosome 18q, identified inactivating mutations in CRCs that had LOH of chromosome 18q providing clues that these genes are tumor suppressor genes inactivated in CRC (Thiagalingam et al. 1996; Riggins et al. 1997).

To identify additional genetic changes in CRCs, studies interrogated a larger number of genes. Some of the efforts focused on identifying mutations in genes that could be targetable, like kinases. In such an effort, activating mutations in *PIK3CA* were identified in 32 % of CRCs and they were associated with late stage disease, adding one more step of clonal expansion to the progression model (Fig. 9.1b) (Samuels et al. 2004). Although *PIK3CA* is mutated in a third of CRCs, overall 40 % have mutations in the PI3K signaling pathway. The cancers without *PIK3CA* mutations have mutations in *PTEN*, *PAK4*, *AKT2,* or *MARK3* (Parsons et al. 2005). Mutual exclusivity of mutations suggests that they provide equivalent growth advantage to cancer cells.

Like with FAP, identification of the genetic basis of Hereditary Non-Polyposis Colon Cancer (HNPCC) provided clues to the understanding of the genetic basis of sporadic cancers. Mismatch-repair genes mutated in the germline of HNPCC patients are also mutated in a subset of CRCs that have very high mutation burden, providing a molecular mechanism for the MSI phenotype (Fishel et al. 1993; Leach et al. 1993; Bronner et al. 1994; Papadopoulos et al. 1994; Nicolaides et al. 1994; Papadopoulos et al. 1995; Liu et al. 1996). About one-third of CRCs have inactivating mutations of *TGFBR2*. In tumors with mismatch-repair defects, TGFBR2 inactivation results from frameshift mutations occurring in a polyadenine repeat within the coding regions of *TGFBR2*. *TGFBR2* was also somatically inactivated in CRCs without mismatch-repair defects (Markowitz et al. 1995; Grady et al. 1999). Mutations in *TGFBR2* coincided with the transition of adenoma to carcinoma, the same stage of progression that *SMAD4*, a component of the TGFβ pathway, was inactivated. In a similar manner, *BRAF* was shown to be mutated in CRCs that do not have mutations in *KRAS* (Rajagopalan et al. 2002). Overall, in a study of 330 CRCs, 51 % had mutations in *KRAS* and 10 % in *BRAF*. However, *BRAF* mutations were more prevalent in cancers with mismatch-repair defects. Fifty nine percent of the mismatch-repair proficient cancers have mutations in *KRAS* and 7 % in *BRAF*. In contrast, 43 % of mismatch-repair deficient CRCs have mutations in *KRAS* and 31 % in *BRAF*. Similarly, mutations in *CTNNB1* and *APC* were mutually exclusive;

however *CTNNB1* showed higher prevalence in the mismatch-repair deficient cancers (Morin et al. 1997; Kitaeva et al. 1997; Ilyas et al. 1997; Iwao et al. 1998; Sparks et al. 1998).

Based on the above studies, the same pathways are dysregulated in mismatch-repair proficient and deficient CRCs. What is different is the prevalence of the main target of mutations in the pathways. The largest distinction between these two types of CRCs is the mode of mutations. Fifteen percent of CRCs are mismatch-repair deficient and exhibit the microsatellite instability genotype (MSI), while 85 % of CRCs exhibit chromosomal instability (CIN) (Lengauer et al. 1998; Thiagalingam et al. 2001). Genes with polyadenine or other tracks in their coding regions, like *TGFBR2*, *BAX*, and *BRAF* are targeted more frequently in tumors with MSI (Fig. 9.1b).

9.3 The Era of Unbiased Studies

The selection of genes for mutation analysis in cancer was based on information from linkage studies, identification of chromosomal abnormalities in cancers, or known functional attributes of certain genes that could fit a model. In the meantime, it was starting to become clear that genetic markers were predictive for hereditary disease, and potentially prognostic for disease progression. Furthermore, the emergence of targeted therapies necessitated a better understanding of the genetic changes contained in a cancer. Questions like the following ones needed answers. How many genes are mutated in a human cancer? How different are cancers of the same type at the genetic level? Could genetics explain the differential response to therapies of patients with seemingly identical disease? Thus, a global approach on understanding the genetic landscape of CRC was necessary.

Improvements in technology and bioinformatics, and the sequence of the human genome made it possible to examine the sequence of the whole cancer genome in an unbiased fashion. In a pair of seminal papers, all of the known protein-coding genes were sequenced in 11 human CRCs, the first ever such analysis in human cancer (Sjoblom et al. 2006; Wood et al. 2007). These studies ushered the era of whole genome sequencing in cancer genetics. The effort was a herculean one. In total, 20,857 transcripts corresponding to 18,191 genes were PCR amplified exon-by-exon in 198,098 reactions per sample and sequenced by Sanger sequencing. An expedition in discovering the genetic makeup of cancer cells, in the absence of previous information and with a "wild type" reference of the human genome assembled from few humans was underway. Samples with high cancer cell content, adequate amounts of DNA and of clonal origin were selected, like xenografts developed from primary tumors. Every somatic mutation identified was tested in the cancer lesion used to develop the xenograft to ensure that the mutation arose in the cancer and not during ex vivo growth. The number of tumors selected provided sufficient statistical power to identify commonly mutated genes in these cancers. These included *TP53*, *APC*, *KRAS*, *SMAD4*, *FBXW7*, *TGFBRII*, and *TCF7L2* (*TCF4*). Overall, mutations

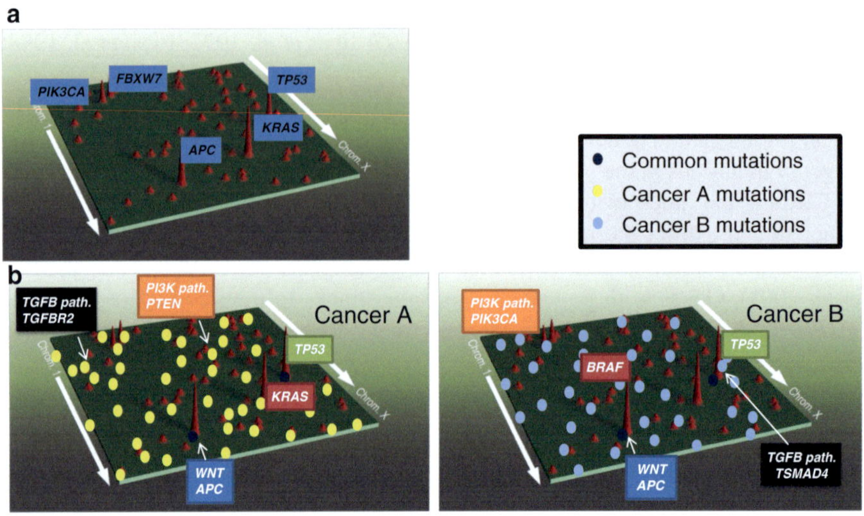

Fig. 9.2 Panel (**a**): Depiction of the colorectal genetic landscape, modified from Wood et al., Science (2007). Panel (**b**): The landscape of two different colorectal cancers. Very few common mutations, however, both have the same main pathways altered

were identified in 848 different genes. In addition to mutations, copy number variation was assessed in these samples, providing a comprehensive picture of the coding region of the genome (Leary et al. 2008). The mutated genes were organized in pathways or cellular processes in an effort to reduce the complexity of the hundreds of different genes that were mutated in CRCs (Lin et al. 2007). It can be argued that this study would have identified all the known oncogenes and tumors suppressor genes that had been discovered in a course of the previous 20 years. The landscape of the CRC genome emerged (Fig. 9.2a). The key features are: (1) there are few genes that are mutated with high prevalence, called "mountains," interspersed with many genes mutated in small fraction of CRCs, called "hills," and a lot of other genes with mutations unique to each cancer. (2) Each cancer has approximately 80 genes mutated. (3) Each cancer is genetically unique (Fig. 9.2b). (4) The most common type of mutations are point mutations of C to T. (5) The number of mutations and the number of genes involved are surprisingly high and therefore, it is not clear which of the mutations drive the processes of tumorigenesis. Based on previous information, some genes were suspected being drivers, like *GNAS*, *RET*, *TCF1* and genes that have been part of translocations, like *DDX10*, and *GLI1*.

Recently, two independent studies aimed at identifying genetic changes in CRCs. The first study was performed by The Cancer Genome Atlas (TCGA) (Cancer Genome Atlas Network 2012). The approach was a comprehensive genome-scale analysis including exome sequencing for the identification of mutations, whole genome sequencing for the identification of translocations and other rearrangements, DNA copy number analysis, promoter methylation, and messenger RNA and microRNA expression. Overall 276 samples were analyzed by at least one of the approaches. The main findings of this study were in concordance with previous findings. In the non-hypermutable tumors, the eight most frequently mutated genes were *APC*, *TP53*, *KRAS*, *PIK3CA*, *FBXW7*, *SMAD4*, *TCF7L2*, and *NRAS*, adding no new "mountains" in the landscape of CRCs. The biggest distinction of this approach compared to the one described earlier was the number of samples analyzed. Large-scale studies including hundreds of samples could identify genes that are mutated in small fractions of cancers. In this light, 7 % of cancers had mutations in *FAM123B* and 4 % in *SOX9*, a gene that has not been associated with cancer before. Not surprising, a small percent of samples had mutations in *CTNNB1* (5 %). Whole genome sequencing was also performed at low depth coverage with the goal to identify rearrangements and copy number variation, instead of exhaustively examining each base. The major regions that were amplified or deleted in this set of samples encompassed previously known loci, as well as new ones. One of the most common focal amplification involved *IGF2*. *IGF2* amplification was mutually exclusive with mutations in genes in the PI3K signaling pathway. Translocation analysis validated the findings of *VTI1A-TCF7L2* fusion present in 12 % of CRCs (Bass et al. 2011). Another recurrent fusion was observed involving *TTC28*, a gene target of *TP53*. Some of the samples analyzed were rectal cancer and allowed comparison between rectal and colon cancers. Their genetic landscapes were identical.

Sixteen percent of the cancers analyzed had a higher number of mutations and they were identified as hypermutable. These tumor samples had mutations in mismatch-repair genes, including *POLE*, or inactivation of *MLH1* by methylation, providing a validation of previous findings (Herman et al. 1998). These samples were the first hypermutable CRCs to undergo genome-scale analyses. Not surprisingly, the hypermutable tumors had more mutations in genes that have repeats within their sequence (*BAX*, *TGFBR2*). So, although it appears that the hypermutable tumors progress through a different sequence of genetic events, the pathways affected are the same as those in the stable cancers.

Integrated analysis defined five common pathways, WNT, TGFB, PIK3CA, p53, and RTK/RAS, all known to be involved in CRC development. Not all cancers from the 195 analysed had mutations in all of the pathways; however, there was not a clear correlation between certain pathways as demarcating a different progression model for colorectal tumorigenesis in a subset of CRCs.

The second study analyzed 72 pairs of primary colon cancers by next-generation sequencing to characterize their exomes, transcriptomes, and copy number alterations (Seshagiri et al. 2012). The main findings were concordant with the previous studies. The most frequently mutated genes were *APC*, *KRAS*, *TP53*, *PIK3CA*, *SMAD4*, and *FBXW7*. As in the previous study, some genes were mutated at low frequency. Mutations in TET gene families have not been reported in CRC before. The mutations reported in this study not only include inactivating mutations but also missense mutations of unclear significance. In other tumor types, inactivating *TET2* mutations have been identified in a mutually exclusive manner with *IDH1* or *IDH2* mutations. Another finding of this study was the identification of R-spondin family genes translocations.

9.4 Interpretation of Genomic Studies

So, how many genes are altered in a human CRC? Integrated analysis of gene mutations, amplifications, and deletions as presented in the three studies discussed above indicates that the number of non-silent mutations is about 80. In hypermutable CRCs the number of mutations is in the hundreds (Akhtar-Zaidi et al. 2012, 2012; Cancer Genome Atlas Network 2012). The number of mutations within a tumor is informative. Tumors of the same histological and clinical type appear to have more or less the same number of mutations. However, there is large variability between tumors of different types. For example, melanomas have a lot more mutations than CRCs, and pediatric tumors have very few mutations. The number of mutations present within a cancer depends on several things. One is the exposure to mutagens; perhaps that is why lung cancers and melanomas have many mutations compared to other tumor types. Another is the age of the tumor; the fewer generations the fewer number of mutations.

Somatic mutations can be used to model tumor development. In one study, the mutations present in the advanced malignant colorectal adenocarcinomas were also tested for their presence in other lesions from the same patient (Jones et al. 2008a). The time separating the birth of each lesion can be determined based on the number of mutations detected in benign, invasive, and metastatic cancer from the same individual, the time between divisions of the cells, and the mutation rate. The main conclusions of the study were: (1) it takes ~17 years for a large advanced adenoma to evolve into a malignant cancer, but only 2 years for cells within that cancer to acquire the ability to metastasize; (2) it requires few, if any, genetic changes to transform a highly invasive cancer to one with capacity to metastasize; (3) the rate in which point mutations develop in advanced cancers are similar to that of normal cells. These results not only have important implications for understanding human tumor pathogenesis, but perhaps more important, they indicate that there is a window between an advanced cancer in situ, which can be curable by surgery, and its metastasis.

Taken together, the compiled genomic data has revealed a complex genetic landscape for CRC. (1) There are many genes mutated in each individual tumor, but only few of them are drivers; (2) There are many genes mutated in each tumor, but only few of them are genes that are mutated at high prevalence; (3) each tumor, even of the same histopathological subtype, is genetically unique. Each CRC has ~80 non-silent mutations, but only very few, three or so, are in common with another cancer, (4) there is heterogeneity within a primary tumor as well among and within metastatic lesions. This high complexity of CRCs was somewhat surprising. However, there are ways to simplify this heterogeneity (Fig. 9.3).

With all the mutations present in a cancer, it is unlikely that each one of them contributes to tumorigenesis. Passenger mutations are best defined as those that do not confer a selective growth advantage to the cells in which they occur, while driver mutations are those that do confer a growth advantage. It is often difficult to distinguish driver mutations from passenger mutations when the mutations occur at low

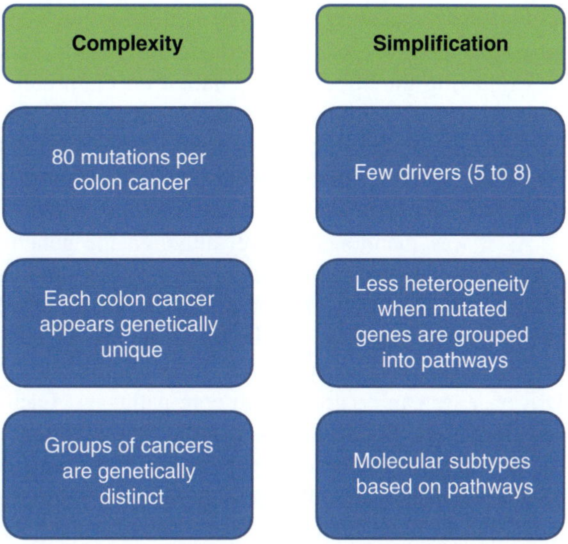

Fig. 9.3 Cancer genome complexities

frequency. From the 80 mutations in a given CRC, few drive the process, while the rest are passenger mutations that happened to be present in a clone that expanded. One way to distinguish drivers from passengers is statistical. Genes with high prevalence of mutations indicate that they are preferentially selected during the processes of tumorigenesis. Genes mutated at low frequnecy, but which have been shown to promote tumorigenesis in other tumor types, may also be drivers. The cancer gene list is constantly evolving and, although the rate of new "mountain" discovery has slowed down, the list of "hills" is increasing mainly by identifying components within pathways dysregulated in cancer (see examples in WNT and PI3K3 signaling above). Without this list complete, it is not always possible to identify a gene as a cancer gene the first time is encountered. One of the best examples of this challenge is provided by *IDH1* mutations. A single mutation of *IDH1*, R132H, was discovered in a whole exomic screen of 11 CRCs (Sjoblom et al. 2006). This mutation was not identified in more than 200 additional CRC samples and was presumed to be a passenger mutation. However, frequent *IDH1* mutations at the identical residue were found when brain tumors, such as lower grade astrocytomas and oligodendrogliomas, were evaluated (Parsons et al. 2008; Yan et al. 2009). Thus, the *IDH1* mutation in that original CRC, in retrospect, was undoubtedly a driver. This example illustrates that, once a genetic alteration is identified as a driver in one tumor type, infrequent mutations of the same type in the same gene in other tumors can be more reliably interpreted. Given that, it is now known that *ARID1A* is a bona fide tumor suppressor gene in ovarian clear cell carcinomas (Jones et al. 2010; Wiegand et al. 2010). Evaluation of *ARID1A* mutations in other tumor types identified mutations in 8 % of CRCs (Jones et al. 2012). Other genomic studies have identified a small number of mutations in other components of the protein complex that ARID1A participates, indicating that there is a subgroup of CRCs that have mutations in genes that regulate chromatin remodeling. Finally, another way to identify drivers is by evaluating the effect of a mutation on the protein of the gene. Genes that are mutated with lower prevalence but the mutations have high probability of altering the function or structure of the protein could turn out to be driver mutations. For this, there are algorithms that have been developed to determine if the mutations have the potential to alter the function of the protein (Carter et al. 2010).

As it was mentioned above, although the genetic landscape of two different CRCs have similar features, there are very few "mountains" and even fewer "hills" that are common between them. This is an unprecedented complexity in human cancers and presents several issues in the management of patients with CRC. One way to reduce this complexity is to group the mutated genes into core pathways. Driver genes can be grouped in core pathways. Each cancer has a subset of these core pathways altered by mutation in one of its components (Jones et al. 2008). These core pathways regulate the hallmark cellular processes involved in tumorigenesis (Hanahan and Weinberg 2011). The cellular pathways involved in cancer development have been described elsewhere (Vogelstein and Kinzler 2004). The theory is that any gene within the pathway, when altered by mutation, affects the pathway in an equivalent way.

There are five pathways known to be either activated or inactivated in CRCs. (1) The WNT pathway is activated in almost 100 % of CRCs. It is very likely that all CRCs have this pathway activated, but that not all of its mutated components have been identified yet, or that some mutations in the known components have been either missed or are cryptic, like promoter mutations that have not been evaluated or difficult to interpret. The main target is *APC* inactivation, but mutations in *CTNNB1*, *AXIN2*, *DKKs*, *SOX9*, *TCF7L2* have also been identified. (2) The second pathway is the *KRAS/NRAS/BRAF* pathway activated in ~50 % of CRCs. It is not yet clear if the rest of the CRCs have mutations in other components of this pathway, or mutations in these genes demarcate a subgroup of CRCs. (3) The third pathway altered in CRCs is the PI3K signaling pathway with *PIK3CA* the main target of mutations. (4) The p53 pathway is inactivated in ~70 % of CRCs. (5) Finally, the TGFβ signaling pathway is inactivated 30–40 % of CRCs. In most of the pathways, there is a main target. It is not clear why this is the case, assuming equivalence of mutations. One explanation is based on the preferred mode of mutagenesis within a cancer. For example, as was mentioned above, genes with poly-nucleotite tracks are better substrates for mutation in MSI CRCs. *TGBR2* is the main target of TGFB signaling inactivation in hypermutable tumors, while loss and inactivating mutations of *SMAD4* is the preferred mechanism in non-hypermutable tumors, which are CIN. The potential of functional differences between mutations is also possible, however, less well understood. Most *KRAS* mutations are in codon 12, but some are in codons 13, 61, and 146. Why is codon 12 more prevalent than the others? Does a codon 12 mutation provide a different advantage than codon 13 or codon 61 under different microenvironments? Interestingly, pancreatic cancers have mutations in codon 12 only. The role of *KRAS* in tumorigenesis has been explored in model systems and with functional studies (Haigis et al. 2008; Yun et al. 2009; Ying et al. 2012).

The vast majority of the new cancer genes identified through whole genome approaches belong to family of genes that are involved in regulating epigenetic processes in the cell, like changes in methylation, chromatin remodeling, and link to telomere integrity (Jones et al. 2010; Ley et al. 2010; Morin et al. 2011; Jiao et al. 2011; Wilson and Roberts 2011). Thus, genetic changes regulate the epigenetic landscapes in cancer cells. These genes are tumor suppressors and their inactivation should result in global epigenetic changes within the cancer cells. Unfortunately, it is not yet clear what the targets of such mutations are. In colon cancer, although there is a decrease in the overall methylation in the genome, the promoters of specific genes are methylated resulting in the silencing of genes that contribute to tumorigenesis (Herman et al. 1995). Comparison of the epigenetic histone mark H3K4me1 for gain and loss of transcription in a series of colon cancer cell lines and colon crypts resulted in a reproducible signature of genomic areas with altered transcription indicative of the tumor state and overlapping with the in vivo transcriptome of CRC (Akhtar-Zaidi et al. 2012). Studies correlating mutations of genes affecting epigenomic changes with expression and epigenetic profiles will be of great value for understanding the functional consequences of such mutations.

The response of cancers to therapies can be very variable and this is due, at least partly, to the underlying genetic changes within the cancer cells. Molecular profiling of cancers has led to the identification of molecular subtypes in other tumor types. Larger studies like the TCGA study on CRC have the potential to identify such subtypes, however, this was not the case for CRCs.

9.5 Clinical Value of Genomic Studies

Genomic studies provide targets for therapeutics and markers for diagnostics.

9.5.1 Targeted Therapies

Targeted therapies have been proven to work in clinical trials, however, patients' responses are heterogeneous and resistance typically ensues. Part of the issue is the lack of druggable targets for the development of new therapies. Designing targeted therapies for tumor suppressor genes is difficult and strategies like synthetic lethality need to be developed. This requires a better understanding of the pathways in which the tumor suppressors function. Oncogenes are rare in most cancers with an average of less than one mutation in an oncogene per cancer genome. *KRAS*, the most commonly mutated oncogene in CRC, has been proven to be untargetable, and it is a negative predictor for other targeted therapies. Inhibitors for *BRAF* that have worked in other tumor types with *BRAF* mutations have not yet been efficacious in CRCs. The development of *PIK3CA* inhibitors is still ongoing.

One of the main limitations of targeted therapies on kinase signaling pathways is the emergence of resistance. One of the therapeutic approaches approved for use in CRC involves the use of monoclonal antibodies against EGFR, like cetuximab or panitumumab (Jonker et al. 2007; Van Cutsem et al. 2007). Overall response to anti-EGFR therapy depends largely on the genotype of *KRAS* (Karapetis et al. 2008; Amado et al. 2008). These therapies are initially effective against cancers with wild-type *KRAS*, but patients with CRCs that carry *KRAS* mutations have intrinsic resistance. However, after the initial response, even *KRAS* wild-type cancers acquire resistance. The basis for this resistance is not well understood. Recently, two studies showed that patients with cancers expressing wild-type *KRAS* who underwent anti-EGFR therapies developed recurrent tumors that were positive for *KRAS* mutations (Misale et al. 2012; Diaz et al. 2012). Furthermore, mathematical modeling indicated that the *KRAS* mutations were not present in the primary tumor; rather they developed in the clones that metastasized (Diaz et al. 2012). This would also explain the presence of more than one *KRAS* mutation arising in the same patient after anti-EGFR therapy, indicating the development of multiple resistant clones. One of the promises of targeted therapy and the development of companion diagnostics is the identification of patients suited for a therapy (Papadopoulos et al. 2006). The information derived from genomic studies provides tools to do exactly that. However, the development of resistance is so frequent that this model is in jeopardy. With the development of sensitive methods for the detection of rare mutations, cells in the primary tumor that carry mutations that could result in resistant clones after treatment with therapies would be detected ahead of time with the hope to inform the management of the patient. However, in the example above, this is not the case as the resistant clones arose after treatment.

The issue of resistance to targeted therapies is rather common. Targeted therapies against the mutant form of *BRAFV600E* were first used in melanoma patients because *BRAF* is commonly mutated in these cancers. *BRAF* inhibitors have shown great responses in melanoma patients (Flaherty et al. 2010). However, these responses are not durable and almost all cancers come back resistant to the initially therapy. Resistant tumors have mutations downstream or in other parts of the same pathway ensuring that it remains activated, a testament to the importance of oncogenic mutations affecting pathways necessary for the growth of cancer cells. In some cases the mutations that confer resistance involve *MEK1* (Wagle et al. 2011). This proposed the use of both *BRAF* and *MEK* inhibitors as a strategy targeting the pathway at multiple points, reminiscent of the approaches of therapies for AIDS.

Recently, another type of innate resistance to *BRAF* inhibitors has been described related to the microenvironment. Growth factors, especially *HGF* that can activate *MET*, can bypass the need for *BRAF*. In a more general view, resistance to anticancer kinase inhibitors can be mediated by growth factors (Wilson et al. 2012).

The lack of response of colon cancer to *BRAF* inhibitors was paradoxical. While 80 % of melanomas of patients respond to the therapy, only 5 % of CRC patients with the same mutation show some response (Tol et al. 2009; Kopetz et al. 2010). Recently, it has been suggested in in vitro systems in melanomas that *BRAF* inhibitors could result in feedback loops that can circumvent the inhibition. In contrast to melanomas, colon cancers express EGFR and this feedback activation results in resistance to the inhibition of *BRAF* (Prahallad et al. 2012).

Targeting multiple members of the same pathway, like MEK and BRAF, is emerging as a strategy to circumvent this issue. Perhaps, an approach that targets a pathway downstream from the genes that get mutated could be more successful, at least in some situations. This will require better understanding of the pathways involved in cancer development. Genomic studies can identify the components of a pathway that are mutated. Then, association of the mutations (the genotype) of the cancer cells with other global changes in the cell (i.e., gene expression profiles) should lead to understanding of the effects of the mutations on the phenotype of the cancer cells.

Because of the issues with targeted therapies, alternative approaches based on the genotype of the cancer cells are been developed. Mutations in cancer cells result in the production of new peptides which when they are present in the circulation could be viewed as foreign antigens by the immune system. Utilizing the data from the sequencing of CRCs and with the help of in silico prediction algorithms, it was determined that colon cancer cells accumulate on average approximately ten epitopes, some on peptides from cancer genes, that are novel and unique HLA-A*0202 epitopes. This raises the possibility that with the appropriate manipulation of the immune system, the cancer cells can be attacked without previous knowledge of the specific mutations (Segal et al. 2008).

9.5.2 Early Detection

There is another way to reduce cancer deaths, early detection and prevention. Close to 50 % of cancers can be prevented by utilizing information that we already know (Colditz et al. 2012). Detection of cancers early, before they metastasize and are still curable by surgery, will have a large impact in decreasing mortality due to cancer. For CRC, colonoscopy can do that. It is the best screening tool available. However, there are issues with patient compliance, especially from healthy individuals. It is an expensive procedure and it is associated with risk of perforation of the intestine. It could miss tumors on the right colon and it is administered later in life, missing a fraction of cancers. For these reasons, an accurate test that is noninvasive, using biological fluids that can be collected in a visit to the doctor's office, can augment the number of people screened for colon cancer. In addition to increasing compliance for screening, the test can be administered earlier in life and it can be easily repeated to increase sensitivity.

Genetic analysis has provided a wealth of information to be used towards this. Mutations are specific to cancer cells, they are not just associated with the disease, they rather cause them, and they are present before metastasis. A test that is based on their detection can be a very specific and sensitive screening tool and improve on the specificity of the methylation-based test currently available. Circulating tumor DNA methods are being developed to do just that. Two issues need to be addressed. One is technical. The test needs to be able to detect a very small number of molecules that represent the mutation in a very large number of molecules that represent the wild-type sequence. The development of digital genomics has addressed this issue to a great extend. Digital genomics allow the simultaneous interrogation of independent molecules (Dressman et al. 2003; Diehl et al. 2005). One mutated molecule can be identified in the presence of at least 10,000 wild-type molecules. The other issue is biological. Studies in patients with different stage CRC and adenomas, as well as normal unaffected individuals, showed that the detection of ctDNA is feasible in a large fraction of localized disease, while is detectable in all advanced cases and not detectable in the controls, indicating that the approach is specific (Traverso et al. 2002; Lecomte et al. 2002). Advances in massively parallel sequencing, which is a digital approach, provide an opportunity to develop high throughput, cheap tests. One key issue that needs to be addressed here is the development of methods that circumvent the innate error rate of these technologies. Recently, two approaches have been developed to address this (Kinde et al. 2011; Schmitt et al. 2012). Both approaches utilize two principles to distinguish real mutations from errors. (1) Each template molecule is assigned a unique identifier, which is a unique sequence introduced in primers that are used for PCR or adaptors ligated to the template, before any manipulation of the template. This unique relationship is examined at the end of the processes. Changes on the DNA that have retained these unique relations between a mutation and a unique identifier called mutations. (2) The mutations examined in both strands of the molecule increasing confidence that the observed alterations are not artifacts.

The mutations identified in genomic projects have further value. For example, they can be used to monitor patient's tumor burden by following mutations in ctDNA. Few years ago, a proof-of-principle study showed that following tumor burden in CRC patients that were managed with different therapies, showed that ctDNA can be more predictive of tumor burden than CEA (Diehl et al. 2008). This is a personalized approach, where mutations are identified in patient's tumor and their presence is followed in the plasma of the patient throughout the clinical management period.

9.6 Perspective

Cancer is a unique disease in that its cure relies on the complete elimination of cells that have only few differences from the other cells in the body. From the differences between the normal and cancer cells, somatic mutations are what have unequivocally distinguished the cancer from the normal cells. Genomic studies are a mean to identify these differences. Towards this, great progress has been made the last few years on the understanding of the complexity of the genetic landscape of cancers. Soon all of the cancer genes will be identified, providing the raw materials for the development of new targeted therapies and novel diagnostics. However, there are still a number of issues that need to be addressed.

The great majority of the mutations that have been identified in solid tumors are passengers. As a consequence, these mutations could be lost in subsequent divisions of the cancer cells and it is important to distinguish them from the drivers. However, the number and nature of mutations can be informative, too. The number of mutations varies based on endogenous processes, for example increased mutation burden in MSI tumors, or exogenous exposures, like UV light or other exposures.

These mutations can arise during the normal development of the tissue, during the normal turnover of the tissue, or during tumorigenesis. It will be informative to identify the number of passenger mutations in different stages of CRC development and compare it with that of normal cells. Furthermore, the global signature of mutations in the genome, in conjunction with the number of mutations, can be revealing of certain exposures to mutagens.

Molecular subtyping and understanding of the interplay of pathways may help understand how to manage patients more effectively. For this to be achieved, a better understanding of the components of the pathways involved in tumorigenesis is needed. The identification of mutually exclusive mutations in different genes that appear at the same progression stage of a cancer type has helped to identify components of intracellular pathways important for cancer. The correlation of a phenotype, clinical or not, with specific alterations in a gene or pathway, coupled with functional studies in model systems can help define the pathways involved in cancer and identify the best point of attack for targeted therapies.

Within each tumor, the heterogeneity is great and understanding it at the genetic level can help us understand the clonal evolution of individual cancers. In turn, this information may prove to be important not only scientifically but also clinically by identifying and monitoring the presence of clones resistant to a therapy. The combination of massive parallel sequencing and sensitive methods for the detection of rare mutations can achieve this.

Finally, the low hanging fruit from the genomic projects is early detection, prognosis and monitoring and which can be developed now and they are agnostic to the type of mutation or function of the gene that is mutated, as long at the mutation is a driver.

These are exiting times as the field of cancer genetics is moving from discovery to clinical implementation.

References

Akhtar-Zaidi B, Cowper-Sal-lari R, Corradin O et al (2012) Epigenomic enhancer profiling defines a signature of colon cancer. Science 336(6082):736–739

Amado RG, Wolf M, Peeters M et al (2008) Wild-type KRAS is required for panitumumab efficacy in patients with metastatic colorectal cancer. J Clin Oncol 26(10):1626–1634

Bass AJ, Lawrence MS, Brace LE et al (2011) Genomic sequencing of colorectal adenocarcinomas identifies a recurrent VTI1A-TCF7L2 fusion. Nat Genet 43(10):964–968

Bos JL, Fearon ER, Hamilton SR et al (1987) Prevalence of ras gene mutations in human colorectal cancers. Nature 327(6120):293–297

Bronner CE, Baker SM, Morrison PT et al (1994) Mutation in the DNA mismatch repair gene homologue hMLH1 is associated with hereditary non-polyposis colon cancer. Nature 368(6468):258–261

Cancer Genome Atlas Network (2012) Comprehensive molecular characterization of human colon and rectal cancer. Nature 487(7407):330–337

Carter H, Samayoa J, Hruban RH, Karchin R (2010) Prioritization of driver mutations in pancreatic cancer using cancer-specific high-throughput annotation of somatic mutations (CHASM). Cancer Biol Ther 10(6):582–587

Colditz GA, Wolin KY, Gehlert S (2012) Applying what we know to accelerate cancer prevention. Sci Transl Med 4(127):127rv4

Diaz LA Jr, Williams RT, Wu J et al (2012) The molecular evolution of acquired resistance to targeted EGFR blockade in colorectal cancers. Nature 486(7404):537–540

Diehl F, Li M, Dressman D et al (2005) Detection and quantification of mutations in the plasma of patients with colorectal tumors. Proc Natl Acad Sci USA 102(45):16368–16373

Diehl F, Schmidt K, Choti MA et al (2008) Circulating mutant DNA to assess tumor dynamics. Nat Med 14(9):985–990

Dressman D, Yan H, Traverso G, Kinzler KW, Vogelstein B (2003) Transforming single DNA molecules into fluorescent magnetic particles for detection and enumeration of genetic variations. Proc Natl Acad Sci USA 100(15):8817–8822

Fearon ER (2011) Molecular genetics of colorectal cancer. Annu Rev Pathol 6:479–507

Fearon ER, Hamilton SR, Vogelstein B (1987) Clonal analysis of human colorectal tumors. Science 238(4824):193–197

Fishel R, Lescoe MK, Rao MR et al (1993) The human mutator gene homolog MSH2 and its association with hereditary nonpolyposis colon cancer. Cell 75(5):1027–1038

Flaherty KT, Puzanov I, Kim KB et al (2010) Inhibition of mutated, activated BRAF in metastatic melanoma. N Engl J Med 363(9):809–819

Forrester K, Almoguera C, Han K, Grizzle WE, Perucho M (1987) Detection of high incidence of K-ras oncogenes during human colon tumorigenesis. Nature 327(6120):298–303

Grady WM, Myeroff LL, Swinler SE et al (1999) Mutational inactivation of transforming growth factor beta receptor type II in microsatellite stable colon cancers. Cancer Res 59(2):320–324

Groden J, Thliveris A, Samowitz W et al (1991) Identification and characterization of the familial adenomatous polyposis coli gene. Cell 66(3):589–600

Hahn SA, Schutte M, Hoque AT et al (1996) DPC4, a candidate tumor suppressor gene at human chromosome 18q21.1. Science 271(5247):350–353

Haigis KM, Kendall KR, Wang Y et al (2008) Differential effects of oncogenic K-Ras and N-Ras on proliferation, differentiation and tumor progression in the colon. Nat Genet 40(5):600–608

Hanahan D, Weinberg RA (2011) Hallmarks of cancer: the next generation. Cell 144(5):646–674

Herman JG, Merlo A, Mao L et al (1995) Inactivation of the CDKN2/p16/MTS1 gene is frequently associated with aberrant DNA methylation in all common human cancers. Cancer Res 55(20):4525–4530

Herman JG, Umar A, Polyak K et al (1998) Incidence and functional consequences of hMLH1 promoter hypermethylation in colorectal carcinoma. Proc Natl Acad Sci USA 95(12): 6870–6875

Ilyas M, Tomlinson IP, Rowan A, Pignatelli M, Bodmer WF (1997) Beta-catenin mutations in cell lines established from human colorectal cancers. Proc Natl Acad Sci USA 94(19): 10330–10334

Iwao K, Nakamori S, Kameyama M et al (1998) Activation of the beta-catenin gene by interstitial deletions involving exon 3 in primary colorectal carcinomas without adenomatous polyposis coli mutations. Cancer Res 58(5):1021–1026

Jiao Y, Shi C, Edil BH et al (2011) DAXX/ATRX, MEN1, and mTOR pathway genes are frequently altered in pancreatic neuroendocrine tumors. Science 331(6021):1199–1203

Jones S, Chen WD, Parmigiani G et al (2008a) Comparative lesion sequencing provides insights into tumor evolution. Proc Natl Acad Sci USA 105(11):4283–4288

Jones S, Zhang X, Parsons DW et al (2008b) Core signaling pathways in human pancreatic cancers revealed by global genomic analyses. Science 321(5897):1801–1806
Jones S, Wang TL, Shih Ie M et al (2010) Frequent mutations of chromatin remodeling gene ARID1A in ovarian clear cell carcinoma. Science 330(6001):228–231
Jones S, Li M, Parsons DW et al (2012) Somatic mutations in the chromatin remodeling gene ARID1A occur in several tumor types. Hum Mutat 33(1):100–103
Jonker DJ, O'Callaghan CJ, Karapetis CS et al (2007) Cetuximab for the treatment of colorectal cancer. N Engl J Med 357(20):2040–2048
Joslyn G, Carlson M, Thliveris A et al (1991) Identification of deletion mutations and three new genes at the familial polyposis locus. Cell 66(3):601–613
Karapetis CS, Khambata-Ford S, Jonker DJ et al (2008) K-ras mutations and benefit from cetuximab in advanced colorectal cancer. N Engl J Med 359(17):1757–1765
Kinde I et al (2011) Detection and quantification of rare mutations with massively parallel sequencing. Proc Natl Acad Sci USA 108:9530–9535
Kinzler KW, Vogelstein B (1996) Lessons from hereditary colorectal cancer. Cell 87(2):159–170
Kinzler KW, Nilbert MC, Su LK et al (1991) Identification of FAP locus genes from chromosome 5q21. Science 253(5020):661–665
Kitaeva MN, Grogan L, Williams JP et al (1997) Mutations in beta-catenin are uncommon in colorectal cancer occurring in occasional replication error-positive tumors. Cancer Res 57(20):4478–4481
Kopetz S et al. (2010) PLX4032 in metastatic colon cancer patients with mutant BRAF tumors. J Clin Oncol 28, abstract 3543
Leach FS, Nicolaides NC, Papadopoulos N et al (1993) Mutations of a mutS homolog in hereditary nonpolyposis colorectal cancer. Cell 75(6):1215–1225
Leary RJ, Lin JC, Cummins J et al (2008) Integrated analysis of homozygous deletions, focal amplifications, and sequence alterations in breast and colorectal cancers. Proc Natl Acad Sci USA 105(42):16224–16229
Lecomte T, Berger A, Zinzindohoue F et al (2002) Detection of free-circulating tumor-associated DNA in plasma of colorectal cancer patients and its association with prognosis. Int J Cancer 100(5):542–548
Lengauer C, Kinzler KW, Vogelstein B (1998) Genetic instabilities in human cancers. Nature 396(6712):643–649
Ley TJ, Ding L, Walter MJ et al (2010) DNMT3A mutations in acute myeloid leukemia. N Engl J Med 363(25):2424–2433
Lin J, Gan CM, Zhang X et al (2007) A multidimensional analysis of genes mutated in breast and colorectal cancers. Genome Res 17(9):1304–1318
Liu B, Parsons R, Papadopoulos N et al (1996) Analysis of mismatch repair genes in hereditary non-polyposis colorectal cancer patients. Nat Med 2(2):169–174
Lynch HT, de la Chapelle A (2003) Hereditary colorectal cancer. N Engl J Med 348(10):919–932
Markowitz SD, Bertagnolli MM (2009) Molecular origins of cancer: molecular basis of colorectal cancer. N Engl J Med 361(25):2449–2460
Markowitz S, Wang J, Myeroff L et al (1995) Inactivation of the type II TGF-beta receptor in colon cancer cells with microsatellite instability. Science 268(5215):1336–1338
Misale S, Yaeger R, Hobor S et al (2012) Emergence of KRAS mutations and acquired resistance to anti-EGFR therapy in colorectal cancer. Nature 486(7404):532–536
Mitelman F, Mark J, Nilsson PG, Dencker H, Norryd C, Tranberg KG (1974) Chromosome banding pattern in human colonic polyps. Hereditas 78(1):63–68
Monpezat JP, Delattre O, Bernard A et al (1988) Loss of alleles on chromosome 18 and on the short arm of chromosome 17 in polyploid colorectal carcinomas. Int J Cancer 41(3):404–408
Morin PJ, Sparks AB, Korinek V et al (1997) Activation of beta-catenin-Tcf signaling in colon cancer by mutations in beta-catenin or APC. Science 275(5307):1787–1790
Morin RD, Mendez-Lago M, Mungall AJ et al (2011) Frequent mutation of histone-modifying genes in non-Hodgkin lymphoma. Nature 476(7360):298–303

Nicolaides NC, Papadopoulos N, Liu B et al (1994) Mutations of two PMS homologues in hereditary nonpolyposis colon cancer. Nature 371(6492):75–80
Nishisho I, Nakamura Y, Miyoshi Y et al (1991) Mutations of chromosome 5q21 genes in FAP and colorectal cancer patients. Science 253(5020):665–669
Nowell PC (1976) The clonal evolution of tumor cell populations. Science 194(4260):23–28
Papadopoulos N, Nicolaides NC, Wei YF et al (1994) Mutation of a mutL homolog in hereditary colon cancer. Science 263(5153):1625–1629
Papadopoulos N, Nicolaides NC, Liu B et al (1995) Mutations of GTBP in genetically unstable cells. Science 268(5219):1915–1917
Papadopoulos N, Kinzler K, Vogelstein B (2006) The role of companion diagnostics in the development and use of mutation-targeted cancer therapies. Nat Biotechnol 24:985–995
Parsons DW, Wang TL, Samuels Y et al (2005) Colorectal cancer: mutations in a signalling pathway. Nature 436(7052):792
Parsons DW, Jones S, Zhang X et al (2008) An integrated genomic analysis of human glioblastoma multiforme. Science 321(5897):1807–1812
Prahallad A, Sun C, Huang S et al (2012) Unresponsiveness of colon cancer to BRAF(V600E) inhibition through feedback activation of EGFR. Nature 483(7387):100–103
Rajagopalan H, Bardelli A, Lengauer C, Kinzler KW, Vogelstein B, Velculescu VE (2002) Tumorigenesis: RAF/RAS oncogenes and mismatch-repair status. Nature 418(6901):934
Reichmann A, Martin P, Levin B (1981) Chromosomal banding patterns in human large bowel cancer. Int J Cancer 28(4):431–440
Riggins GJ, Kinzler KW, Vogelstein B, Thiagalingam S (1997) Frequency of Smad gene mutations in human cancers. Cancer Res 57(13):2578–2580
Samuels Y, Wang Z, Bardelli A et al (2004) High frequency of mutations of the PIK3CA gene in human cancers. Science 304(5670):554
Schmitt MW et al (2012) Detection of ultra-rare mutations by next-generation sequencing. Proc Natl Acad Sci USA 36:14508–14513
Segal NH, Parsons DW, Peggs KS et al (2008) Epitope landscape in breast and colorectal cancer. Cancer Res 68(3):889–892
Seshagiri S, Stawiski EW, Durinck S et al (2012) Recurrent R-spondin fusions in colon cancer. Nature 488(7413):660–664
Siegel R, Naishadham D, Jemal A (2012) Cancer statistics, 2012. CA Cancer J Clin 62(1):10–29
Sjoblom T, Jones S, Wood LD et al (2006) The consensus coding sequences of human breast and colorectal cancers. Science 314(5797):268–274
Solomon E, Voss R, Hall V et al (1987) Chromosome 5 allele loss in human colorectal carcinomas. Nature 328(6131):616–619
Sparks AB, Morin PJ, Vogelstein B, Kinzler KW (1998) Mutational analysis of the APC/beta-catenin/Tcf pathway in colorectal cancer. Cancer Res 58(6):1130–1134
Sugarbaker JP, Gunderson LL, Wittes RE (1985) Colorectal cacner. In: DeVita VT, Hellman S, Rosenberg SA (eds) Cancer: principles and practices of oncology. JB Lippincott, Philadelphia, pp 800–803
Thiagalingam S, Lengauer C, Leach FS et al (1996) Evaluation of candidate tumour suppressor genes on chromosome 18 in colorectal cancers. Nat Genet 13(3):343–346
Thiagalingam S, Laken S, Willson JK et al (2001) Mechanisms underlying losses of heterozygosity in human colorectal cancers. Proc Natl Acad Sci USA 98(5):2698–2702
Tol J, Nagtegaal ID, Punt CJ (2009) BRAF mutation in metastatic colorectal cancer. N Engl J Med 361(1):98–99
Traverso G, Shuber A, Levin B et al (2002) Detection of APC mutations in fecal DNA from patients with colorectal tumors. N Engl J Med 346(5):311–320
Van Cutsem E, Peeters M, Siena S et al (2007) Open-label phase III trial of panitumumab plus best supportive care compared with best supportive care alone in patients with chemotherapy-refractory metastatic colorectal cancer. J Clin Oncol 25(13):1658–1664
Vogelstein B, Kinzler KW (2004) Cancer genes and the pathways they control. Nat Med 10(8):789–799

Vogelstein B, Fearon ER, Hamilton SR, Feinberg AP (1985) Use of restriction fragment length polymorphisms to determine the clonal origin of human tumors. Science 227(4687):642–645
Vogelstein B, Fearon ER, Hamilton SR et al (1987) Clonal analysis using recombinant DNA probes from the X-chromosome. Cancer Res 47(18):4806–4813
Vogelstein B, Fearon ER, Hamilton SR et al (1988) Genetic alterations during colorectal-tumor development. N Engl J Med 319(9):525–532
Vogelstein B, Fearon ER, Kern SE et al (1989) Allelotype of colorectal carcinomas. Science 244:207
Wagle N, Emery C, Berger MF et al (2011) Dissecting therapeutic resistance to RAF inhibition in melanoma by tumor genomic profiling. J Clin Oncol 29(22):3085–3096
Wiegand KC, Shah SP, Al-Agha OM et al (2010) ARID1A mutation sin endometriosis-associated ovarian carcinomas. N Engl J Med 16:1532–1543
Wilson BG, Roberts CW (2011) SWI/SNF nucleosome remodellers and cancer. Nat Rev Cancer 11(7):481–492
Wilson TR et al (2012) Widespread potential for growth factor-driven resistance to anticancer kinase inhibitors. Nature 487:505–509
Wood LD, Parsons DW, Jones S et al (2007) The genomic landscapes of human breast and colorectal cancers. Science 318(5853):1108–1113
Yan H, Parsons DW, Jin G et al (2009) IDH1 and IDH2 mutations in gliomas. N Engl J Med 360(8):765–773
Ying H, Kimmelman AC, Lyssiotis CA et al (2012) Oncogenic Kras maintains pancreatic tumors through regulation of anabolic glucose metabolism. Cell 149(3):656–670
Yun J, Rago C, Cheong I et al (2009) Glucose deprivation contributes to the development of KRAS pathway mutations in tumor cells. Science 325(5947):1555–1559

Chapter 10
Copy-Number Alterations in the Colorectal Cancer Genome

Jihun Kim and Adam J. Bass

Abstract Colorectal cancer (CRC) has classically been divided into two main genetic/molecular subtypes; tumors characterized by chromosomal instability (CIN) and those with microsatellite instability (MSI). Although cases with MSI often have relatively bland copy-number profiles, cases characterized by CIN typically possess many somatic copy-number alterations (SCNAs). Thanks to the remarkable progress in copy-number profiling techniques with both increased resolution and sample throughput, the landscape of the SCNAs in CRC has increasingly begun to be revealed. Many of the arm-level SCNAs of CRC are shared by many epithelial cancers but some of them are unique to gut epithelial cancers or to CRC. Gain of 8q, 20p/q and loss of 17p are commonly observed across the gut adenocarcinomas. More unique to CRC are highly recurrent chromosomal gains of 13q. Important focal SCNAs include the amplifications of 8q at *MYC*, 20q around *BCL2L1*, 11p at *IGF2*, and *miR-483*, and 17q at *ERBB2*. The amplification of *ERBB2* is particularly important because it is clinically targetable. Focal loss of tumor suppressor genes such as *TP53* and *SMAD4* reflects the selective advantage of loss of these factors. Although we began to reveal the landscape of SCNA in CRC, we have yet to fully appreciate the biologic rationale and significance for this spectrum of recurrent structural alterations in the genomes of these cancers.

J. Kim
Department of Medical Oncology, Dana-Farber Cancer Institute,
Boston, MA, USA

Department of Pathology, University of Ulsan College of Medicine,
Asan Medical Center, Seoul, Korea

A.J. Bass (✉)
Department of Medical Oncology, Dana-Farber Cancer Institute,
Boston, MA, USA
e-mail: Adam_Bass@dfci.harvard.edu

K.M. Haigis (ed.), *Molecular Pathogenesis of Colorectal Cancer*,
DOI 10.1007/978-1-4614-8412-7_10,

10.1 Colorectal Cancer as a Complex Genomic Disease

Colorectal cancer (CRC) is the third leading cause of cancer death in the United States (Jemal et al. 2010). The development of CRC is recognized to follow a multistep process starting from the initial precursor lesion (adenoma), which then progresses towards dysplasia and, ultimately, adenocarcinoma. That carcinomas arise from adenomas was initially suggested by the observation that colorectal adenoma, if untreated, significantly increased the risk of CRC (Murakami et al. 1990) and that carcinomatous foci were frequently detected in colorectal adenomas, particularly in those with large size, high-grade dysplasia, or villous morphology.

This progression series that leads the normal epithelium of the colorectum to develop into cancer is driven by a number of critical alterations in the genome of colorectal epithelial cells. These genomic alterations (or mutations) are thought to cause cancer through their ability to activate certain oncogenes, imparting constitutive pro-growth functions, and to inactivate tumor suppressor genes, which normal function to restrain cellular growth. While the emergence of cancer depends upon the acquisition of a sufficient combination of these genomic alterations, individual tumors vary in both the exact genomic alterations they have and the manner in which these genomic lesions were acquired. An investigation of CRC genomes should be prefaced by a consideration of three characteristic patterns of genomic instability that occur in these cancers: chromosomal instability (CIN), microsatellite instability (MSI), and the CpG island methylator phenotype (CIMP). Among these three forms of general genomic instability, the MSI and CIMP groups often overlap and are largely (but not completely) exclusive with the CIN tumors (Goel et al. 2007; Cheng et al. 2008; Geiersbach and Samowitz 2011).

Here, we focus on the contribution of larger-scale structural alterations of the genome to the development of CRC, forms of genomic instability falling under the CIN category. The term "chromosomal instability" refers to an accelerated gains or losses of whole or large portions of chromosomes that results in karyotypic variability from cell to cell (Lengauer et al. 1998). In contrast to mutations that alter as little as a single nucleotide, i.e., those in MSI, CIN-induced events are larger-scale alterations, ranging between just several kilobases of DNA to alterations impacting an entire chromosome. In other cases, CIN can induce genomic alterations that are copy-neutral yet still grossly alter the structure of the cancer genome, through formation of rearrangements in which distinct loci in the genome are stitched together aberrantly. Within the framework of CRC, the salient point is that these genomic aberrations can enable changes to genes whose normal function is to maintain homeostatic balance in colorectal epithelial cells, thus propelling the progression of cancer.

Indeed, genomic alterations such as those caused by the CIN phenotype are well recognized as contributing to some of the steps in the multistep genetic model of colorectal carcinogenesis. While activation of certain oncogenes (e.g., *KRAS*) typically results from a point mutation, genomic amplification can also serve as a means of oncogene activation. For example, oncogenic amplification of *ERBB2*, *MYC*, and *MYB* has been observed in primary colorectal tumors and their derived cell lines (Alitalo et al. 1983, 1984; D'Emilia et al. 1989; Finley et al. 1989).

For the loss of tumor suppressor genes, copy-number loss of the corresponding chromosomal loci is a frequent mechanism of inactivation. Unlike oncogenes, genomic alterations to tumor suppressors must typically involve both copies of the gene, thus leading to the commonly observed co-occurrence of larger-scale deletions of one copy with a focal deletion, point mutation, or epigenetic alteration of the other allele. In some cases, these chromosomal alterations can involve a loss of the chromosome with the normal allele of a tumor suppressor, coupled to the acquisition of two distinct copies of the sister chromosome with the mutant copy. These cases of "copy-neutral" deletions would thus be a subset of the phenomenon of loss of heterozygosity (LOH), as the tumor's DNA would no longer have heterozygous alleles on the chromosome. Indeed, many classic tumor suppressors in colorectal adenocarcinoma are found in regions of the genome often subject to LOH in CRC. For example, *APC* is subject to allelic loss when chromosome 5q is deleted (Ashton-Rickardt et al. 1989). At the later stages of carcinogenesis, deletions of chromosome 17p and 18q are often reported (Fearon and Vogelstein 1990). Within chromosome 17p is *TP53*, the most important tumor suppressor gene and chromosome 18q harbors several candidate genes such as *SMAD4*, *DCC*, *SMAD2*, and *CABLES1* (Pino and Chung 2010).

In this chapter, genomic copy-number changes resulting from the CIN in CRCs will be discussed in depth in terms of previous findings and recent progress. Although we knew several cancer-related genes important for CRC biology, the technological advances since the completion of human genome project greatly improved our ability to detect novel cancer-related genes. The contents comprise three subtopics: (1) the molecular mechanism of CIN known so far, (2) technological advancements in investigating copy-number alterations with an emphasis on the increased resolution that greatly enable de novo discovery of focal events, (3) a recently uncovered genome-wide landscape of copy-number alterations in CRC and its clinical implications.

10.2 Mechanisms Underlying Somatic Copy-Number Alterations in Cancer

Normal somatic cells usually have two sets of chromosomes, and these two sets of chromosomes are nonidentical, as one chromosome is of paternal origin and one is of material origin. Normal cells maintain chromosomal integrity, despite the process of recurrent cellular divisions, by using several failsafe mechanisms including mitotic checkpoint, telomere end protection, and DNA damage response (Pino and Chung 2010). In cancer, mutations or gene expression changes in the genes involved in those failsafe mechanisms can lead to a permissive cellular environment in which gross chromosomal alterations can occur and are tolerated.

Proper segregation of chromatids during mitosis requires the highly coordinated action of several kinetochore and associated proteins as well as centrosome-associated kinases and this process is compromised in CRC in several ways. For example, mutations in genes involved in the mitotic spindle checkpoint such as *hZw10*, *hZwilch/FLJ10036*, and *hRod/KNTC* kinetochore proteins and in *Ding*, which is essential for proper chromosome disjunction, have been reported in CRC (Wang et al. 2004). It has been controversial whether the CIN is the cause or the consequence of cancers. However, several lines of evidence support a causal role of CIN in cancer development. The introduction of mutant *BUB1* gene to microsatellite unstable tumor cells with normal mitotic checkpoint resulted in the disruption of mitotic checkpoint leading to CIN (Cahill et al. 1998). Moreover, targeted disruption of hCDC4, a regulator of cyclin E, in karyotypically stable CRC cells caused CIN (Rajagopalan et al. 2004). In addition, the hemizygous loss of CENP-E, a centromere protein and the overexpression of MAD2 facilitated tumor formation in mice (Pellman 2007). Abnormal centrosome number has also been proposed to cause CIN and, accordingly, abnormal centrosome alignment resulted in improper attachment of mitotic spindle and chromosomal mis-segregation in some CRC cell lines (Ganem et al. 2009). Recently, micronuclei, which result from chromosome mis-segregations, reportedly undergo defective asynchronous replication and profound fragmentation followed by integration to daughter cell nuclei (Crasta et al. 2012). This mechanism can explain the recently recognized catastrophic genomic rearrangement called "chromothripsis." Finally, a potential role of centrosome-associated Aurora kinase and Polo-like kinase in CIN development has been suggested. In CRC, amplification of *AURKA* was correlated with the degree of CIN (Anand et al. 2003; Ewart-Toland et al. 2003) and overexpression of Aurora B was associated with advanced stage (Katayama et al. 1999). Overexpression of PLK1, a Polo-like kinase, correlated with an advanced clinical stage (Takahashi et al. 2003).

Telomere dysfunction has been proposed as another important cause of CIN. Telomeres are DNA-protein complexes that protect the ends of eukaryotic chromosomes from fusing and breaking during segregation. In somatic cells, telomeres usually are shortened after each round of cell division because DNA polymerase fails to completely synthesize the lagging strand at the end of each chromosome. When the length of telomeres is shortened to a critical level, the DNA damage checkpoint triggers senescence or apoptosis. Cancer cells often overcome the "telomere crisis" by activating telomerase or using other mechanisms to elongate their telomeres. If the telomere end protection is compromised, chromosomal ends are subject to breakage-fusion-bridge cycles, leading to CIN (Maser and DePinho 2002). The breakage-fusion-bridge cycles reportedly lead to genomic amplification of fragile sites (O'Hagan et al. 2002) and mice lacking *Terc*, the RNA component of telomerase, exhibited an increased frequency of aberrant crypt foci and microadenomas in the gastrointestinal tract (Rudolph et al. 2001). Accumulated evidence suggests that telomere shortening is associated with early stage CRC and precursor lesions and that telomerase activation is associated with advanced CRC (Engelhardt et al. 1997; Takagi et al. 1999; Plentz et al. 2003; Gertler et al. 2004).

Abnormalities in DNA damage response have also been implicated in the CIN of CRCs. Normal somatic cells protect themselves from genotoxic stresses by arresting the cell cycle and repairing damages. In case of irreversible damage, cells normally induce senescence or apoptosis. Several genes such as *ATM*, *TP53*, *BRCA1*, and *BRCA2* have important roles in the DNA damage responses and their mutations have been associated with well-characterized cancer syndromes. In CRCs, mutations of *TP53* are thought to have at least a permissive role in the development of chromosomal instabilities (Pino and Chung 2010) and *MRE11*, which is involved in DNA double-strand break repair, was shown to be mutated in some CRC samples (Wang et al. 2004).

Chromosomal abnormalities can be roughly divided into those broad events that lead to alterations of an entire chromosome or chromosome arm (i.e., an event which breaks a chromosome at the centromere) and the more focal events comprising less than a chromosome arm. Among the various mechanisms described above, abnormalities in mitotic checkpoint are thought to result in arm-level alterations, whereas defects in telomere maintenance or DNA damage response are associated with more focal alterations. Focal amplifications have been proposed to arise from episome formation with subsequent extra-chromosomal replication or from breakage-fusion-bridge cycles as described previously. The breakage-fusion-bridge cycle hypothesis might be supported by the observation that luminal B subtype and ERBB2-amplified subtype of breast cancers show tandem inverted repeats of the focal chromosomal region spanning *ERBB2* locus (Kwei et al. 2010). It is conceivable that if a particular region next to the *ERBB2* locus undergoes breakage-fusion-bridge cycles repeatedly after DNA replication, the region will take a form of inverted repeats of *ERBB2* locus. Meanwhile, frequent low-amplitude focal gains or losses are observed in BRCA1-associated or basal-like subtype of breast cancers. Because those cancer types are associated with mutations or deletions of *TP53* and *BRCA1* genes, this type of focal copy-number alterations might be attributed to the defects in the DNA damage repair system (Kwei et al. 2010).

Although some of the molecular mechanisms of somatic copy-number alterations (SCNAs) in cancer genomes have been proposed, much remains to be discovered. Although we do not fully understand how these events occur, we can still build off the elucidation of highly recurrent SCNAs to derive insights into the genes responsible for driving the progression of these cancers. While such a focus on the ultimate results of CIN in CRC can lead one to look past the mechanisms leading to this instability, ultimately better insights into the machinery behind these alterations will help us identify those alterations of greatest pathogenic importance to cancer. As discussed further below, we typically view the generation of SCNAs as a random process that ultimately results in recurrent genomic alterations through positive selection of aberrations leading to advantageous changes in critical oncogenes and tumor suppressors. While such selection clearly occurs, the assumption that the underlying genomic instability is random is likely incorrect. Nevertheless, statistical models that seek to identify key genomic aberrations typically assume these events are random. Should we have the capacity to more faithfully model the process of genomic instability in specific cancers, we will better be able to identify those recurrent alterations more likely targeting key cancer genes from those that are merely a byproduct of the process of instability. Despite these caveats, we can still use the study of SCNAs in CRC to derive insights into the pathophysiology and therapeutic vulnerabilities in CRC. In the next section, methods of discovering SCNAs in cancer cells will be briefly reviewed.

10.3 Methods to Study Copy-Number Alterations: Technical and Statistical Considerations

The first identification of chromosomal number abnormalities dates back to 1958, when metaphase spread analysis revealed trisomy of chromosome 21 to be a pathognomonic feature in patients with Down syndrome. The most prominent cornerstone for aneuploidy as a feature of cancer was the identification of a novel minute chromosome, an abnormally small-sized extra-chromosome, in patients with chronic myeloid leukemia (Rudkin et al. 1964). After the development of chromosomal banding techniques, more detailed chromosomal analysis became feasible. Thanks to the banding technique, the novel minute chromosome, Philadelphia chromosome, turned out to be a translocation between chromosome 9 and 22 (Rowley 1973). The spatial resolution metaphase chromosome banding at this time was approximately 10 Mb, thus representing a large area given the average chromosome size of 128 Mb. With the advancement of molecular cytogenetic techniques, the resolution became higher, allowing identification of smaller alterations and greater refinement of the potential targets of genomic alterations.

The subsequent generation of approaches for SCNA characterization was based upon the hybridization of DNA from the tumor to specific bait sequences and then quantification of the tumor DNA signal present. The most popular method among them was fluorescent in situ hybridization (FISH), an approach allowing focused measurement of a specific locus in the genome. FISH is limited, however both by the inability to look across multiple genomic loci in an unbiased fashion and by the requirement for high-quality metaphase chromosome spreads. To overcome these limitations, comparative genomic hybridization (CGH) was developed. In CGH, DNA is directly isolated from test samples and normal reference samples and the two DNA samples are labeled by different fluorescent dyes. In its original variation, two differently labeled DNA samples were pooled and allowed to competitively hybridize to target normal cell metaphase chromosomes. The detected color ratio indicated the amount of test sample DNA relative to that of reference sample DNA across each chromosomal locus, but provided only relatively low resolution.

New approaches to SCNA characterization followed the success of the human genome project. With this new knowledge of the genome sequence, it became possible to design microarrays with specific nucleotide probes scattered across the genome thus enabling quantification of the DNA content of the tumor at much higher resolution than with prior approaches. CGH approaches were adapted to microarrays (array-CGH) whereby the mixture of labeled test and reference DNA samples are hybridized to microarrays onto which many predesigned genomic DNA clones were spotted. An additional approach using a different class of DNA microarrays, single nucleotide polymorphism (SNP) arrays, has also emerged. As the SNP arrays quantify DNA content at the site of polymorphic regions of the genome, they enable measurement of LOH as well as copy-number (Bignell et al. 2004; Zhao et al. 2004). For both SNP arrays and array CGH, the resolution of analysis is limited only by the size and density of spotted DNA clones. Recently, arrays that contain approximately

Table 10.1 Methods to study copy-number alterations

Method	Scan coverage	Resolution (kb)	Translocation/ inversion detection	LOH detection	Sample throughput
Karyotyping	Genome wide	~10,000	Yes	No	Low
FISH	Targeted	~1	Yes	No	High
Array CGH	Genome wide	Up to 10	No	No	High
SNP array	Genome wide	Up to 1 (depends on the actual probe location)	No	Yes	High
Massively parallel sequencing	Genome wide	Unlimited	Yes	Yes	Low

LOH loss of heterozygosity, *FISH* fluorescence in situ hybridization, *CGH* comparative genomic hybridization, *SNP* single nucleotide polymorphism

two million oligonucleotides probes have allowed these approaches to define SCNAs comprising 10's of Kb of DNA, thus providing >100-fold increase in resolution relative to chromosomal banding. Despite all the improvements mentioned above, however, those array-based techniques cannot detect copy-neutral structural alterations such as balanced translocation and inversion.

Finally, SCNAs can also be studied with next-generation sequencing technologies that involve massive parallel sequencing of short genomic reads. These individual short-read sequences, typically obtained not from direct primer-based sequencing but from "shotgun" sequencing of genomic DNA, can be computationally aligned to the reference genome sequence. By counting the number of reads aligned to specific chromosomal regions and normalizing against background coverage, researchers can detect SCNA at finer resolution than with array-based platforms (Chiang et al. 2009; Meyerson et al. 2010). In addition to the benefit of greater resolution, genomic characterization with these newest technologies also provides the benefit of identifying translocations, inversions, and other copy-neutral structural alterations in addition to, of course, point mutations found through genomic sequencing. The techniques for the detection of SCNAs are summarized in Table 10.1 in terms of their resolution, utility, and limitations.

While the importance of developing technologies to identify SCNAs in cancer-derived DNA samples is clear, it is equally important to develop the analytic tools used to identify regions of the genome likely to harbor genes contributing to the pathophysiology of cancer. The need for these analytic techniques follows the substantial degree of background genomic alteration in many cancers. The bulk of the genomic alterations in cancer likely represent "passenger" events, which do not confer any selective advantage. Approaches for statistically analyzing these data typically are premised upon the idea that, although the development of SCNAs is random, those SCNAs that lead to the amplification of oncogenes or deletions of tumor suppressors will be preferentially selected for when they occur. Thus, considering a population of cancers, those SCNAs that contribute to cancer will occur more often than expected by chance given the background rate of genomic alterations. Following those principles, several statistical algorithms have been developed and applied to successfully identify biologically relevant targets. For example, the functional importance of *HGF* and *MET* genes in glioma harboring broad chromosome 7 amplification was identified by applying Genomic Identification of Significant Targets in Cancer, GISTIC (Beroukhim et al. 2007).

10.4 Arm-Level Copy-Number Alterations in CRC: Both Epithelial Cancer-Common and CRC-Specific Types

There are many recurrent genomic alterations affecting genes critical to the development and maintenance of CRC (Hermsen et al. 2002; Rajagopalan et al. 2003). Several studies performed using chromosome banding techniques or CGH repeatedly showed that CRCs harbored gains of 7p, 7q, 8q, 13q, and 20q and losses of 4p, 4q, 5q, 8p, 15q, 17p, and 18q (Ried et al. 1996; Meijer et al. 1998; Douglas et al. 2004; Nakao et al. 2004). Among them, gain of 7p/q and losses of 4q and 8p were observed in adenomas, whereas gains of 8q, 13q, and 20q and losses of 18q were preferentially found in carcinomas, suggesting their involvement in the progression from adenoma to carcinoma (Ried et al. 1996; Meijer et al. 1998). In addition, gains of 7p/q and 8q and losses of 4pq, 8p, and 18q were associated with advanced Dukes' stage or liver metastases (Diep et al. 2006; Hughes et al. 2006). From these data it is apparent that arm-level copy-number alterations accumulate during the course of colorectal carcinogenesis. However, no arm-level event seems to be specific for any particular stage of colorectal carcinogenesis.

More recent large-scale genome-wide studies with array-based technologies linked to statistical approaches to analyze SCNA data have built upon our understanding of the chromosome arm-level events (Martin et al. 2007; Wood et al. 2007; Beroukhim et al. 2010). Largely overlapping with data derived from non-array-based techniques for copy-number profiling, copy-number gain was observed in 1q, 7p/q, 8q, 12q, 13q, 19q, and 20p/q and copy-number loss was detected in 1p, 4q, 5q, 8q, 14q, 15q, 17p/q, 18p/q, 20p, and 22q (Fig. 10.1) (Martin et al. 2007). Among them, significantly deleted 18p/q and 17p/q loci harbored *SMAD4* and *TP53*, respectively. The detailed frequencies and q-values (levels of significance after correction for multiple hypothesis testing) of those arm-level SCNAs from a recent study (Cancer Genome Atlas Network 2012) are summarized in Table 10.2.

Among the arm-level alterations that are commonly seen in CRC, some of them are commonly seen in many cancers, while others are more characteristic of CRC in particular. Some efforts to cluster arm-level SCNA profiles have shown that there are clear similarities across many epithelial cancers, including events such as gains of 8q, 12p, 1q, 3q, 20p/q and losses of 17p, 19p, 4p/q, and 8p (Beroukhim et al. 2010). Additionally, certain events are more characteristic only of adenocarcinomas emerging from the gut (including gastric and esophageal adenocarcinomas), such as gain of 8q, 20p/q and loss of 17p. Interestingly, chromosome13q gain, a common feature of CRC, is more unique to this disease. Many other tumors have strong

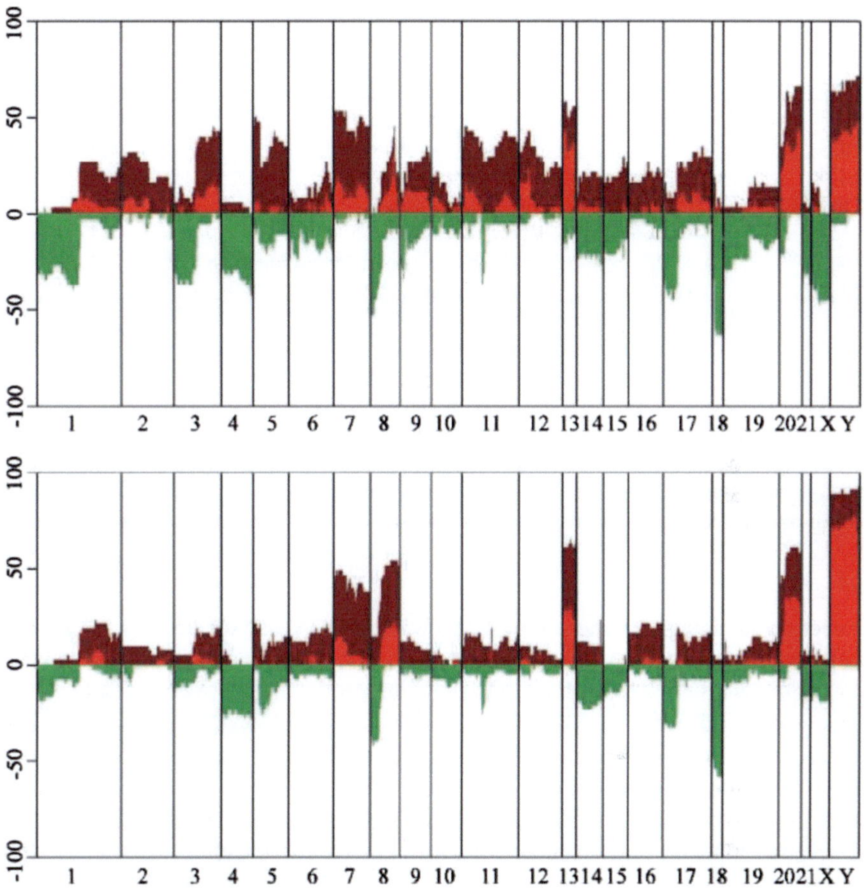

Fig. 10.1 Recurrent arm-level copy-number alterations in colorectal cancer cell lines and primary colorectal cancer samples. Recurrent copy-number alterations in 39 colorectal cancer cell lines (*top*) and in 42 primary colorectal cancer samples (*bottom*). *Red color* represents gained regions and *green color* represents lost regions. Data from cell lines and primary cancer samples correlate well with each other. Although focal events cannot be recognized mainly due to the limitations in resolution, recurrent broad gain or loss is clearly seen. Note the gains of 7pq, 8q, 13q, and 20q and the losses of 4pq, 5q, 8p, 15q, 17p, and 18q (reprinted by the permission from the American Association for Cancer Research: Martin ES et al. *Cancer Res* 2007;67(22):10736–43. DOI: 10.1158/0008-5472.CAN-07-2742)

selective pressures to delete chromosome 13, a finding often attributed to the presence of the tumor suppressor gene *RB* on this chromosome. However, CRCs, which largely depend on canonical β-catenin pathway for their survival, might require gain of chromosome 13q to keep *RB* gene intact and/or to increase the levels of *CDK8* as *RB* and *CDK8* may activate canonical β-catenin pathway by suppressing E2F1 transcription factor that would otherwise repress β-catenin (Morris et al. 2008).

Given the recurrent nature of arm-level or chromosome-level alterations in the development of colorectal adenocarcinoma, these events must carry a significant

Table 10.2 Major arm-level copy-number alterations and important genes included in the peaks

Amplification peak	Potential targets	Biological significance	References
8p11	FGFR1	Known oncogene	1
8p12	WHSC1L1	Chromatin modifier, NE	1
8q24	MYC	Known oncogene	1, 2
11p15	INS, IGF2	Insulin signaling, IC	1
13q12	CDK8	Known oncogene	1, 3
13q22	KLF5	Involved in gut epithelial proliferation, NE	1, 2
13q33	IRS2	Insulin signaling, IC	2
	FGF14	Growth factor, NE	2
17q21	ERBB2	Known oncogene	1
20q11	BCL2L1	Antiapoptotic function, IC	1, 4, 5
20q13	HNF4A	Genetic variants are associated with diabetes, A few evidences as a potential oncogene	1

1, colon tcga; 2, Martin et al. Cancer Res 2007; 3, Firestein et al. Nature 2008; 4, Beroukhim et al. Nature 2010; 5, Sillars-Hardebol et al. J Pathol 2011

NE Its role in colorectal cancer has not been established yet, *IC* Implicated in colorectal cancer

selection advantage for these tumors. However, given that such events impact the copy-number of so many genes, it often difficult to determine the biologic rationale for these events. Indeed regions impacted recurrently by SCNAs may harbor multiple genes whose altered expression contributes to neoplasia (Solimini et al. 2012). Efforts to identify key individual genes targeted by these broad recurrent genomic alterations often involve looking for the overlap of these events with other alterations such as somatic mutation of specific genes or more focal SCNAs. Indeed, the frequently mutated tumor suppressor *TP53* lies on chromosome 17q. On the frequently lost chromosome arm 18q, it has been noted that the gene *SMAD4* on this arm is both subject to recurrent focal deletion and also mutation arguing for this gene to be a rationale for these losses. However, other potential tumor suppressors, *DCC*, *SMAD2*, and *CABLES*, also fall on this arm (Pino and Chung 2010). Chromosome 8q gains are often attributed to the presence of oncogene *MYC*. Recently, pinpointing amplified regions in chromosome 20q by statistical analysis of many adenoma and CRC cases coupled with expression data analysis came up with several putative oncogenes attributable to 20q gain such as *AUKRA* and *TCFL5* (Carvalho et al. 2009).

10.5 Focal Copy-Number Alterations: A Key to the Discovery of Putative Oncogenes or Tumor Suppressor Genes

When evaluating the SCNAs in the cancer genome, those events that are focal, i.e., less than a chromosome arm, it becomes much more feasible to identify the specific gene (or genes) likely to be the biologic target. This process is greatly assisted both by the new technologies allowing more resolution in the definition of small SCNAs, the ability to interrogate larger sample collections and, of course, statistical approaches to analyze these data. The study of focal SCNAs is typically divided into the evaluation of focal amplifications, presumably targeting oncogenic factors, and the presence of focal deletions, typically inactivating tumor suppressors. These rules are not absolute as there are instances where focal deletions can be activating, such as with the sub-genic deletions within *EGFR* that lead to an activated form of this gene in glioblastoma and lung squamous cell carcinoma. Additionally, like other somatic genomic analyses, the majority of focal alterations in a specific cancer are likely merely passenger events related to the background instability. Still, evaluation of highly recurrent alterations can provide great insight into specific genes active in cancer.

While focal genomic amplifications are reported in CRC, the frequency of such events are lower than has often been reported for other epithelial cancers such as lung, breast, or ovary cancers. Indeed, an analysis of SNCA profiles across ~3,000 epithelial cancers identified that CRC genomes have lower rates of focal SCNAs than many other comparable tumor (Beroukhim et al. 2010). By contrast, the commonly gained arm-level events in CRC are much more frequent, demonstrating much more inter-tumor heterogeneity in the spectrums of focal SCNAs. Early efforts to define the focal SCNAs in CRC followed hypothesis-driven methods using fluorescence in situ hybridization (FISH), to query amplification in strongly suspect cancer genes. Through these approaches, the amplification of *MYC* gene locus was found with the frequency ranging from 9 to 14 % of the studied population (Al-Kuraya et al. 2007) and the amplification of *ERBB2* and *EGFR* was reported in a small set of the CRCs (Ooi et al. 2004).

With the advancement of high-density array-based methods such as array CGH and SNP array, the entire landscape of focal amplifications and deletions is now being revealed. These approaches are allowing identification of focal amplifications or deletions that could not be detected by the previous low-resolution techniques. Although focal amplification peaks are not fully concordant among the published studies, some reliable studies came up with potential therapeutic targets in common.

Some focal SCNAs occur on top of arm-level SCNAs. A striking set of examples is the focal amplifications superimposed upon chromosome 13q gain in CRC. These events include a set of amplifications at 13q12 near *CDK8* and *CDX2*, another focal amplification locus at *KLF5*, and also some amplifications distally at the locus of *IRS2* (Martin et al. 2007; Firestein et al. 2008; Cancer Genome Atlas Network 2012). *CDK8* was shown to act as a pro-survival factor in β-catenin-dependent CRCs as we discussed in the previous section. *IRS2* has been implicated in the survival of CRC as a messenger downstream of insulin signaling. However, the role of *KLF5* in CRC is not clear, although *KLF5* transcription factor has been implicated in the mutant KRAS-driven intestinal tumorigenesis (Nandan et al. 2008). Chromosome 8, which shows recurrent arm-level gains, also has interesting focal amplification peaks on 8q24 and 8p12 containing *MYC* and *WHSC1L1*, respectively (Cancer Genome Atlas Network 2012). *MYC* is a known oncogene but the role of *WHSC1L*, a histone methyltransferase, in CRC is unclear. The 8p12 focal amplification is further interesting because the tyrosine kinase *FGFR1*, a potential therapeutic target, is the immediate neighbor of *WHSC1L1*, although it does not fall in the computationally defined peak region. Additionally, a broad peak on chromosome 20q has some focal amplification peaks around *BCL2L1* and *HNF4A*. The pro-survival function of BCL2L1 in chromosome 20q amplified CRCs (Beroukhim et al. 2010; Sillars-Hardebol et al. 2012) and the tumor-promoting role of HNF4A in the mouse model of colorectal tumorigenesis (Darsigny et al. 2010) have been shown.

Meanwhile, some recurrent focal amplifications are distinct from the arm-level gains. An isolated focal amplification peak on chromosome 11p, which has been found in 7 % of CRC, contains insulin (*INS*), insulin-like growth factor (*IGF2*), and tyrosine hydroxylase (*TH*) as well as *miR-483*, which is located within an intron of *IGF2* (Cancer Genome Atlas Network 2012). When the expression levels of those potential targets were considered, only *IGF2* and *miR-483* were biologically relevant amplified targets corroborating the previous studies (Nakagawa et al. 2001; Cui et al. 2003; Veronese et al. 2010).

Another amplified peak on chromosome 17q containing *ERBB2* gene has been found in a significant proportion, i.e., approximately 5 %, of the CRCs as well as in several other cancer types (Fig. 10.1a) (Beroukhim et al. 2010; Cancer Genome Atlas Network 2012). *ERBB2* amplifications are well described in breast, gastro-esophageal, and CRCs and cancers bearing this amplification can be treated with trastuzumab, a monoclonal antibody against ERBB2 (Burstein et al. 2003; Bang et al. 2010). In addition, genomic amplification of *ERBB2* locus correlates well with its elevated expression, suggesting that ERBB2 is a real target in CRC (Camps et al. 2009). Regarding EGFR, which is currently a therapeutic target in CRC, high-level focal genomic amplifications at the *EGFR* locus seems to be rare (Ooi et al. 2004; Al-Kuraya et al. 2007). Rather, most gains of *EGFR* locus are arm-level events (Fig. 10.2b) (Cancer Genome Atlas Network 2012).

The study of focal deletions in the cancer genome reveals highly recurrent events, some of which are attributed to *bona fide* tumor suppressors, while others are of less clear cancer relevance. While the presence of a highly recurrent focal deletion that recurrently targets as little as a single gene would seem to be solid support for a pathogenic role of such a gene in cancer, there are concerns about the relevance of some of these targets. Many genes subject to such highly focal deletion are genes whose coding exons are spread over a large genomic footprint, typically spanning >1 Mb of the genome. The concern has thus emerged that these sequences may be subject to breakage in the genome due to marked structural fragility rather than the presence of a functional tumor suppressor. Indeed, many of these loci of recurrent focal are putative “fragile sites,” sites that exhibit nonrandom gaps or breaks when chromosomes are exposed to specific cell culture conditions. Genes targeted by such putative fragile site deletions include *FHIT*, *A2BP1*, *WWOX*, *NAALAD2*, *FAM190B*, *GMDS*, and *PDE4D*, among others. The potential tumor suppressive roles of genes such as *FHIT* and *WWOX* have been suggested in several animal models (Drusco et al. 2011), but their roles in colorectal tumorigenesis are still questionable. The incidence of deletions in these events is quite striking, as these genes are often deleted in the majority of samples. In some cases, there is stronger evidence for a pathogenic role of a frequently deleted gene that may also

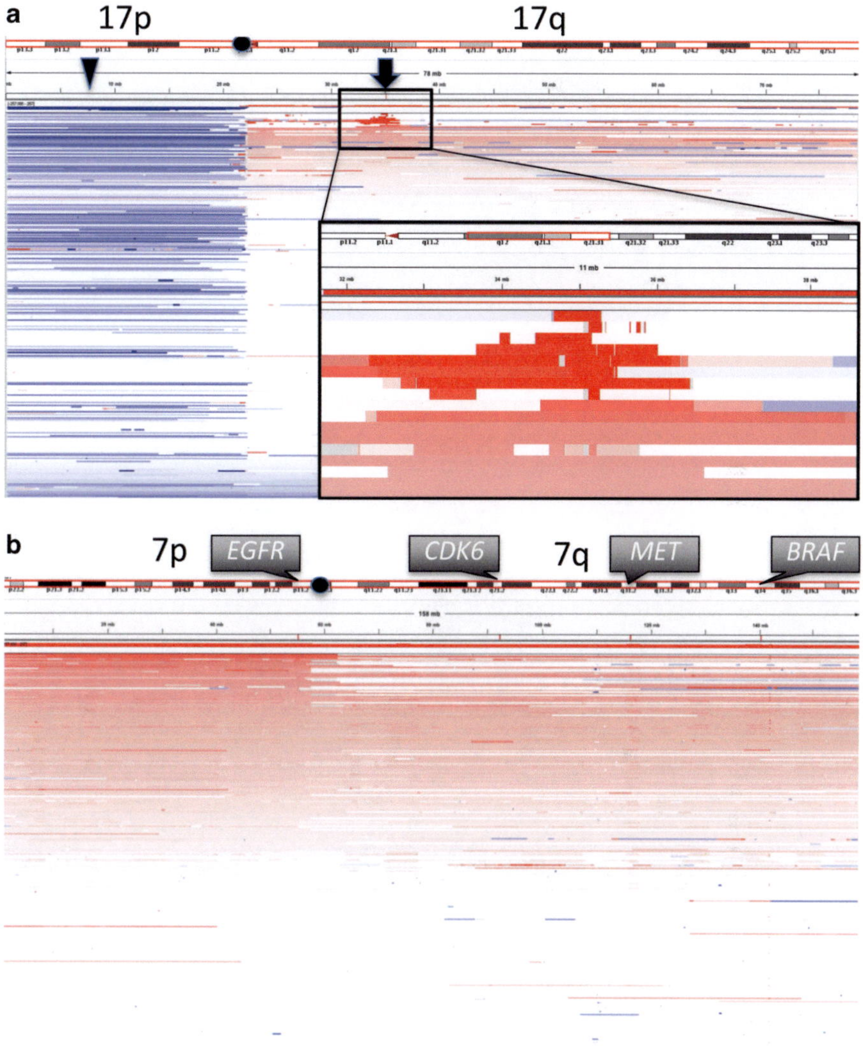

Fig. 10.2 Visualized examples of representative copy-number alterations in colorectal cancers. Representative copy-number alteration patterns are presented by the integrative genomics viewer (IGV). *Horizontal axis* represents chromosomal location and *vertical axis* represents individual cases. *Red color* depicts gain and *blue color* depicts loss of genetic material. (**a**) A representative example of focal amplification involving *ERBB2* locus. Focal regions around 17q21 show marked gains of genetic material in a subset of cases. Those regions share the *ERBB2* locus (*arrow*) in common. It seems clear that eight cases have this amplification (*inset*). Note that approximately half of cases harbor recurrent arm-level losses on the short arm of the Chromosome 17 where *TP53* locus (*arrowhead*) sits. (**b**) Arm-level gain of the chromosome 7. Approximately a half of the cases show a broad gain of both arms. Several major oncogenes such as *EGFR*, *CDK6*, *MET*, and *BRAF* are located in this chromosome (The location of each gene is marked along the chromosome). Note that

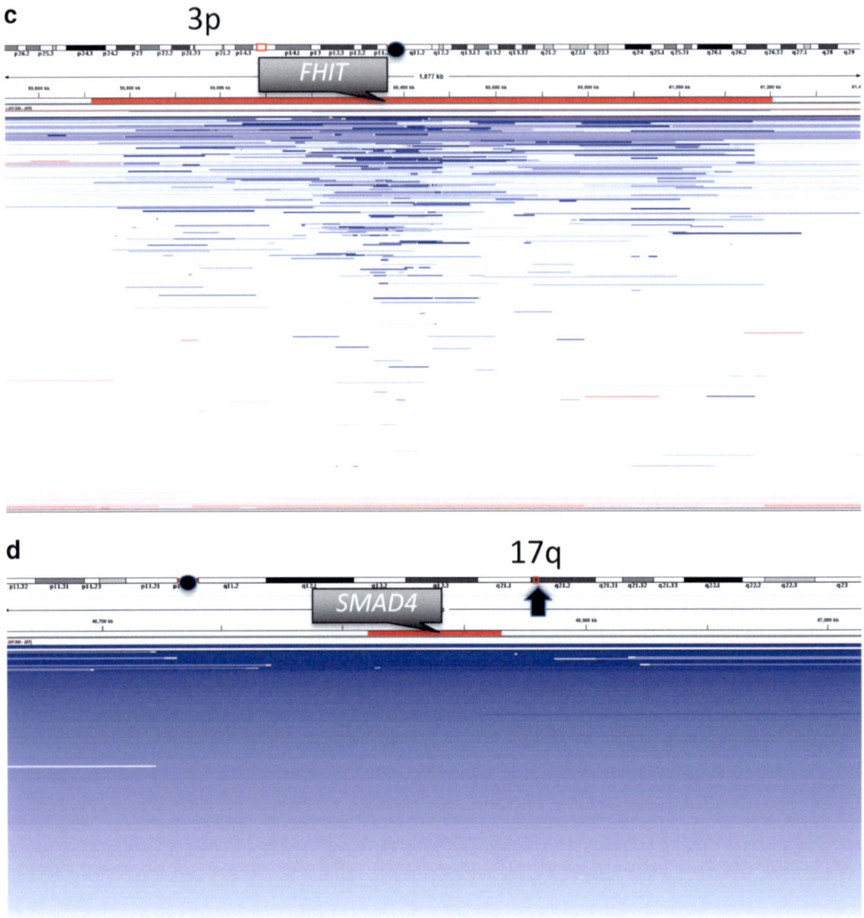

Fig. 10.2 (continued) the gain of *EGFR* locus is not a focal event. (**c**) A typical pattern of common fragile sites. Genomic region around *FHIT* locus (marked with a *red bar*) shows irregularly fragmented loss. Note that some segments of surrounding region are gained. (**d**) A typical pattern of genomic loss involving tumor suppressor gene. The chromosomal segment surrounding the *SMAD4* tumor suppressor is lost in a continuous fashion. Note that the *SMAD4* locus (*arrow* in 17q, marked with a *red bar*) falls within the most overlapping region. All figures were generated with SNP array data from the colorectal TCGA project (The Cancer Genome Atlas, Nature 487:330–7. DOI:10.1038/nature11252)

happen to be at a fragile site. For example, a focal deletion at the gene *PARK2* is found on chromosome 16. Despite the potential that deletions may occur due to fragility, frequent mutations of this gene have been found (Veeriah et al. 2010) and deletion of *PARK2* gene has been shown to drive CRCs in cooperation with APC deficiency in a mouse model (Poulogiannis et al. 2010).

Interestingly, recent high-resolution SCNA profiling techniques show the different patterns of loss between common fragile sites and real tumor suppressor loci.

In common fragile sites, genetic losses are fragmented and include deletions often falling entirely within a single intron (Fig. 10.2c). By contrast, in deletions at *bona fide* tumor suppressor loci, genetic losses occur in continuous fashion and are centered at the target tumor suppressor genes (Fig. 10.2d) (Cancer Genome Atlas Network 2012).

Other significant focal deletions affect established tumor suppressor gene loci such as *SMAD4*, *APC*, *PTEN*, and *SMAD3*. The functional impact of SMAD4, APC, and PTEN in colorectal tumorigenesis has been indirectly shown by a clever in vivo insertional mutagenesis approach (Starr et al. 2009) and germ-line mutations or deletions of them are responsible for juvenile polyposis syndrome, familial adenomatous polyposis, and Cowden disease, respectively. A tumor suppressive role for SMAD3 has also been demonstrated in mouse model of colorectal tumorigenesis (Zhu et al. 1998; Sodir et al. 2006). In addition to the previously known tumor suppressors, genome-wide interrogation of significantly deleted regions revealed novel functionally relevant targets. One good example is *TCF7L2*, which was found in a significantly deleted region on chromosome 10p25.2. The oncogenic role of TCF7L2 is supported by the fact that it has been known as an essential β-catenin cofactor (Clevers 2004) and was frequently mutated in CRCs (Cancer Genome Atlas Network 2012). Additionally, some of the deletions were shown to cause an in-frame fusion with neighboring *VTI1A* gene and the resulting *VTI1A-TCF7L2* fusion gene was required for survival of CRC cells bearing the translocation (Bass et al. 2011). As we can see in this example, the better we understand the genome structure, the more likely we identify other ways in which deletions have unexpected functional impact, in some cases, activation.

10.6 Bedside Applications of Somatic Copy-Number Alterations in CRCs

In addition to the cancer genome serving as a window into the underlying pathophysiology of CRC, genomic alterations also can serve as potential clinical biomarkers to help guide therapeutic decision-making. SCNAs themselves can point to a specific oncogene active in a cancer, thus directly informing the selection of targeted therapeutics. A classical example of linkage between SCNAs and therapeutic implication is *ERBB2* amplification in breast cancers, where *ERBB2* gene amplification is the primary biomarker guiding the use of the ERBB2-targeting antibody trastuzumab. Interestingly, *ERBB2* amplification has been noted in CRC in ~5 % of cases. When linked with new data suggesting a role for ERBB2-targeting agents in CRC (Camps et al. 2009), there may be a role for using *ERBB2* SCNA status as a biomarker in CRC. The target that receives substantial attention in CRC, however, is the epidermal growth factor receptor (EGFR). Antibodies targeting ERFR have led to modest improvements in survival of patients with CRC when used in the second-line treatment of metastatic disease. However, although clear data support the decision to not provide EGFR-targeted agents in patients with tumors harboring KRAS mutations, no clear positive biomarkers exist to guide the use of EGFR-targeting agents. Although EGFR copy-number gain has been reported to predict the response to anti-EGFR therapy, their reproducibility was limited (Personeni et al. 2008). Thus, it would be worthwhile to study additional biomarkers for unresponsiveness to EGFR-targeted therapies.

Beyond informing the selection of individual targeted agents, chromosomal aberrations can also serve as biomarkers of prognosis or of the potential response to conventional chemotherapeutics. LOH at chromosome 18q has been proposed as a marker of poor prognosis and also a predictor of unfavorable outcomes after adjuvant 5-fluorouracil-based chemotherapy (Martinez-Lopez et al. 1998; Ogunbiyi et al. 1998; Watanabe et al. 2001). The association between 18q LOH and poor prognosis may seem reasonable as this event is known to occur late in colorectal carcinogenesis and the chromosomal region 18q contains many tumor suppressor candidates such as *DCC*, *SMAD4*, *SMAD2*, and *CABLES1*. However, the possibility also exists that there could be confounding in these analyses as 18q loss is frequently seen in microsatellite stable tumors, tumors with worse prognosis compared to those with MSS. Indeed, a couple of recent prospective studies found that 18q LOH has no impact on prognosis when the analysis was done in a large set of microsatellite stable CRCs (Ogino et al. 2009; Bertagnolli et al. 2011).

10.7 The Future of SCNA Study in Colorectal Adenocarcinoma

CRCs can be divided into microsatellite stable and unstable tumors and microsatellite stable tumors have SCNAs far more frequently than microsatellite unstable tumors. The most common SCNAs in CRC that have been identified are also those that we have known about the longest, arm-level gains and losses such as gain of 13q and 20q and loss of 18q. The advent of new genome-scale profiling techniques allowing the measurement of SCNAs, gene expression, epigenetics, and sequencing of the same samples will allow integrated bioinformatics studies to help elucidate the key genes responsible for these highly recurrent events. Given the commonality of these events, efforts to leverage these lesions to define new therapeutics could have great impact on the disease.

Compared to several other epithelial cancers, highly focal amplifications are less common. Nonetheless, focal alterations at genes such as *ERBB2* and *IGF2* may be able to be leveraged to define new targets for subsets of patients with CRC. Additionally, as genomic technology continues to improve, we have the potential to define even more focal alterations that may be missed by current array-based approaches. This improvement may follow the emerging use of next-generation DNA sequencing to profile SCNAs. This approach has already been implemented in some pilot studies demonstrating how it is feasible not only to define SCNAs using these approaches but also to build upon elucidation of SCNAs by better understanding the detail of the genome structure generated by these alterations. SCNAs not only lead to changes in copy-number, they also lead to pieces of the genome being cut apart and pieced together to form aberrant connections between segments of the genome. Next-generation sequencing approaches can directly identify these

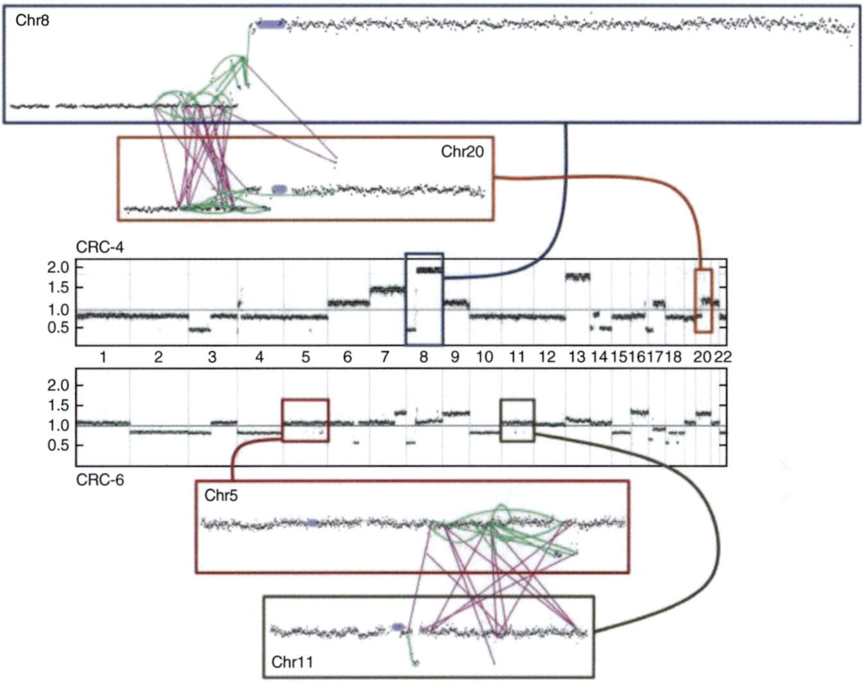

Fig. 10.3 Complex rearrangements between different chromosomes in two colorectal carcinomas. The central portion of the figure shows global copy-number profiles in two primary colorectal carcinomas. The chromosome identity is labeled across the *x*-axis and the $\log_2$ copy-number ratio is depicted along the *y*-axis. The *upper inset boxes* show detailed views of the copy numbers and rearrangements of chromosome 8 (*dark blue*) and 20 (*ochre*) for a colorectal cancer sample with the centromere labeled as a *purple circle*. Rearrangements detected by dRanger are shown in *green* (intrachromosomal) and *purple* (interchromosomal). The *lower inset boxes* show detailed copy numbers and rearrangements for another colorectal cancer sample. The *lower inset boxes* show chromosome 5 (*red*) and 11 (*gray*), with lines marking positions of genomic rearrangements. The rearrangements shown in both cases are very complex and they are thought to result from a catastrophic genomic event called "chromothripsis" (reprinted from Bass AJ et al. *Nature Genetics* 2011;43(10):964–8. DOI:10.1038/ng.936)

aberrant connections generated as the unstable colorectal adenocarcinoma genome is weaved together. These approaches are beginning to reveal previously unrecognized structural complexity of cancer genomes. By the whole-genome sequencing approach, a recurrent VTI1A-TCF7L2 fusion was found in a small subset of CRCs, and complex rearrangements between two different chromosomes, which probably resulted from chromothripsis, could also be detected (Fig. 10.3) (Bass et al. 2011). Now that we have this new window into the genomes of colorectal and other cancer, we will be able to make new generations of discoveries and, perhaps, gain greater insight into genomic events that are well recognized but whose impact on cancer pathophysiology is not fully appreciated. Together, these growing insights should provide the research community information on novel therapeutic targets, candidate biomarkers, and pathways that may be active in these deadly diseases.

References

Alitalo K, Schwab M et al (1983) Homogeneously staining chromosomal regions contain amplified copies of an abundantly expressed cellular oncogene (c-myc) in malignant neuroendocrine cells from a human colon carcinoma. Proc Natl Acad Sci USA 80(6):1707–1711

Alitalo K, Winqvist R et al (1984) Aberrant expression of an amplified c-myb oncogene in two cell lines from a colon carcinoma. Proc Natl Acad Sci USA 81(14):4534–4538

Al-Kuraya K, Novotny H et al (2007) HER2, TOP2A, CCND1, EGFR and C-MYC oncogene amplification in colorectal cancer. J Clin Pathol 60(7):768–772

Anand S, Penrhyn-Lowe S et al (2003) AURORA-A amplification overrides the mitotic spindle assembly checkpoint, inducing resistance to Taxol. Cancer Cell 3(1):51–62

Ashton-Rickardt PG, Dunlop MG et al (1989) High frequency of APC loss in sporadic colorectal carcinoma due to breaks clustered in 5q21-22

Bang YJ, Van Cutsem E et al (2010) Trastuzumab in combination with chemotherapy versus chemotherapy alone for treatment of HER2-positive advanced gastric or gastro-oesophageal junction cancer (ToGA): a phase 3, open-label, randomised controlled trial. Lancet 376(9742):687–697

Bass AJ, Lawrence MS et al (2011) Genomic sequencing of colorectal adenocarcinomas identifies a recurrent VTI1A-TCF7L2 fusion. Nat Genet 43(10):964–968

Bean J, Brennan C et al (2007) MET amplification occurs with or without T790M mutations in EGFR mutant lung tumors with acquired resistance to gefitinib or erlotinib. Proc Natl Acad Sci USA 104(52):20932–20937

Beroukhim R, Getz G et al (2007) Assessing the significance of chromosomal aberrations in cancer: methodology and application to glioma. Proc Natl Acad Sci 104(50):20007–20012

Beroukhim R, Mermel CH et al (2010) The landscape of somatic copy-number alteration across human cancers. Nature 463(7283):899–905

Bertagnolli MM, Redston M et al (2011) Microsatellite instability and loss of heterozygosity at chromosomal location 18q: prospective evaluation of biomarkers for stages II and III colon cancer—a study of CALGB 9581 and 89803. J Clin Oncol 29(23):3153–3162

Bignell GR, Huang J et al (2004) High-resolution analysis of DNA copy number using oligonucleotide microarrays. Genome Res 14(2):287–295

Burstein HJ, Harris LN et al (2003) Trastuzumab and vinorelbine as first-line therapy for HER2-overexpressing metastatic breast cancer: multicenter phase II trial with clinical outcomes, analysis of serum tumor markers as predictive factors, and cardiac surveillance algorithm. J Clin Oncol 21(15):2889–2895

Cahill DP, Lengauer C et al (1998) Mutations of mitotic checkpoint genes in human cancers. Nature 392(6673):300–303

Camps J, Nguyen QT et al (2009) Integrative genomics reveals mechanisms of copy number alterations responsible for transcriptional deregulation in colorectal cancer. Genes Chromosomes Cancer 48(11):1002–1017

Cancer Genome Atlas Network (2012) Comprehensive molecular characterization of human colon and rectal cancer. Nature 487(7407):330–337

Carvalho B, Postma C et al (2009) Multiple putative oncogenes at the chromosome 20q amplicon contribute to colorectal adenoma to carcinoma progression. Gut 58(1):79–89

Cheng YW, Pincas H et al (2008) CpG island methylator phenotype associates with low-degree chromosomal abnormalities in colorectal cancer. Clin Cancer Res 14(19):6005–6013

Chiang DY, Getz G et al (2009) High-resolution mapping of copy-number alterations with massively parallel sequencing. Nat Methods 6(1):99–103

Clevers H (2004) Wnt breakers in colon cancer. Cancer Cell 5(1):5–6

Crasta K, Ganem NJ et al (2012) DNA breaks and chromosome pulverization from errors in mitosis. Nature 482(7383):53–58

Cui H, Cruz-Correa M et al (2003) Loss of IGF2 imprinting: a potential marker of colorectal cancer risk. Science 299(5613):1753–1755

D'Emilia J, Bulovas K et al (1989) Expression of the c-erbB-2 gene product (p185) at different stages of neoplastic progression in the colon. Oncogene 4(10):1233–1239

Darsigny M, Babeu JP et al (2010) Hepatocyte nuclear factor-4alpha promotes gut neoplasia in mice and protects against the production of reactive oxygen species. Cancer Res 70(22):9423–9433

Diep CB, Kleivi K et al (2006) The order of genetic events associated with colorectal cancer progression inferred from meta-analysis of copy number changes. Genes Chromosomes Cancer 45(1):31–41

Douglas EJ, Fiegler H et al (2004) Array comparative genomic hybridization analysis of colorectal cancer cell lines and primary carcinomas. Cancer Res 64(14):4817–4825

Drusco A, Pekarsky Y et al (2011) Common fragile site tumor suppressor genes and corresponding mouse models of cancer. J Biomed Biotechnol 2011:984505

Engelhardt M, Drullinsky P et al (1997) Telomerase and telomere length in the development and progression of premalignant lesions to colorectal cancer. Clin Cancer Res 3(11):1931–1941

Ewart-Toland A, Briassouli P et al (2003) Identification of Stk6/STK15 as a candidate low-penetrance tumor-susceptibility gene in mouse and human. Nat Genet 34(4):403–412

Fearon ER, Vogelstein B (1990) A genetic model for colorectal tumorigenesis. Cell 61(5): 759–767

Finley GG, Schulz NT et al (1989) Expression of the myc gene family in different stages of human colorectal cancer. Oncogene 4(8):963–971

Firestein R, Bass AJ et al (2008) CDK8 is a colorectal cancer oncogene that regulates beta-catenin activity. Nature 455(7212):547–551

Ganem NJ, Godinho SA et al (2009) A mechanism linking extra centrosomes to chromosomal instability. Nature 460(7252):278–282

Geiersbach KB, Samowitz WS (2011) Microsatellite instability and colorectal cancer. Arch Pathol Lab Med 135(10):1269–1277

Gertler R, Rosenberg R et al (2004) Telomere length and human telomerase reverse transcriptase expression as markers for progression and prognosis of colorectal carcinoma. J Clin Oncol 22(10):1807–1814

Goel A, Nagasaka T et al (2007) The CpG island methylator phenotype and chromosomal instability are inversely correlated in sporadic colorectal cancer. Gastroenterology 132(1):127–138

Hermsen M, Postma C et al (2002) Colorectal adenoma to carcinoma progression follows multiple pathways of chromosomal instability. Gastroenterology 123(4):1109–1119

Hughes S, Williams RD et al (2006) Meta-analysis and pooled re-analysis of copy number changes in colorectal cancer detected by comparative genomic hybridization. Anticancer Res 26(5A):3439–3444

Jemal A, Siegel R et al (2010) Cancer statistics, 2010. CA Cancer J Clin 60(5):277–300

Katayama H, Ota T et al (1999) Mitotic kinase expression and colorectal cancer progression. J Natl Cancer Inst 91(13):1160–1162

Kwei KA, Kung Y et al (2010) Genomic instability in breast cancer: pathogenesis and clinical implications. Mol Oncol 4(3):255–266

Lengauer C, Kinzler KW et al (1998) Genetic instabilities in human cancers. Nature 396(6712): 643–649

Martin ES, Tonon G et al (2007) Common and distinct genomic events in sporadic colorectal cancer and diverse cancer types. Cancer Res 67(22):10736–10743

Martinez-Lopez E, Abad A et al (1998) Allelic loss on chromosome 18q as a prognostic marker in stage II colorectal cancer. Gastroenterology 114(6):1180–1187

Maser RS, DePinho RA (2002) Connecting chromosomes, crisis, and cancer. Science 297(5581):565–569

Meijer GA, Hermsen MA et al (1998) Progression from colorectal adenoma to carcinoma is associated with non-random chromosomal gains as detected by comparative genomic hybridisation. J Clin Pathol 51(12):901–909

Meyerson M, Gabriel S et al (2010) Advances in understanding cancer genomes through second-generation sequencing. Nat Rev Genet 11(10):685–696

Morris EJ, Ji JY et al (2008) E2F1 represses beta-catenin transcription and is antagonized by both pRB and CDK8. Nature 455(7212):552–556
Murakami R, Tsukuma H et al (1990) Natural history of colorectal polyps and the effect of polypectomy on occurrence of subsequent cancer. Int J Cancer 46(2):159–164
Nakagawa H, Chadwick RB et al (2001) Loss of imprinting of the insulin-like growth factor II gene occurs by biallelic methylation in a core region of H19-associated CTCF-binding sites in colorectal cancer. Proc Natl Acad Sci USA 98(2):591–596
Nakao K, Mehta KR et al (2004) High-resolution analysis of DNA copy number alterations in colorectal cancer by array-based comparative genomic hybridization. Carcinogenesis 25(8):1345–1357
Nandan MO, McConnell BB et al (2008) Kruppel-like factor 5 mediates cellular transformation during oncogenic KRAS-induced intestinal tumorigenesis. Gastroenterology 134(1):120–130
O'Hagan RC, Chang S et al (2002) Telomere dysfunction provokes regional amplification and deletion in cancer genomes. Cancer Cell 2(2):149–155
Ogino S, Nosho K et al (2009) Prognostic significance and molecular associations of 18q loss of heterozygosity: a cohort study of microsatellite stable colorectal cancers. J Clin Oncol 27(27):4591–4598
Ogunbiyi OA, Goodfellow PJ et al (1998) Confirmation that chromosome 18q allelic loss in colon cancer is a prognostic indicator. J Clin Oncol 16(2):427–433
Olshen AB, Venkatraman ES et al (2004) Circular binary segmentation for the analysis of array-based DNA copy number data. Biostatistics 5(4):557–572
Ooi A, Takehana T et al (2004) Protein overexpression and gene amplification of HER-2 and EGFR in colorectal cancers: an immunohistochemical and fluorescent in situ hybridization study. Mod Pathol 17(8):895–904
Pellman D (2007) Cell biology: aneuploidy and cancer. Nature 446(7131):38–39
Personeni N, Fieuws S et al (2008) Clinical usefulness of EGFR gene copy number as a predictive marker in colorectal cancer patients treated with cetuximab: a fluorescent in situ hybridization study. Clin Cancer Res 14(18):5869–5876
Pino MS, Chung DC (2010) The chromosomal instability pathway in colon cancer. Gastroenterology 138(6):2059–2072
Plentz RR, Wiemann SU et al (2003) Telomere shortening of epithelial cells characterises the adenoma-carcinoma transition of human colorectal cancer. Gut 52(9):1304–1307
Poulogiannis G, McIntyre RE et al (2010) PARK2 deletions occur frequently in sporadic colorectal cancer and accelerate adenoma development in Apc mutant mice. Proc Natl Acad Sci USA 107(34):15145–15150
Rajagopalan H, Nowak MA et al (2003) The significance of unstable chromosomes in colorectal cancer. Nat Rev Cancer 3(9):695–701
Rajagopalan H, Jallepalli PV et al (2004) Inactivation of hCDC4 can cause chromosomal instability. Nature 428(6978):77–81
Ried T, Knutzen R et al (1996) Comparative genomic hybridization reveals a specific pattern of chromosomal gains and losses during the genesis of colorectal tumors. Genes Chromosomes Cancer 15(4):234–245
Rowley JD (1973) Letter: a new consistent chromosomal abnormality in chronic myelogenous leukaemia identified by quinacrine fluorescence and Giemsa staining. Nature 243(5405): 290–293
Rudkin CT, Hungerford DA et al (1964) DNA contents of chromosome ph1 and chromosome 21 in human chronic granulocytic leukemia. Science 144:1229–1231
Rudolph KL, Millard M et al (2001) Telomere dysfunction and evolution of intestinal carcinoma in mice and humans. Nat Genet 28(2):155–159
Sillars-Hardebol AH, Carvalho B et al (2012) BCL2L1 has a functional role in colorectal cancer and its protein expression is associated with chromosome 20q gain. J Pathol 226(3):442–450
Sodir NM, Chen X et al (2006) Smad3 deficiency promotes tumorigenesis in the distal colon of ApcMin/+ mice. Cancer Res 66(17):8430–8438
Solimini NL, Xu Q et al (2012) Recurrent hemizygous deletions in cancers may optimize proliferative potential. Science 337(6090):104–109

Starr TK, Allaei R et al (2009) A transposon-based genetic screen in mice identifies genes altered in colorectal cancer. Science 323(5922):1747–1750
Takagi S, Kinouchi Y et al (1999) Telomere shortening and the clinicopathologic characteristics of human colorectal carcinomas. Cancer 86(8):1431–1436
Takahashi T, Sano B et al (2003) Polo-like kinase 1 (PLK1) is overexpressed in primary colorectal cancers. Cancer Sci 94(2):148–152
Veeriah S, Taylor BS et al (2010) Somatic mutations of the Parkinson's disease-associated gene PARK2 in glioblastoma and other human malignancies. Nat Genet 42(1):77–82
Veronese A, Lupini L et al (2010) Oncogenic role of miR-483-3p at the IGF2/483 locus. Cancer Res 70(8):3140–3149
Wang Z, Cummins JM et al (2004) Three classes of genes mutated in colorectal cancers with chromosomal instability. Cancer Res 64(9):2998–3001
Watanabe T, Wu TT et al (2001) Molecular predictors of survival after adjuvant chemotherapy for colon cancer. N Engl J Med 344(16):1196–1206
Wood LD, Parsons DW et al (2007) The genomic landscapes of human breast and colorectal cancers. Science 318(5853):1108–1113
Zhao X, Li C et al (2004) An integrated view of copy number and allelic alterations in the cancer genome using single nucleotide polymorphism arrays. Cancer Res 64(9):3060–3071
Zhu Y, Richardson JA et al (1998) Smad3 mutant mice develop metastatic colorectal cancer. Cell 94(6):703–714

Chapter 11
Genome-Wide Association Studies in Colorectal Cancer

Ian Tomlinson

Abstract The genome-wide association study (GWAS) generally involves the molecular and statistical analysis of thousands of individuals with a disease and similar numbers of unaffected or population controls. In short, a search is made for polymorphic alleles that occur more or less often in cases compared with controls, using stringent thresholds of statistical significance. GWAS have principally aimed to identify disease risk alleles that are common in the general population and that have modest effects on susceptibility. Over 20 common polymorphisms that influence bowel cancer risk have been identified by GWAS. Many of these polymorphisms act in the bone morphogenetic protein (BMP) signaling pathway, providing unexpected and important insights that could potentially be used to prevent adenomas and carcinoma of the colorectum, either using novel molecular agents or by stratifying the population for tailored screening by colonoscopy, sigmoidoscopy or fecal occult blood testing.

11.1 Background: The Genetic Contribution to CRC

Cancer families have been known for hundreds of years. Sometimes these families have had unusual clinical features or combinations of features that have immediately marked them as genetic. In colorectal cancer, families with hundreds of polyps or the pigmentation of Peutz–Jeghers syndrome are particular examples (Westerman et al. 1999; Lynch et al. 2009). There has, however, been an unspoken assumption, by both the general population and those involved in medicine, that a family history of colorectal cancer, even in the absence of highly unusual features, increases the risk to that patient's relatives. In fact, whilst there is good evidence that the risk of cancer

I. Tomlinson (✉)
Wellcome Trust Centre for Human Genetics, University of Oxford, Oxford, UK
e-mail: iant@well.ox.ac.uk

K.M. Haigis (ed.), *Molecular Pathogenesis of Colorectal Cancer*, DOI 10.1007/978-1-4614-8412-7_11, © Springer Science+Business Media New York 2013

is indeed increased in the presence of a family history of the disease, data generally show that this tendency is not strong. Most of the family clusters of cancer probably have a strong chance component, but there is also evidence that genetics plays an important role.

There are several measures of the genetic contribution to the common cancers, but these have taken two predominant forms: (1) the relative risk compared with the general population in those with an affected family member; and (2) trait heritability. Familial relative risk, especially sibling relative risk (λs), is a commonly used measure of the tendency for a disease to occur within families and it is often used as a proxy measure for the genetic contribution to disease. λs is relatively easily measured, although problems of positive and negative ascertainment exist. For example, if the data set is based on a screening program, self-referral or unconfirmed patient reports without full medical records, familial risks are likely to be overestimated. Other factors may cause underestimation of risks. Another difficult issue is a tendency for non-genetic factors (or, more accurately, complex factors with a non-genetic component) to be influenced by their presence in family members. For colorectal cancer, smoking, diet and weight are all examples of factors with clear familial tendencies and potential effects on CRC risk. However, it has not yet been possible to accurately estimate the proportion of familial risk that might be "environmental". In general, the most reliable estimates suggest that λs is about 2.0–2.5 for the common cancers. The risk is quite similar if a parent is affected, suggesting that the susceptibility genes involved mostly act dominantly. In other words, if your brother, sister, mother or father has had colorectal cancer during their lifetime, you yourself are at about double the population risk of disease, equating to a lifetime risk of about 10 % in the western world. The more family members affected, the higher is the relative risk. For this reason, some recommend regular colonoscopy for anyone with a blood relative who has had CRC, although often intensive screening is reserved for those with multiple affected family members or specific mutations in high-risk (Mendelian) predisposition genes.

Heritability is another, more direct measure of the genetic component of disease. In humans, twin studies probably provide the best measure of heritability, since monozygous and dizygous twins can be compared. In colorectal cancer, not many such studies have been performed, but a typical estimate is that about 30 % of the variance in disease risk has a genetic origin (Lichtenstein et al. 2000). In general, it is unlikely that shared environments and Mendelian cancer syndromes can account for all the excess familial risk and heritability. On this basis, searches for cancer-causing genetic variants of low or moderate penetrance have been justified, funded and undertaken.

11.2 Approaches Based on Candidate Genes

In the late twentieth century, the major activity in the genetics of cancer predisposition was the identification of susceptibility genes for the Mendelian cancer syndromes. In colorectal cancer, this led to the identification of the genes for familial adenomatous polyposis (*APC*), juvenile polyposis (*SMAD4*, *BMPR1A*), Peutz–Jeghers syndrome (*LKB1/STK11*) and Lynch syndrome (*MSH2*, *MLH1*, *MSH6*, *PMS2*). *MUTYH*-associated polyposis was characterised later. Some of these studies relied on linkage analysis and positional cloning, although there were also important contributions to the successful cloning efforts from the study of somatic mutation spectra in tumors candidate gene analysis and the identification of rare patients with cytogenetically identifiable mutations. However, despite the primacy of Mendelian syndromes, the possibility that there existed non-Mendelian predisposition genes was already fully recognised and numerous small-scale association studies were performed on that basis.

Unfortunately, at that time methods did not exist for reliably genotyping very large numbers of samples. Furthermore, given the large effect sizes of Mendelian predisposition genes, relative risks of twofold or more were considered quite reasonable assumptions for the effect sizes of more common, lower-penetrance genes. The typical association study in the year 2000 might therefore consist of a couple of hundred cases and controls typed at a handful of candidate common single nucleotide polymorphisms (SNPs) using a method such as restriction fragment length polymorphism analysis. Typical candidate polymorphisms for colorectal cancer were within genes involved in carcinogen metabolism (cytochrome P450 family, glutathione-S-tranferase), DNA repair, the folate pathway or iron metabolism (haemochromatosis, methylene tetrahydrofolate reductase) or with a role in the somatic pathways of colorectal tumorigenesis (E-cadherin, p53, RAS) (Houlston and Tomlinson 2001). It turned out that almost all of the individual studies were far too small, because the relative risks of common cancer predisposition variants are very rarely as high as two. Moreover, consistent global genetic differences between cancers and controls (population stratification) led to false associations being reported. Despite this, meta-analyses did suggest that one or two of the variants assessed might be associated with CRC predisposition. Clearly, the choice of candidates was not that bad. However, by the start of the twenty-first century there was a gradually developing consensus that candidate gene-based association studies for cancer predisposition had, by-and-large, failed.

In fact, the ability to move beyond the candidate gene association study was driven partly by disillusionment on the part of the research community, but mostly by technological advances that allowed rapid, high-quality genotyping of thousands of DNA samples at a relatively low price per sample per polymorphism. These two factors, together with large-scale human variation discovery programmes such as HapMap, drove the emergence of the genome-wide association study, or GWAS.

11.3 Planning and Evolution of the GWAS

GWAS typically involves association tests between genotypes in disease cases and controls at many thousands of SNPs throughout the genome. Although there have been focused GWAS, for example, based on protein coding variants, most have taken a hypothesis-free approach based on linkage disequilibrium (LD) mapping, an established method that had mostly been used in the past in attempts to fine-map the locations of Mendelian cancer predisposition genes. The ability to use LD mapping on a genomic scale came about largely as a result of efforts to catalogue common human polymorphisms, the HapMap project being the principal early exponent (International HapMap Consortium 2003). By discovering millions of common SNPs and comparing genetic and physical maps, it was confirmed that LD tended to occur in "blocks" because meiotic recombination events were not random, but clustered at "hotspots" in the genome. Therefore, a subset of SNPs—so-called tagging SNPs—could be used as proxies for many other SNPs with which they had strong LD. In practice, several hundred thousand SNPs were required to capture the majority of common genetic variants in the European population. Typically, assays for these SNPs were developed on commercial microarrays that permitted typing of several hundred or more samples genome-wide in a few months.

GWAS strategy has developed over time, as the sensitivity of the methods to design and technical issues have become apparent. Most of the initial cancer GWAS involved about 1,000 cases with disease and a similar number of controls (Easton et al. 2007; Tomlinson et al. 2007). The patients were generally derived from specific prevalent or incident case collections rather than cohorts. Sometimes, patients were selected to be more "genetic" by virtue of family history, early onset of disease, multiplicity of tumors (such as bilateral breast cancer) and exclusion of the known Mendelian cancer syndromes. Controls were sometimes obtained through the cases under study, often being spouses, partners or friends, thus providing a modicum of control for environmental and lifestyle factors. For some studies, controls were selected for an absence of a personal or family history of cancer, although increasing use has been made of population-based controls such as those from the Wellcome Trust-funded genotyping of the UK 1958 Birth Cohort and National Blood Service collections.

Although the early GWAS based on one or two thousand patients did yield some associations (or "hits"), it was clear from an early stage that the risks conferred by the common SNP alleles were at the low end of expectations—or, more accurately, smaller than the hopes or hype. It followed that the typical cancer GWAS would require many thousands of cases and controls to be successful. A means had to be found to balance type I and type II errors, in other words, to genotype sufficient cases and controls to demonstrate an association convincingly, whilst not spending the budget on attempting in vain to validate chance associations. Almost all GWAS for the common cancers therefore evolved a multiphase design with the aim of identifying as many true associations as possible at minimal expense. In short, the design involved an initial genome-wide SNP typing discovery phase (sometimes a meta-analysis of more than one study), followed by one or often more phases during which the SNPs with the strongest association signal by P value were genotyped in additional samples. Those SNPs that continued to show evidence of association were then genotyped in additional samples until validation was deemed to have failed, or there was convincing evidence of association (generally $P<5\times10^{-8}$, by consensus, based on an approximate Bonferroni correction of $P=[0.05/1{,}000{,}000$

SNPs]). However, ignorance of the underlying genetic architecture of disease, and the fragmented sample sets and funding streams, meant that study design was in reality more opportunistic or ad hoc than carefully planned. There were particular problems in combining data derived from different large-scale SNP genotyping platforms. Increasingly, however, large GWAS consortia have developed—such as COGENT (Houlston 2012) and GECCO for colorectal cancer—in which a more co-ordinated approach is being taken, although the concerted GWAS efforts for the "hormonally driven" trio of breast, ovarian and prostate cancer have been particularly impressive.

One early realization was that suboptimal sample selection, poor-quality sample preparation and errors in genotyping could all be fatal for a GWAS, especially—but not exclusively—at the full-genome genotyping stage. Various quality control checks have been adopted for the genome-wide phases of GWAS. These include technical assessments such as genotyping calls, deviation from Hardy–Weinberg equilibrium and sample duplication, relatedness or contamination. Biological checks include assessment of systematic inflation of the test statistics, identification of population substructure by principal component analysis or multidimensional scaling and incorporation of some measure of prior probability, such as eliminating SNPs with low minor allele frequency or using Bayesian methods of analysis. Although some of these measures can be applied to GWAS validation phases, there is no easy way in the absence of large-scale genotyping data to detect unsuspected differences arising from, say, systematic ethnic or sub-population differences between cases and controls. Even relatively robust genotyping assays based on widely used methods such as Sequenom/Massarray, ABI Taqman and KASPar are often not centralised and can vary considerably among the groups who contribute to a multiphase multicentre GWAS. In general, these latter technical risks are conservative, in that they are more likely to prevent reports of true associations in an otherwise sound GWAS.

The statistical analysis of GWAS has spawned a large number of methods for dealing with the data, but it remains true that the simple frequentist test of association is the most commonly used, whether by allele counting in a 2×2 table design, Cochran-Armitage trend test or logistic regression analysis of allelic dosage. The results from these two methods are generally equivalent, the former potentially providing a little more power, with the latter capable of flexibility, such as the incorporation of co-variates, such as sex, age, environment, other genotypes and measures of inter-sample differences, such as the eigenvalues derived from principal component analysis. Other basic analyses are often performed alongside the allele-based test, in order to examine dominant, recessive and genotype-specific models of SNP action. In addition, Bayesian tests have been developed to take account of prior expectations, such as the probability that a test result is a true positive given allele frequency and even the possible function of the SNP concerned. Despite these and other more sophisticated methods, most cancer GWAS rely on testing allele counts in cases and controls or conditional logistic regression with case/control status as the outcome variable. The multiple stages of GWAS are correspondingly combined using size-weighted meta-analysis of allele counts or of odds ratios (or beta coefficients) and 95 % confidence intervals (or standard errors) from logistic regression.

If an overall $P < 5 \times 10^{-8}$ is achieved, significance is declared. Alternatives, such reporting of separate results for the discovery and validation phases—regarding the former as generating a number of hypothetical "hits" that can be tested with less stringent *P* values in the latter—seem not to have found favor.

An issue with GWAS is that very few studies of a sufficient sample size have been performed to re-test the reported statistically-significant association signals. It is not difficult to appreciate why this is the case when the various scientific rewards for "discovery" are greater than those for "failure to replicate". An associated problem is that even for true associations, effect sizes from GWAS are likely to overestimate those in the general population owing to the use of selected cases and controls and the "Winner's curse". As some very large longitudinal/cohort studies start to genotype cancer SNPs to estimate the true effects, it is likely that a few associations will be shown to be errors. On the other hand, the limited amount of independent replication testing performed so far has supported the great majority of cancer GWAS findings.

11.4 Timeline and Findings of GWAS in Colorectal Cancer

Colorectal cancer GWAS first discovered an association between disease and a SNP, rs6983267 (chr8:128,413,305, Human Genome Build 37), in 2007. The odds ratio per allele was about 1.25 and a sample of about 2,000 selected cases and a similar number of controls was sufficient to demonstrate the association convincingly. The same allele (G) at this SNP had already been shown to be associated with prostate cancer risk (Yeager et al. 2007) and was subsequently shown to be associated with ovarian cancer risk (White et al. 2010). Other, independent SNPs at the same genetic locus are associated with prostate and breast cancer. The best candidate gene in the region—albeit 140 kb distal—is the c-myc oncogene, which plays a central role in the maintenance of normal cell proliferation and is amplified and/or over-expressed in many cancer types. Fine mapping (see below) showed no known SNP to have a stronger association signal than rs6983267, and *in silico* prediction, followed by chromatin immunoprecipitation, showed that rs6983267 itself changes a binding site for the transcription factor TCF4. Although other functional variation may be in linkage disequilibrium, the region around rs6983267 can itself act as a transcriptional enhancer in the gastrointestinal tract. However, not all studies have successfully linked the alleles at rs6983267 to differences in c-myc transcription, and other candidate genes exist in the region, closer to rs6983267. One is *POU5F1P1*, a transcribed so-called pseudogene of the *OCT1* stem cell transcription factor; in fact, POU5F1P1 protein may exist. In addition, various long non-coding RNAs near rs6983267 have been found, some with potential effects on c-myc and other genes. At the time of writing, several groups are trying to pin down the functional variants at rs6983267, their efforts illustrating the many difficulties of this type of work, including our fundamental ignorance of the organs, tissues and cells in which the functional variation acts, and the stage or time at which it has its effects (Pomerantz et al. 2009; Tuupanen et al. 2009; Wasserman et al. 2010).

A detailed description of the discovery of the other 20 or so tagSNPs now known to be associated with colorectal cancer risk (Dunlop et al. 2012a) is neither possible nor desirable here. Nevertheless, the identification of these SNPs has been more than a train- or bus-spotting exercise, however much it may have seemed like that at times! Some of the lessons that have been learned are as follows.

1. The notion that common alleles affect cancer risk in the general population is almost certainly correct. Fine mapping to identify functional variation has not, in general, detected that the tagSNP signals result from, say, rare variants of stronger effect size.
2. The effects per allele are small, typically explaining a minimum of 10 % differential risk, as the SNP set currently stands. Presumably, there exists a larger number of as-yet unknown CRC susceptibility SNPs of weaker effects. These findings are consistent with predictions that, even for relatively late-onset diseases like cancer, variants with even moderately strong effects on risk would provide a non-trivial selective disadvantage and hence their frequencies would be low.
3. Many of the best candidate functional pathways—such as Wnt signaling and DNA repair—for colorectal cancer predisposition are absent or under-represented in the set of CRC SNPs.
4. Several CRC SNPs are at genes that encode proteins that act in the bone morphogenetic protein (BMP) pathway. In fact, this was a good candidate pathway, but one that was previously thought to be involved only in predisposition to rare intestinal polyp syndromes.
5. All risks appear to be log-additive, with no good evidence of epistasis (gene × gene interactions) to date.
6. Some genes harbor more than one, independent risk variant.
7. A variety of molecular processes and pathways seem capable to influencing CRC risk. It is likely that some effects are cell-autonomous (e.g., eukaryotic translation initiation factor 3 subunit H, the BMP inhibitor SMAD7, telomerase RNA component) and others are microenvironmental (e.g., BMP2, BMP4).
8. For all variation identified to date, there are plausible reasons (but no more than that) to suspect that the critical influences are on the colorectal mucosa rather than other organs or tissues.
9. In most cases, the underlying functional variation probably affects gene transcription through altering enhancer, insulator or repressor function. However, some CRC SNPs may tag variants with more direct influences on protein function (for example, in rhophilin2).

It must be emphasised that the underlying functional genetic variation has not been fully characterised for any of the identified CRC SNPs. the identified CRC SNPs. In some cases, moreover, one or more genes in the region may be affected by the functional variant, and the possibility that *in trans* effects influence distant genes or genes on other chromosomes cannot be discounted. Nevertheless, the above tentative conclusions appear reasonable and are generally supported by GWAS in other types of cancer.

11.5 The BMP Pathway and Colorectal Cancer Susceptibility

One of the most interesting findings from CRC GWAS has been the seven disease-associated SNPs close to BMP pathway genes. These principally include the secreted BMP antagonist gremlin (GREM1), the secreted BMP ligands BMP4 and BMP2 and the BMP-specific signal-inhibitory protein SMAD7. In addition, an eighth SNP lies near laminin A5 (*LAMA5*), which is required for BMP antagonist production. The SNPs near SMAD7 (rs4939827) and *GREM1* (rs4779584) were the second and third to be shown to associate with CRC susceptibility (Broderick et al. 2007; Jaeger et al. 2008), with relatively large differential risks of about 20 % per allele. Since then, the signal at rs4939827 has been fine-mapped to another SNP within an intron of SMAD7 that may differentially bind transcription factors (Pittman et al. 2009). The rs4779584 signal has been shown to result from two independent (non-correlated) SNPs represented by rs16969861 and rs11632715, both a few tens of kb upstream of *GREM1* (Tomlinson et al. 2011). *GREM1* thus represents an example of a synthetic association (Dickson et al. 2010)—a somewhat abused term, but essentially one that denotes a GWAS signal resulting from tagging of two independent underlying signals—and is a gene that, somewhat unusually, harbors both high- and low-penetrance risk variants for colorectal cancer (Jaeger et al. 2012). The CRC risk SNPs near *BMP4* (rs4444235 and rs1957636) lie on either side of the gene. Their association with *BMP4* expression remains controversial. The *BMP2* SNPs (rs4813802 and rs962153) both lie upstream of the gene, although effects on gene expression are currently unknown.

The precise role of BMP signaling in the gastrointestinal tract is not known (Hardwick et al. 2008). It probably has an important role in development, but may also contribute to the maintenance of the adult colorectal epithelium. One model posits that the BMP4 and BMP2 ligands, which are largely produced by mesenchymal cells near the colorectal crypt tops in response to epithelial Hedgehog signaling, act to modulate and attenuate Wnt signaling to ensure that colorectal epithelial cells differentiate as they move upwards from the stem cell niche at the bottom of the crypt. The BMP ligands bind to receptors such as type I and type II BMPR/activin receptors to activate downstream signaling though SMADs 1, 5 and 8 and the "common SMAD" SMAD4. The known BMP signaling targets include the ID1, 2, 3 genes, which are thought to function principally as transcriptional repressors, although their repertoire of target genes is not well known. Secreted BMP antagonists include Gremlin1, Gremlin2 and Noggin. These are produced by sub-epithelial myofibroblasts and smooth muscle cells at the crypt bottoms and adjacent to the basement membrane. One aspect of their function may be to aid Wnt signaling in maintaining the stem cell population and niche. This raises the possibility that the BMP pathway SNPs affect CRC risk by increasing the number of potentially cancer-forming cells. However, recent discoveries, such as the fact that Gremlin1 can also act as a VEGF receptor agonist, suggest that the underlying mechanism of raised cancer risk may be more complex. A particularly fascinating aspect of BMP signaling is its relationship to TGF-beta signaling, which is attenuated in some sporadic CRCs. The two pathways are related in their effects and overlap at points (although their target genes appear to be largely non-overlapping), but there is also evidence to show that TGF-beta signaling can suppress BMP signaling.

In conclusion, the identification of common, risk-associated polymorphisms in the BMP genes has highlighted the importance of this somewhat-neglected signaling pathway in the normal colorectum and in CRC risk. Although much remains unknown, it is entirely possible that other, perhaps less common, CRC predisposition variants exist within BMP pathway genes in addition to the SNPs and rare mutations already known. It is not clear why other important signaling pathways—Wnt, Delta-Notch, Ephrins and TGF-beta itself—harbor far fewer CRC predisposition variants, but we can surmise either that the BMP pathway is much more important than we have supposed or that the other pathways are too important for selection to tolerate polymorphisms of even small functional effect.

11.6 Criticisms of GWAS

Criticisms of cancer GWAS usually fall into one of the following categories:

1. GWAS have failed to explain more than a tiny part of the heritability of cancer.
2. GWAS alleles have small effects on risk that cannot be used in the clinic—this fact was predictable since alleles with stronger (and useable) effects sizes would be rendered rare by natural selection.
3. GWAS have been very expensive.
4. GWAS have been "handle-turning" exercises with very little intellectual content.

It must be acknowledged by GWAS proponents that while many of these criticisms were made prior to the first reports of GWAS findings: all have some validity. However, the criticisms also have a "straw man" element. GWAS is not the apotheosis of human genetics, but a contribution to a full understanding of the inherited basis of disease. Moreover, GWAS has clearly had not just a discovery role, but also an exploratory element, since neither experimental nor theoretical data could have predicted that so few common cancer predisposition alleles would have had relative risks as high as, say, twofold. Nevertheless, the possibility always existed that most common cancer alleles modulated risks by 10 % or less, most GWAS protagonists made no claims otherwise and a critic would have had to have true clairvoyancy to know this in advance. The notion that GWAS were wasteful is impossible to disprove without having some idea of what other projects would have been funded by the money, but at the time of writing, most GWAS have actually been remarkably efficient considering their scale, and they have delivered a great deal of scientifically fascinating data. In addition, the common criticism that GWAS results have no clinical application is not necessarily correct, as we shall see below. Since cancer develops to a large extent as a result of truly chance mutations and usually presents relatively late in life, individual risk prediction is always going to be difficult, since the variance in risk will only form a proportion of the variance in cancer development.

GWAS scientists have, perhaps, been unhelpful to themselves by emphasising how little of the sibling relative risk or heritability can be explained by common polymorphisms. Many explanations are proposed for this "missing heritability" of cancer, including our failure to date to find many truly "causal" variants, the existence of undiscovered rare risk alleles of larger effects or many common alleles of smaller effects, copy number variation, epigenetic variation, etc. Sibling relative risk is, moreover a useful, but rather crude measure, that may capture myriad genetic and non-genetic factors that cause disease to run in families. Instead, arguably more important measures for common diseases are the absolute and relative risks that can be explained by GWAS-discovered variants. Upper and lower quartiles of risk for the CRC SNPs, for example, differ by about twofold or more.

11.7 Current and Future Prospects

The list of current or potential follow-on experiments in the "Post-GWAS" world is long. It includes:

1. searching for additional risk variants, by using even larger sample sets and/or examining variants, including rare alleles, that are not captured by the available SNP arrays.
2. exploring risk in subgroups, such as microsatellite-unstable cancers (breast cancer GWAS has shown that different histological sub-types have different genetic risk profiles).
3. searching for risk loci in ethnic groups other than the white northern Europeans who have been the focus of CRC GWAS to date.
4. using additional strategies, such as admixture mapping.
5. exploring intermediate phenotypes, such as colorectal adenomas or serrated polyps, or even normal traits, such as weight and height, that are associated with CRC risk; these intermediate phenotype studies may be better powered than cancer-based studies in some circumstances.
6. undertaking a deeper analysis of existing data by imputation, haplotype-based tests, epistasis tests, gene–environment interaction searches, tests based on groups or sets of variants, etc.
7. searching for functional variation by methods ranging from genetic fine mapping though characterisation of the regulatory landscape and mRNA expression studies to animal models.

11.7.1 Clinical Application

It is common practice in CRC Genetics Clinics to offer enhanced cancer prevention measures to those whose family history predicts a mean increased risk of two–three-fold over the general population. Although counseling is delivered on an individual level, this enhanced screening is effectively applied on a cohort level to all "moderate-risk" patients in this situation (unlike those who carry mutations in one of the Mendelian cancer predisposition genes, who have intensive screening). The true CRC risks of the "moderate-risk" individuals are likely to vary greatly around the mean increased risk, which is itself highly likely to be an inaccurate estimate. If SNP typing can predict a twofold risk difference between top and bottom genetic risk quartiles (see above), it could be argued that SNP typing could be performed as well as, or even instead of, taking detailed family histories, and screening modulated on that basis, since both methods are poor at accurately predicting individual risks (Dunlop et al. 2012b). The fact that SNP typing has not yet happened in this context in part reflects the historical development of clinical genetics services around highly skilled diagnoses of rare pediatric conditions and its associated counseling, rather than as a molecular specialty.

If SNP typing to predict CRC risk is currently unlikely in the Genetics Clinic, might it be undertaken in the context of the entire population? Since population-based cancer screening by colonoscopy or fecal occult blood testing is in place in many countries, it would, in principle, be possible to modulate screening frequency and/or modality according to SNP genotype. At the level of the population, this would use resources more efficiently. However, there are potential problems wit this approach, including the following: a detailed economic assessment has not been performed to determine whether this could be cost-neutral; take-up rate of the genetic test might be low and might even reduce the numbers using the tumor screening test; and there could be political problems in stratifying population screening by population genetic risk. Nevertheless, in most countries, cancer screening has been introduced with little objection and special screening for those at very high risk of cancer is intrinsically unobjectionable. If the economic assessment were favorable, the principle of SNP-based population cancer screening could be established.

11.8 Conclusions

The history of cancer GWAS is a short, successful one. A panel of common, CRC-predisposition tagSNPs has been established, important molecular pathways identified and the basis established for functional studies to determine how the common variants have their effects. Like all other large undertakings, GWAS has suffered criticism, some fair and some not. The fact that common alleles have weaker effects on risk than hoped for is a disappointment, but this is actually an important GWAS finding that could not have been predicted in advance, especially for a relatively late-onset disease like cancer. It is unclear as to whether GWAS findings can be used in clinical practice for risk stratification for screening, or even for chemoprophylaxis. In the near future, GWAS activity is likely to be less than in the boom years of the late 2000s, but although GWAS may be more focused in the future, it is unlikely to die away. Colorectal cancer GWAS have added 20 low-penetrance tagSNPs to the 10 genes known to predispose to CRC in the Mendelian setting. Experience from GWAS and the patient cohorts obtained will allow much more efficient and better-planned future studies to identify the hoped-for class of CRC predisposition genes with allele frequencies and effects intermediate between Mendelian alleles and common SNPs. To undertake searches for these intermediate alleles without first undertaking GWAS would have been a mistake, and it remains to be seen whether the "glass half-full or half-empty" view of GWAS is replicated in the results of the intermediate allele searches.

References

Broderick P, Carvajal-Carmona L, Pittman AM, Webb E, Howarth K, Rowan A, Lubbe S, Spain S, Sullivan K, Fielding S, Jaeger E, Vijayakrishnan J, Kemp Z, Gorman M, Chandler I, Papaemmanuil E, Penegar S, Wood W, Sellick G, Qureshi M, Teixeira A, Domingo E, Barclay E, Martin L, Sieber O, Kerr D, Gray R, Peto J, Cazier JB, Tomlinson I, Houlston RS (2007) A genome-wide association study shows that common alleles of SMAD7 influence colorectal cancer risk. Nat Genet 39(11):1315–1317

International HapMap Consortium (2003) The International HapMap Project. Nature 426(6968):789–796

Dickson SP, Wang K, Krantz I, Hakonarson H, Goldstein DB (2010) Rare variants create synthetic genome-wide associations. PLoS Biol 8(1):e1000294

Dunlop MG, Dobbins SE, Farrington SM, Jones AM, Palles C, Whiffin N, Tenesa A, Spain S, Broderick P, Ooi LY, Domingo E, Smillie C, Henrion M, Frampton M, Martin L, Grimes G, Gorman M, Semple C, Ma YP, Barclay E, Prendergast J, Cazier JB, Olver B, Penegar S, Lubbe S, Chander I, Carvajal-Carmona LG, Ballereau S, Lloyd A, Vijayakrishnan J, Zgaga L, Rudan I, Theodoratou E, Thomas H, Maher E, Evans G, Walker L, Halliday D, Lucassen A, Paterson J, Hodgson S, Homfray T, Side L, Izatt L, Donaldson A, Tomkins S, Morrison P, Brewer C, Henderson A, Davidson R, Murday V, Cook J, Haites N, Bishop T, Sheridan E, Green A, Marks C, Carpenter S, Broughton M, Greenhalge L, Suri M, Starr JM, Deary I, Kirac I, Kovacevic D, Aaltonen LA, Renkonen-Sinisalo L, Mecklin JP, Matsuda K, Nakamura Y, Okada Y, Gallinger S, Duggan DJ, Conti D, Newcomb P, Hopper J, Jenkins MA, Schumacher F, Casey G, Easton D, Shah M, Pharoah P, Lindblom A, Liu T, Edler D, Lenander C, Dalen J, Hjern F, Lundqvist N, Lindforss U, Pahlman L, Smedh K, Tornqvist A, Holm J, Janson M, Andersson M, Ekelund S, Olsson L, Smith CG, West H, Cheadle JP, Macdonald G, Samuel LM, Ahmad A, Corrie P, Jodrell D, Palmer C, Wilson C, O'Hagan J, Smith D, McDermott R, Walshe J, Cassidy J, McDonald A, Mohammed N, White J, Yosef H, Breathnach O, Grogan L, Thomas R, Eatock M, Henry P, Houston R, Johnston P, Wilson R, Geh I, Danwata F, Hindley A, Susnerwala S, Bradley C, Conn A, Raine A, Twelves C, Falk S, Hopkins K, Tahir S, Dhadda A, Maraveyas A, Sgouros J, Teo M, Ahmad R, Cleator S, Creak A, Lowdell C, Riddle P, Benstead K, Farrugia D, Reed N, Shepherd S, Levine E, Mullamitha S, Saunders M, Valle J, Wilson G, Jones A, Weaver A, Clark PI, Haylock B, Iqbal MI, Myint AS, Beesley S, Sevitt T, Nicoll J, Daniel F, Ford V, Talbot T, Butt M, Hamid A, Mack P, Roy R, Osborne R, McKinna F, Alsab H, Basu D, Murray P, Sizer B, Azam FA, Neupane R, Waterston A, Glaholm J, Blesing C, Lowndes S, Medisetti A, Gaya A, Leslie M, Maisey N, Ross P, Dunn G, Al-Salihi O, Wasan HS, Tan LT, Dent J, Hofmann U, Joffe JK, Sherwin E, Soomal R, Chakrabarti A, Joseph S, Van der Voet J, Wadd NJ, Wilson D, Anjarwalia S, Hall J, Hughes R, Polychronis A, Scarffe JH, Hill M, James RD, Shah R, Summers J, Hartley A, Carney D, McCaffrey J, Bystricky B, O'Reilly S, Gupta R, Al-Mishlab T, Gidden F, O'Hara R, Stewart J, Ashford R, Glynne-Jones R, Harrison M, Mawdsley S, Barlow H, Tighe M, Walther J, Neal J, Rees C, Bridgewater J, Karp S, McGovern U, Atherton PJ, El-Deeb H, Macmillan C, Patel K, Bessell EM, Dickinson PD, Potter V, Jephcott C, McAdam K, Wrigley J, Muthuramalingam S, O'Callaghan A, Melcher L, Braconi C, Geh JI, Palmer D, Narayana P, Steven N, Rudman S, Chakraborti P, Kelly K, Macgregor C,

Whillis D, Freebairn A, Gildersleve J, Sharif S, Astras G, Hickish T, Beech D, Ellis R, Kulkarni R, Shankland K, Begent R, Mayer A, Meyer T, Strauss S, Hall V, Raj S, Chau I, Cunningham D, Birtle A, Biswas A, Wise M, Cummins S, Essapen S, Middleton G, Topham C, Langley R, Webb A, Wilkins M, Iveson TJ, Askill C, Wagstaff J, Azzabi A, Bateman A, Prejbisz J, Tsang D, Ali N, O'Neill P, Cottrill C, Propper D, Lofts FJ, Kennedy J, Anthoney DA, Cooper R, Crellin A, Melcher A, Seymour M, Baughan C, Alexander E, Crown J, Fennelly D, Adab F, Giridharan S, Pedley I, Wright K, Bliss P, Cogill G, Lo N, Toy E, Hochhauser D, Ledermann J, Brewster A, Maughan T, Mort D, Mukherjee S, Dobrowsky W, Calvert P, Leonard G, Ford H, Moody AM, Goriah S, Clive S, Dawson L, McLean C, Phillips HA, Gopi K, Tomlinson M, Clenton S, Furniss D, Hornbuckle J, Pledge S, Wadsley J, Abbas M, Marshall E, Harper-Wynne C, Barnes A, Kumar S, Vigneswaran V, Gollins S, Genton M, Sparrow G, Bale C, Fuller C, Mullard A, Stuart N, Williams R, Keane M, Wasan H, Adams R, Madi A, Hodgkinson E, Rogers P, Pope M, Kaplan R, Meade A, Parmar M, Kenny S, Fisher D, Harper L, Mitchell J, Nichols L, Sydes B, Clement L, Kay E, Courtney C, Gallagher M, Murphy C, Thompson L, Beall S, Hassan S, Gracie R, Griffiths G, Mason M, Parker C, Rudd R, Johnson P, Whelan J, Northover J, Brown J, Aapro M, Stout R, Midgley R, Kerr DJ, Campbell H, Tomlinson IP, Houlston RS (2012a) Common variation near CDKN1A, POLD3 and SHROOM2 influences colorectal cancer risk. Nat Genet 44(7):770–776

Dunlop MG, Tenesa A, Farrington SM, Ballereau S, Brewster DH, Koessler T, Pharoah P, Schafmayer C, Hampe J, Volzke H, Chang-Claude J, Hoffmeister M, Brenner H, von Holst S, Picelli S, Lindblom A, Jenkins MA, Hopper JL, Casey G, Duggan D, Newcomb PA, Abuli A, Bessa X, Ruiz-Ponte C, Castellvi-Bel S, Niittymaki I, Tuupanen S, Karhu A, Aaltonen L, Zanke B, Hudson T, Gallinger S, Barclay E, Martin L, Gorman M, Carvajal-Carmona L, Walther A, Kerr D, Lubbe S, Broderick P, Chandler I, Pittman A, Penegar S, Campbell H, Tomlinson I, Houlston RS (2012b) Cumulative impact of common genetic variants and other risk factors on colorectal cancer risk in 42 103 individuals. Gut 62(6):871–881

Easton DF, Pooley KA, Dunning AM, Pharoah PD, Thompson D, Ballinger DG, Struewing JP, Morrison J, Field H, Luben R, Wareham N, Ahmed S, Healey CS, Bowman R, Meyer KB, Haiman CA, Kolonel LK, Henderson BE, Le Marchand L, Brennan P, Sangrajrang S, Gaborieau V, Odefrey F, Shen CY, Wu PE, Wang HC, Eccles D, Evans DG, Peto J, Fletcher O, Johnson N, Seal S, Stratton MR, Rahman N, Chenevix-Trench G, Bojesen SE, Nordestgaard BG, Axelsson CK, Garcia-Closas M, Brinton L, Chanock S, Lissowska J, Peplonska B, Nevanlinna H, Fagerholm R, Eerola H, Kang D, Yoo KY, Noh DY, Ahn SH, Hunter DJ, Hankinson SE, Cox DG, Hall P, Wedren S, Liu J, Low YL, Bogdanova N, Schurmann P, Dork T, Tollenaar RA, Jacobi CE, Devilee P, Klijn JG, Sigurdson AJ, Doody MM, Alexander BH, Zhang J, Cox A, Brock IW, MacPherson G, Reed MW, Couch FJ, Goode EL, Olson JE, Meijers-Heijboer H, van den Ouweland A, Uitterlinden A, Rivadeneira F, Milne RL, Ribas G, Gonzalez-Neira A, Benitez J, Hopper JL, McCredie M, Southey M, Giles GG, Schroen C, Justenhoven C, Brauch H, Hamann U, Ko YD, Spurdle AB, Beesley J, Chen X, Mannermaa A, Kosma VM, Kataja V, Hartikainen J, Day NE, Cox DR, Ponder BA (2007) Genome-wide association study identifies novel breast cancer susceptibility loci. Nature 447(7148):1087–1093

Hardwick JC, Kodach LL, Offerhaus GJ, van den Brink GR (2008) Bone morphogenetic protein signalling in colorectal cancer. Nat Rev Cancer 8(10):806–812

Houlston RS (2012) COGENT (COlorectal cancer GENeTics) revisited. Mutagenesis 27(2):143–151

Houlston RS, Tomlinson IP (2001) Polymorphisms and colorectal tumor risk. Gastroenterology 121(2):282–301

Jaeger E, Webb E, Howarth K, Carvajal-Carmona L, Rowan A, Broderick P, Walther A, Spain S, Pittman A, Kemp Z, Sullivan K, Heinimann K, Lubbe S, Domingo E, Barclay E, Martin L, Gorman M, Chandler I, Vijayakrishnan J, Wood W, Papaemmanuil E, Penegar S, Qureshi M, Farrington S, Tenesa A, Cazier JB, Kerr D, Gray R, Peto J, Dunlop M, Campbell H, Thomas H, Houlston R, Tomlinson I (2008) Common genetic variants at the CRAC1 (HMPS) locus on chromosome 15q13.3 influence colorectal cancer risk. Nat Genet 40(1):26–28

Jaeger E, Leedham S, Lewis A, Segditsas S, Becker M, Cuadrado PR, Davis H, Kaur K, Heinimann K, Howarth K, East J, Taylor J, Thomas H, Tomlinson I (2012) Hereditary mixed polyposis syndrome is caused by a 40-kb upstream duplication that leads to increased and ectopic expression of the BMP antagonist GREM1. Nat Genet 44(6):699–703

Lichtenstein P, Holm NV, Verkasalo PK, Iliadou A, Kaprio J, Koskenvuo M, Pukkala E, Skytthe A, Hemminki K (2000) Environmental and heritable factors in the causation of cancer—analyses of cohorts of twins from Sweden, Denmark, and Finland. N Engl J Med 343(2):78–85

Lynch HT, Lynch PM, Lanspa SJ, Snyder CL, Lynch JF, Boland CR (2009) Review of the Lynch syndrome: history, molecular genetics, screening, differential diagnosis, and medicolegal ramifications. Clin Genet 76(1):1–18

Pittman AM, Naranjo S, Webb E, Broderick P, Lips EH, van Wezel T, Morreau H, Sullivan K, Fielding S, Twiss P, Vijayakrishnan J, Casares F, Qureshi M, Gomez-Skarmeta JL, Houlston RS (2009) The colorectal cancer risk at 18q21 is caused by a novel variant altering SMAD7 expression. Genome Res 19(6):987–993

Pomerantz MM, Ahmadiyeh N, Jia L, Herman P, Verzi MP, Doddapaneni H, Beckwith CA, Chan JA, Hills A, Davis M, Yao K, Kehoe SM, Lenz HJ, Haiman CA, Yan C, Henderson BE, Frenkel B, Barretina J, Bass A, Tabernero J, Baselga J, Regan MM, Manak JR, Shivdasani R, Coetzee GA, Freedman ML (2009) The 8q24 cancer risk variant rs6983267 shows long-range interaction with MYC in colorectal cancer. Nat Genet 41(8):882–884

Tomlinson I, Webb E, Carvajal-Carmona L, Broderick P, Kemp Z, Spain S, Penegar S, Chandler I, Gorman M, Wood W, Barclay E, Lubbe S, Martin L, Sellick G, Jaeger E, Hubner R, Wild R, Rowan A, Fielding S, Howarth K, Silver A, Atkin W, Muir K, Logan R, Kerr D, Johnstone E, Sieber O, Gray R, Thomas H, Peto J, Cazier JB, Houlston R (2007) A genome-wide association scan of tag SNPs identifies a susceptibility variant for colorectal cancer at 8q24.21. Nat Genet 39(8):984–988

Tomlinson IP, Carvajal-Carmona LG, Dobbins SE, Tenesa A, Jones AM, Howarth K, Palles C, Broderick P, Jaeger EE, Farrington S, Lewis A, Prendergast JG, Pittman AM, Theodoratou E, Olver B, Walker M, Penegar S, Barclay E, Whiffin N, Martin L, Ballereau S, Lloyd A, Gorman M, Lubbe S, Howie B, Marchini J, Ruiz-Ponte C, Fernandez-Rozadilla C, Castells A, Carracedo A, Castellvi-Bel S, Duggan D, Conti D, Cazier JB, Campbell H, Sieber O, Lipton L, Gibbs P, Martin NG, Montgomery GW, Young J, Baird PN, Gallinger S, Newcomb P, Hopper J, Jenkins MA, Aaltonen LA, Kerr DJ, Cheadle J, Pharoah P, Casey G, Houlston RS, Dunlop MG (2011) Multiple common susceptibility variants near BMP pathway loci GREM1, BMP4, and BMP2 explain part of the missing heritability of colorectal cancer. PLoS Genet 7(6):e1002105

Tuupanen S, Turunen M, Lehtonen R, Hallikas O, Vanharanta S, Kivioja T, Bjorklund M, Wei G, Yan J, Niittymaki I, Mecklin JP, Jarvinen H, Ristimaki A, Di-Bernardo M, East P, Carvajal-Carmona L, Houlston RS, Tomlinson I, Palin K, Ukkonen E, Karhu A, Taipale J, Aaltonen LA (2009) The common colorectal cancer predisposition SNP rs6983267 at chromosome 8q24 confers potential to enhanced Wnt signaling. Nat Genet 41(8):885–890

Wasserman NF, Aneas I, Nobrega MA (2010) An 8q24 gene desert variant associated with prostate cancer risk confers differential in vivo activity to a MYC enhancer. Genome Res 20(9):1191–1197

Westerman AM, Entius MM, de Baar E, Boor PP, Koole R, van Velthuysen ML, Offerhaus GJ, Lindhout D, de Rooij FW, Wilson JH (1999) Peutz-Jeghers syndrome: 78-year follow-up of the original family. Lancet 353(9160):1211–1215

White KL, Sellers TA, Fridley BL, Vierkant RA, Phelan CM, Tsai YY, Kalli KR, Berchuck A, Iversen ES, Hartmann LC, Liebow M, Armasu S, Fredericksen Z, Larson MC, Duggan D, Couch FJ, Schildkraut JM, Cunningham JM, Goode EL (2010) Variation at 8q24 and 9p24 and risk of epithelial ovarian cancer. Twin Res Hum Genet 13(1):43–56

Yeager M, Orr N, Hayes RB, Jacobs KB, Kraft P, Wacholder S, Minichiello MJ, Fearnhead P, Yu K, Chatterjee N, Wang Z, Welch R, Staats BJ, Calle EE, Feigelson HS, Thun MJ, Rodriguez C, Albanes D, Virtamo J, Weinstein S, Schumacher FR, Giovannucci E, Willett WC, Cancel-Tassin G, Cussenot O, Valeri A, Andriole GL, Gelmann EP, Tucker M, Gerhard DS, Fraumeni JF Jr, Hoover R, Hunter DJ, Chanock SJ, Thomas G (2007) Genome-wide association study of prostate cancer identifies a second risk locus at 8q24. Nat Genet 39(5):645–649

Chapter 12
Future Prospects for Leveraging Molecular Information in the Fight Against Colorectal Cancer

Laura B. Kleiman and Kevin M. Haigis

Abstract Our understanding of the somatic genetic events that contribute to the initiation and progression of colorectal cancer (CRC) has increased in leaps and bounds over the past 25 years. And while this knowledge has dramatically changed the way we think about disease pathogenesis, and in the process led directly to new ideas about therapeutic approaches, questions remain as to how we can leverage this knowledge to cure the disease. This chapter explores two questions that arise from the study of the molecular pathogenesis of CRC. First, can we exploit the multi-step nature of CRC to develop effective therapeutic strategies? Second, do the genetic events driving CRC progression provide a means to develop molecularly targeted assays for early detection?

12.1 Targeted Therapies for Colorectal Cancers

The overarching goal of this book is to highlight the changing landscape of colorectal cancer (CRC) therapy as a result of our increased understanding of the molecular pathogenesis of the disease, focusing primarily on the somatic genetics of CRC. The promise of understanding CRC genetics is that it will lead to the development of patient-specific therapeutic strategies geared toward the exact molecular perturbations that are driving their cancer. In many instances, it is clear to see how gene-level knowledge of an individual's tumor could be useful, for example, using B-RAF inhibitors for cancers expressing mutant B-RAF, although even this concept is not as straightforward as one would hope because treatment of B-RAF mutant CRC cells induces an EGFR-mediated pro-survival response (Corcoran et al. 2012; Prahallad et al. 2012). In other situations, for example in the case of *KRAS*

L.B. Kleiman • K.M. Haigis (✉)
Molecular Pathology Unit, Center for Cancer Research, and Center for Systems Biology, Massachusetts General Hospital, Charlestown, MA 02129, USA
e-mail: khaigis@partners.org

K.M. Haigis (ed.), *Molecular Pathogenesis of Colorectal Cancer*,
DOI 10.1007/978-1-4614-8412-7_12, © Springer Science+Business Media New York 2013

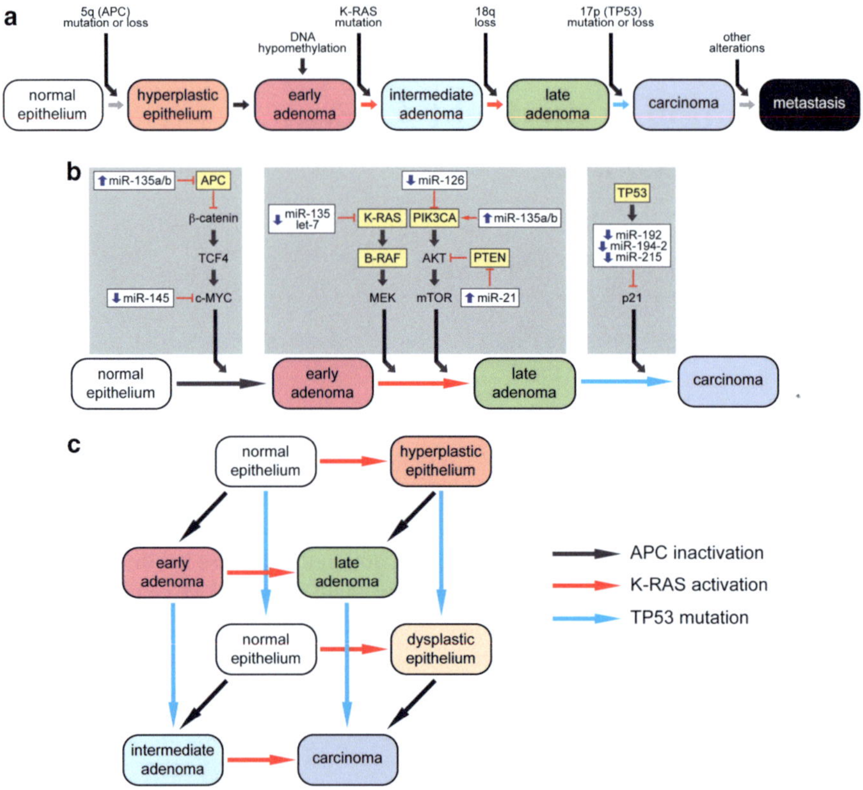

Fig. 12.1 Schematic representations of CRC progression. (**a**) The original Fearon and Vogelstein diagram of CRC progression. This representation is based largely on targeted sequencing efforts that detected mutations in lesions of various histologic states (adapted from Fearon and Vogelstein 1990). (**b**) Updated progression diagram. This representation includes genetic data from large scale sequencing efforts and also analyses of gene and miRNA expression. Note that this is not a comprehensive survey of the molecular pathogenesis of the disease, but rather an updated version of the original linear representation adapted from (Goel and Boland 2012). (**c**) Three-dimensional representation of CRC progression assuming that the accumulation of mutations, rather than the order of mutations, determines the histologic state of tumor. The nonneoplastic endpoints have been revealed through the analysis of genetically engineered mouse models

mutations (where the mutant protein cannot be targeted directly), an obvious therapeutic strategy is not clear.

While many of the chapters in this book focus on a specific pathway that plays a role in CRC, unanswered questions arise from the observation that CRC is a multistep process. In their seminal paper on the genetics of CRC, Fearon and Vogelstein implicated four genomic regions that were altered in the transition from normal colonic epithelium to CRC (Fig. 12.1a) (Fearon and Vogelstein 1990). Over the past 2 decades, many more genes (both coding and noncoding) have been implicated in the pathogenesis of CRC through mutation studies and through gene expression analysis (Fig. 12.1b). To translate this knowledge into therapeutic strategies, we must now determine the relative contribution of each gene to establishing and maintaining the malignant state. If all mutations in a cancer are equally responsible for maintaining the malignant state, an effective therapy might require a cocktail of drugs targeting the entire repertoire of mutations, increasing the likelihood of toxic side effects. Alternatively, if only a subset of mutations is required, fewer drugs would be required for efficacy. This concept is analogous to a pyramidal house of cards. All of the individual cards contribute to the establishment of the final structure. Removal of a single card near the top of the pyramid will leave the bottom of the structure intact. By contrast, removal of a single card at the bottom of the pyramid will bring down the entire structure. Moving forward, the goal for the field will be to identify, and to develop drugs for, those mutations that are most important for CRC progression.

One way that specific genes might contribute disproportionately to malignancy is if the order in which mutations arise matters to the establishment of the endpoint. The original Fearon and Vogelstein diagram (Fig. 12.1a) is often interpreted to mean that mutations arise in a linear order, with APC mutations arising first, followed by K-RAS mutations, etc. Nevertheless, Fearon and Vogelstein indicate that "the total accumulation of changes, rather than their order with respect to one another, is responsible for determining the tumor's biologic properties." Part of the confusion might arise from the fact that the original model did not include nonneoplastic endpoints, for example, aberrant crypt foci. If one considers nonneoplastic endpoints, which have been revealed by studying genetically engineered mouse models, a nonlinear model can describe the step-wise transition from normal colon to malignant cancer that is independent of the order in which mutations are acquired (Fig. 12.1c).

It is likely that order does matter in some genotypic contexts, however. For example, activating mutations in K-RAS and its downstream effector B-RAF occur commonly in CRC. In mouse models, mutational activation of K-RAS in the intestinal epithelium leads to hyperplasia (Haigis et al. 2008), but mutational activation of B-RAF leads to senescence (Carragher et al. 2010). B-RAF mutation also leads to senescence in the lung epithelium and in melanocytes, and this can be overcome with concomitant mutation of tumor suppressor genes, for example, TP53 or PTEN (Dankort et al. 2007, 2009). Thus, in the context of a developing CRC, it is likely that TP53 mutations must precede B-RAF mutation in order to avoid senescence-associated growth arrest. Perhaps therapies that can overcome the loss of tumor suppressor activity would push B-RAF-mutant cells back into senescence.

While the order of mutations might be relevant in certain cases and not in others, genetic data from mice and humans suggest that certain mutations are more critical than others for the formation of CRC. Phenotypic analyses of mouse models carrying mutations in genes that play a role in CRC progression (K-RAS, B-RAF, TP53, SMAD4, etc.) indicate that most of them are incapable of inducing neoplastic growth on their own (Fig. 12.1c) (Carragher et al. 2010; Haigis et al. 2008; Halberg et al. 2000; Takaku et al. 1998). By contrast, mutation of APC is largely sufficient for neoplasia. Since loss of APC initiates neoplastic growth, this event appears at the beginning of the linear progression model (Fig. 12.1a). This observation raises the intriguing possibility that therapies targeting the APC pathway would be capable of transforming CRC back into a nonneoplastic state. The tumor suppressor activity of APC is thought to derive predominantly from its activity as a negative regulator of WNT signaling and WNT pathway inhibitors have emerged over the past several years (Huang et al. 2009; Thorne et al. 2010). Time will tell whether these, or other, pathway inhibitors are effective against benign and malignant CRCs with APC mutations.

12.2 Molecular Methods for CRC Screening

In addition to aiding in the development of therapeutic strategies, the identification of somatic alterations may be important for CRC screening and monitoring. Clinical symptoms of CRC typically arise at an advanced stage and CRCs are often not detected until the cancer has already spread. Nevertheless, if diagnosed at an early stage—while the tumor is still localized—CRC can be cured by surgery. Mathematical models suggest that progression from a benign adenoma to an adenocarcinoma is slow (~17 years) and that the transition from a localized adenocarcinoma to metastatic disease is much faster (~2 years) (Jones et al. 2008), providing a window of opportunity to find and remove pre-metastatic lesions, provided an accurate screening protocol is in place.

Various screening methods are in clinical use and have decreased CRC incidence and mortality. Colonoscopy is clearly the most sensitive way to screen for CRC, but it is invasive and costly. Noninvasive screening methods, for example, detection of biomarkers in serum, would be a major step forward for the field. Carcinoembryonic antigen (CEA) is the most investigated proteomic marker for CRC. CEA is normally expressed only during fetal development, but elevated levels can be detected in the serum of individuals with CRC. Additional proteomic markers under investigation include cytokines, antigens, antibodies, and other mutated or aberrantly expressed proteins (Tjalsma 2010). The major limitation in developing molecular markers is their lack of specificity and sensitivity. Increased expression of CEA, for example, is also detected in heavy smokers and individuals with a variety of malignancies, including gastric and pancreatic cancer. The CEA test is therefore unreliable for CRC screening or diagnosis. Moreover, tumors often induce inflammatory reactions, which can lead to global proteomic changes mirroring inflammatory diseases.

Fecal testing, perhaps, holds the most promise for integrating molecular insights into CRC screening. Currently, the fecal occult blood test (FOBT) is the most commonly used noninvasive method for CRC screening. This assay detects bleeding associated with large adenomas and CRCs by guaiac-based or immunochemical assays with stool samples. The FOBT is inexpensive, but suffers from a low specificity rate. Since it tests for the presence of blood and not a tumor-specific marker, positive results can indicate, in addition to CRC, many non-tumor lesions such as inflammatory bowel disease, celiac disease, endometriosis, or trauma.

Efforts are ongoing to improve upon fecal screening methods in order to detect changes more directly related to the CRC disease process. For example, DNA isolated from fecal samples can be surveyed for mutations associated with CRC. Using a single nucleotide extension assay, mutations in APC, K-RAS, TP53, and PIK3CA were detectable in fecal samples from patients with CRC and these mutations matched those found in the corresponding primary tumors (Diehl et al. 2008). It remains to be seen whether such an assay is sensitive enough to be used in a clinical setting. In the simplest view, increased sensitivity of an APC mutation test may be the most promising as a screening tool to detect the earliest adenomas and the largest number of cancers. Ideally, however, a screening assay would be able to detect the entire catalogue of CRC-associated mutations in a given sample. Given the wealth of knowledge pertaining to the molecular pathogenesis of CRC, a new generation of therapies and screening methods are sure to arise over the coming years.

References

Carragher LA, Snell KR, Giblett SM, Aldridge VS, Patel B, Cook SJ, Winton DJ, Marais R, Pritchard CA (2010) V600EBraf induces gastrointestinal crypt senescence and promotes tumour progression through enhanced CpG methylation of p16INK4a. EMBO Mol Med 2:458–471

Corcoran RB, Ebi H, Turke AB, Coffee EM, Nishino M, Cogdill AP, Brown RD, Della Pelle P, Dias-Santagata D, Hung KE et al (2012) EGFR-mediated re-activation of MAPK signaling contributes to insensitivity of BRAF mutant colorectal cancers to RAF inhibition with vemurafenib. Cancer Discov 2:227–235

Dankort D, Filenova E, Collado M, Serrano M, Jones K, McMahon M (2007) A new mouse model to explore the initiation, progression, and therapy of BRAFV600E-induced lung tumors. Genes Dev 21:379–384

Dankort D, Curley DP, Cartlidge RA, Nelson B, Karnezis AN, Damsky WE Jr, You MJ, DePinho RA, McMahon M, Bosenberg M (2009) Braf(V600E) cooperates with Pten loss to induce metastatic melanoma. Nat Genet 41:544–552

Diehl F, Schmidt K, Durkee KH, Moore KJ, Goodman SN, Shuber AP, Kinzler KW, Vogelstein B (2008) Analysis of mutations in DNA isolated from plasma and stool of colorectal cancer patients. Gastroenterology 135:489–498

Fearon ER, Vogelstein B (1990) A genetic model for colorectal tumorigenesis. Cell 61:759–767

Goel A, Boland CR (2012) Epigenetics of colorectal cancer. Gastroenterology 143:1442–1460 e1441

Haigis KM, Kendall KR, Wang Y, Cheung A, Haigis MC, Glickman JN, Niwa-Kawakita M, Sweet-Cordero A, Sebolt-Leopold J, Shannon KM et al (2008) Differential effects of oncogenic K-Ras and N-Ras on proliferation, differentiation and tumor progression in the colon. Nat Genet 40:600–608

Halberg RB, Katzung DS, Hoff PD, Moser AR, Cole CE, Lubet RA, Donehower LA, Jacoby RF, Dove WF (2000) Tumorigenesis in the multiple intestinal neoplasia mouse: redundancy of negative regulators and specificity of modifiers. Proc Natl Acad Sci U S A 97:3461–3466

Huang SM, Mishina YM, Liu S, Cheung A, Stegmeier F, Michaud GA, Charlat O, Wiellette E, Zhang Y, Wiessner S et al (2009) Tankyrase inhibition stabilizes axin and antagonizes Wnt signalling. Nature 461:614–620

Jones S, Chen WD, Parmigiani G, Diehl F, Beerenwinkel N, Antal T, Traulsen A, Nowak MA, Siegel C, Velculescu VE et al (2008) Comparative lesion sequencing provides insights into tumor evolution. Proc Natl Acad Sci U S A 105:4283–4288

Prahallad A, Sun C, Huang S, Di Nicolantonio F, Salazar R, Zecchin D, Beijersbergen RL, Bardelli A, Bernards R (2012) Unresponsiveness of colon cancer to BRAF(V600E) inhibition through feedback activation of EGFR. Nature 483:100–103

Takaku K, Oshima M, Miyoshi H, Matsui M, Seldin MF, Taketo MM (1998) Intestinal tumorigenesis in compound mutant mice of both Dpc4 (Smad4) and Apc genes. Cell 92:645–656

Thorne CA, Hanson AJ, Schneider J, Tahinci E, Orton D, Cselenyi CS, Jernigan KK, Meyers KC, Hang BI, Waterson AG et al (2010) Small-molecule inhibition of Wnt signaling through activation of casein kinase 1alpha. Nat Chem Biol 6:829–836

Tjalsma H (2010) Identification of biomarkers for colorectal cancer through proteomics-based approaches. Expert Rev Proteomics 7:879–895

Index

A

- Aberrant crypt foci (ACF), 27
- Activin signal, 219–220
- *ACVR2*, 206
- *ACVR2A*, 42
- Adenoma–carcinoma sequence, 26
  - aberrant crypt foci, 27
  - *APC* and *KRAS* gene, 27
  - somatic mutations, 27, 28
- Amado, R.G., 236, 237
- Aneuploidy
  - CDC4 and chromosome 1p deletion, 36
  - 18q loss, 35–36
- Angiogenesis, 77, 88–89
- APC
  - mouse models
    - allele, 113
    - ISC marker genes, 115
    - mutation cluster region, 114
    - sequence, 114
    - Wnt signaling, 113
  - protein, 31–33
- *APC*
  - alleles, 32
  - β-catenin, 32, 112
  - familial adenomatous polyposis
    - *CTNNB1*, 7
    - Gardner's syndrome, 4–5
    - partial karyotype, 5
    - 5q21, 5
    - Wnt signaling, 6
  - genome and implications, 253
  - tumor suppressor gene loci, 280
- Aspirin
  - blocking inflammation, 89–91
  - serrated polyps, 52
- *AXIN2*, 33

B

- Barker, N., 111
- Bass, A.J., 267–283
- Bevacizumab, 139
- Biswas, S., 213
- BKM120 (Novartis), 177
- Blobe, G.C., 206
- B lymphocytes, 87
- *BMPR1A*, 12–14, 206
- Bokemeyer, C., 236
- Bone morphogenetic proteins (BMP)
  - gastro intestinal tract, 294
  - signaling and cancer
    - BMPR1A, 219
    - GWAS, 294–295
    - targets, 294
- *BRAF*
  - clinical significance, 238–239
  - and metastatic CRC, 239
  - mutation
    - B-Raf, 136
    - ERK MAPK pathway, 137
    - V600E, 136
  - serrated pathway, 48–49
- $BRAF^{V600E}$, 39, 49
- Brower, E.T., 157–187

C

- CAC. *See* Colitis-associated cancer (CAC)
- Cameron, D.J., 89
- Cantley, L.C., 157–187
- Carcinoembryonic antigen (CEA) test, 304
- β-Catenin
  - APC, 32, 114
  - familial adenomatous polyposis, 6
  - Wnt signaling, 110

K.M. Haigis (ed.), *Molecular Pathogenesis of Colorectal Cancer*,
DOI 10.1007/978-1-4614-8412-7, © Springer Science+Business Media New York 2013

CCNE1, 37
Centromere dysfunction, 30
Cetuximab, 39, 139, 234–235
Chan, AO-O., 49
Chemokines, 77
Chen, T., 206
Chromoendoscopy, 53
Chromosomal instability pathway (CIN)
  adenoma–carcinoma sequence, 26
    aberrant crypt foci, 27
    APC and *KRAS* gene, 27
    somatic mutations, 27, 28
  centromere dysfunction, 30
  chromosomal segregation, 29
  clinical implications, 38–39
  DNA damage response, 31
  genetic abnormalities
    APC, 31–33
    CCNE1, 37
    CDK8, 37
    CMYC, 36
    COX2, 37
    FBXW7, 37
    KRAS, 34
    LKB1, 38
    PIK3CA and PTEN, 34–35
    18q loss, 35–36
    TGF-β type II receptor, 36
    TP53, 35
    Wnt pathway, 31–33
  genomic instability and cancer, 27, 29
  loss of heterozygosity, 30–31
  sporadic colorectal carcinoma, 78
  telomere dysfunction, 30
  timing, 38
Chu, T.Y., 206
Chung, D.C., 32
Churchill, W.H., 89
CKD8, 37
CMYC, 36
Colitis-associated cancer (CAC)
  angiogenesis, 77
  chemokine expression, 77
  dextran sodium sulfate, 71
  genetic and epigenetic alteration, 71–73
  IL-21, 76
  IL-6/STAT3, 74–75
  NFκB, 73–74
  vs. sporadic colorectal cancer, 77
    clinical and pathologic features, 78
    molecular alterations, 78–81
  TNF- α, 75–76
Complex genomic disease, 267–269
Copy-number alterations
  arm-level, 273–276
  complex genomic disease, 267–269
  mechanisms, somatic cells, 269–271
  SCNA study, in adenocarcinoma, 281–283
  somatic copy-number applications, 281
  technical and statistical methods, 272–273
  tumor suppressor genes
    chromosome 17q, 277
    focal amplifications, 277
    focal deletions, 277, 280
    focal genomic amplifications, 276
    fragile site deletions, 280
    high-density array-methods, 276
    SCNA profiling techniques, 276, 280
    SMAD3 role, 280
    visualized examples, 277–279
Cowden syndrome, 14–15
COX2, 37
CpG island methylator phenotype (CIMP)
  microsatellite instability, 41
  serrated pathway, 48–49
Cripps, W.H., 3, 4
Crohn, B.B., 70
CTNNB1, 7
Cytochrome P450cre, 112

D
DCC, 35
Deacu, E., 206
Dendritic cells (DCs), 84–85
Der, C.J., 121–149
Dextran sodium sulfate (DSS), 71
DNA damage response, 31
DNA mismatch repair (MMR) system
  EXO1, 9
  Lynch syndrome
    HNPCC, 8
    MSH2, 8
    PMS2, 9
  microsatellite instability, 39–40
  PMS1, 9
Douillard, J.Y., 236
Dove, W., 32
Dysplastic serrated polyps, 47

E
Eg5, 29
Elliott, R.L., 206
EMAST, 45
Engle, S.J., 213

Epidermal growth factor receptor (EGFR)
  clinical significance, of mutations, 232
  inhibitors, 131
  KRAS mutations, 235–236
Eppert, K., 206
ERBB2, 281, 282

F
Familial adenomatous polyposis (FAP)
  β-catenin, 6
  CTNNB1, 7
  Gardner's syndrome, 4–5
  MUTYH, 7
  partial karyotype, 5
  5q21, 5
  Wnt signaling, 6
Familial colorectal cancer type X (FCCTX), 11–12
Familial juvenile polyposis, 13
Farina-Sarasqueta, A., 238
Farnesyltransferase inhibitors (FTIs), 142–143
FBXW7
  CIN, 37
  genome and implications, 253
Fearon, E.R., 28, 302
Fecal test, 304
FHIT, 280
5-Fluorouracil in combination with irinotecan (FOLFIRI), 139
5-Fluorouracil in combination with oxaliplatin (FOLFOX), 139

G
Gardner, E.J., 4
Gardner's syndrome, 4–5
GDP-GTP regulation, 125–127
Genetics
  vs. environmental components, 2
  history, 3
  microsatellite stable, 124
  mismatch-repair, 123
  RAS mutational activation, 122–123
Genome and implications
  clinical value
    detection, 259–260
    targeted therapies, 257–259
  era of unbiased studies
    colorectal genetic landscape, depiction, 252
    integrated analysis, 253
    key features, 252
    mutation analysis, 251
    somatic mutation identification, 251
    translocation analysis, 253
  interpretation
    ARID1A, 255–256
    cancer gene list, 255
    cancer genome complexities, 254–255
    epigenetic processes, 257
    integrated analysis, 254
    pathways, 256
    somatic mutations, 254
  molecular genetics, 248
  progression model
    adenomas, 250
    chromosomal losses, 248
    clonal nature, 248
    HNPCC, 250
    KRAS mutations, 249
    mismatch-repair deficient, 250–251
Genome-wide association studies (GWAS)
  BMP pathway, 294–295
  candidate genes approaches, 288–289
  criticisms, 295–296
  current and future prospects
    clinical application, 296–297
    post-GWAS world, 296
  genetic contribution
    cancer families, 287
    heritability, 288
    measurement, 288
  planning and evolution
    involves, 289–290
    SNP typing discovery phase, 290–291
    stages, 291
    statistical analysis, 291
    strategy, 290
    sufficient sample size, 292
    technical assessments, 291
  timeline and findings
    identification, of SNPs, 293
    POU5F1P1, 292
    SNPs, 292
Glick, A., 210
Glucose transporter 4 (GLUT4), 168
Glycogen synthase kinase-3β (GSK3β), 110
Goggins, M., 206
Goldberg, P.M., 237
Grady, W.M., 201–220
Grivennikov, S., 74
Groden, J., 1–17
Gross, P., 3

H
Hahm, K.B., 213
Hahn, S.A., 206
Hamamoto, T., 213

Hankey, W., 1–17
Harada, N., 213
Hecht, J.R., 236
Hempen, P.M., 206
Henry, T., 7
Hereditary non-polyposis colorectal cancer syndrome (HNPCC), 8, 250. *See also* Lynch syndrome (LS)
Herrera, L., 5
Hippocrates, 2
Hohenstein, P., 213
Hong, T.S., 231–242
Howe, J.R., 206
Huang, F.W., 231–242
Hung, K.E., 25–54
Hutchins, G., 237
Hyperplastic polyps (HPs), 47

I
IGF2, 282
IL-6/STAT3, 74–75
Immune cell modulation, 87–88
Inflammation
  acute and chronic, 68
  in angiogenesis, 88–89
  aspirin, 89–91
  fine-tune regulation, 81, 82
  immune cell modulation, 87–88
  immunotherapy, 91–92
  inflammatory bowel disease (*see* Inflammatory bowel disease (IBD))
  invasion and metastasis, 88–89
  non-steroidal anti-inflammatory drugs, 89–91
  polymorphonucleated cells, 81
  transcription factors, 69
  tumor immune and inflammatory cell infiltration
    B lymphocytes, 87
    dendritic cells, 84–85
    macrophages, 82–84
    myeloid-derived suppressor cells, 87
    natural killer cells, 86
    T lymphocytes, 85–86
Inflammatory bowel disease (IBD)
  colitis-associated cancer
    angiogenesis, 77
    chemokine expression, 77
    dextran sodium sulfate, 71
    genetic and epigenetic alteration, 71–73
    IL-21, 76
    IL-6/STAT3, 74–75
    NFκB, 73–74
    TNF-α, 75–76
  MSI, 43, 45
  polymorphisms, 70
Interleukin (IL)-21 (IL-21), 76
Intestinal stem cells
  β-catenin, 110
  gut epithelium, 108
  intestinal crypt, 108, 109
  villus, 109

J
Jeghers, H., 15, 16
Juvenile polyposis syndrome (JPS), 35

K
Karapetis, C.S., 234, 236
Kim, J., 267–283
Kim, S.J., 206
Kinzler, W., 32
Kleiman, K.M., 301–305
Kleiman, L.B., 231–242, 301–305
Klempner, S., 157–187
Knudson, A.G., 10
KRAS
  adenoma–carcinoma sequence, 27
  anti-K-Ras therapeutics, 147–148
  mutations
    EGFR therapy, 235–236
    genome and implications, 253
    interaction and radiation therapy, 241
    leveraging molecular information, 301–302
    location, 233
    predictive role, 234–235
    prognostic factor, 237–238
    progression model, 249
  role of mutant, 135–136
  treatment
    and *BRAF*, 139–140
    EGFR inhibitor treatment, 140–141
    FOLFIRI and FOLFOX, 139
    XELOX, 139
KRAS4A, 136

L
Leveraging molecular information
  molecular methods, for screening, 304–305
  targeted therapies
    APC pathway, 304
    Fearon and Vogelstein diagram, 302, 303
    overarching goal, 301

pathogenesis, 302–303
phenotypic analyses, 303
schematic representations, 302
Lewis, J., 113
Lgr5, 111
LKB1, 38
Loss of heterozygosity (LOH), 30–31
Lynch, H.T., 8, 40
Lynch syndrome (LS)
dominant inheritance, 9–10
FCCTX, 11–12
heterogeneity, 10–11
HNPCC, 8
microsatellite instability, 40, 171
MSH2, 8
PMS2, 9

M
Macrophages
cytotoxic molecules, 84
fine-tune regulation, 82, 83
ROS and NOS, 84
TAMs, 83
Markowitz, S., 206
Maughan, T.S., 236, 237
Mehlen, P., 35
MEK inhibitors, 145–146
MicroRNA, 17
Microsatellite instability (MSI)
CIMP, 41
clinical diagnosis, 45
DNA MMR system, 39–40
inflammatory bowel disease, 43, 45
Lynch syndrome, 40
pathophysiology, 42–44
response to chemotherapy, 45
serrated pathway (*see* Serrated pathway)
sporadic colorectal carcinoma, 78
sporadic MSI, 40–41
Mitogen-activated protein kinase (MAPK), 130
Miyaki, M., 206
Mori, Y., 206
Mouse model, 33, 39
MSI. *See* Microsatellite instability (MSI)
Munoz, N., 213
Mutations
BRAF, 238–239
BRAF and metastatic CRC, 239
combinations of chemotherapies, 231
EGFR, 232
ErbB signaling, 232
genetic alterations, 233
KRAS
EGFR therapy, 235–236
interaction and radiation therapy, 241
location, 233
predictive role, 234–235
prognostic factor, 237–238
multiple signaling pathways, 242
NRAS, 238
PIK3CA, 240
targeted therapies, 232
MUTYH, 7
Myeloid-derived suppressor cells (MDSCs), 87

N
Näthke, I., 107–117
Natural killer (NK) cells, 86
Non-small cell lung cancer (NSCLC), 133
Non-steroidal anti-inflammatory drugs (NSAIDs)
AOM/DSS, 91
COX-2, 89–90
WNT signaling pathway, 90
Nowak, M.A., 35
NRAS, 253
Nuclear factor-κB (NFκB), 73–74

O
O'Brien, M.J., 48
Ogino, S., 237
Oliveira, M.J., 67–92
Overexpression of cyclooxygenase-2 (COX2), 37

P
Paneth cells, 111
Panitumumab, 235–236
Papadopoulos, N., 247–261
Parsons, R., 206
Pasche, B., 206
Pedone, K.H., 121–149
Peutz, J., 15, 16
Peutz–Jeghers syndrome, 15–16
Phosphatidylinositols (PtdIns), 159
Phosphatidylinositol (3,4,5)-trisphosphate (PIP_3)
AKT, 163–164
pivotal role, 162
Ras effector signaling, 132
SHIP phosphatases, 162
structure and generation, 162, 163

Phosphoinositide 3-kinase (PI3K) pathway
AKT activation, biological effects
growth, 166–167
metabolism, 167–168
proliferation, 166
survival, 165–166
angiogenesis
AKT1, 185
physiological process, 185
rapamycin, 186
RTK receptors, 185
class IA and cancer, 159–161
class IB, 161–162
components, 158
cytotoxic fluorouracil, 158
effector signaling
Akt, 146
MEK inhibitor, 147
PIK3CA, 146
PTEN, 147
EGFR, 158
genome and implications, 256
inhibitors
AKT, 179–180
clinical development, 174, 175
clinically effective targeted therapies, 184
dual PI3K/mTOR inhibitors, 182
first generation Pan-PI3K, 175–176
isoforms, 177–179
KRAS, 184
MEK inhibitor combinations, 185
mTOR, 180–181
second generation Pan-PI3K, 177
therapy strategies, 183
mutations
cellular transformation, 168
coexistence and mutual exclusivity, 170–171
EGFR inhibitors, 173
hotspot regions, 169
KRAS mutations, 173
palliative chemotherapy, 173
PIK3CA activation, 169
PIK3CA identification, 169, 170
prognostic significance, 172
PTEN expression, 172
RTK KIT, 169
PIP_3
AKT, 163–164
phosphatases, 162–163
RTK inhibition, 184
VEGFR, 158
PIK3CA
clinical significance of mutations, 240
genome and implications, 253
mutation
E545K, 137
H1047R, 137
PI3K, 138
and PTEN, 34–35
Pino, M.S., 32
Poulogiannis, G., 157–187
p85 regulatory subunit. *See* Phosphoinositide 3-kinase (PI3K) pathway
PTEN
mutations, 15
PIK3CA, 34–35

R

Raf inhibitors
GDC-0879, 144
MEK, 145–146
sorafenib, 143
vemurafenib, 144
Rajagopalan, H., 49
Ras proteins
BRAF
B-Raf, 136
ERK MAPK pathway, 137
V600E, 136
effector signaling
GEF, 138
HRAS, 130
MEK1/2 and ERK1/2, 130
PIP_3, 132
Rac1, 132
Raf-MEK-ERK mitogen-activated protein kinase cascade, 143–146
farnesyltransferase inhibitors, 142–143
gene transcription, 124–125
intracellular signaling, 127
CAAX box, 128
HV sequence, 128
K-Ras4B, 129
KRAS
anti-K-Ras therapeutics, 147–148
and *BRAF*, 139–140
EGFR inhibitor treatment, 140–141
FOLFIRI and FOLFOX, 139
XELOX, 139
KRAS4A, 136
mutational activation
aberrant crypt foci, 135
COSMIC database, 133

KRAS mutation, 135
NSCLC, 133
PIK3CA
E545K, 137
H1047R, 137
PI3K, 138
PI3K effector signaling
Akt, 146
MEK inhibitor, 147
PIK3CA, 146
PTEN, 147
structure and GDP-GTP regulation, 125–127
superfamily, 129
undruggable, 141–142
Reactive nitrogen species (RNS), 71, 72
Reactive oxygen species (ROS), 71, 72
Richman, S.D., 237
Roper, J., 25–54
Rosenberg, H., 70
Roth, A.D., 237

S

Sansom, O.J., 107–117
Sells, J.L., 121–149
Serrated pathway
alternates, 50
CIMP and BRAF, 48–49
clinical characteristics, 52
detection and surveillance, 52–53
dysplastic serrated polyps, 47
epidemiology, 48
HPS, 46
hyperplastic polyps, 47
initiation and progression, 49–51
models, 53
MSI, 48
risk factors, 50, 52
sessile serrated adenomas, 47
Sessile serrated adenomas (SSAs), 47
Shih, I.-M., 35
Single nuclear polymorphism (SNP)
BMP4, 294
CRC risk, 297
planning and evolution, typing discovery phase, 290–291
timeline and findings, 292
SMAD3, 280
SMAD4
familial juvenile polyposis, 12–14
genome and implications, 253
tumor suppressor gene loci, 280
SMAD2/MADH2, 206
SMAD4/MADH4, 206
SMADS, 216–218
Somatic cells
applications, 281
mechanisms, 269–271
Somatic copy-number alterations (SCNAs)
copy-number alterations, 281–283
tumor suppressor genes, 280
Spindle checkpoint regulators, 29
Sporadic colorectal carcinoma (SCRC)
CIMP, 81
CIN, 78
clinical and pathologic features, 78
DNA methylation, 81
IBD, 80
microsatellite instability, 78, 80
molecular alterations, 78–81
pathways, 79
TP53 mutations, 80
Src-homology 2 (SH2), 159
Src-homology 3 (SH3), 159, 160
Stem cells
crypt, 110
Drosophila, 111
LGR-5 positive cells, 110, 111
Paneth cells, 111
Wnt signaling, 115–116
STK11/LKB1, 15–16

T

Takaku, K., 213
Takenoshita, S., 206
TCF7L2, 253
Telomere dysfunction, 30
TGFBR1, 206, 216–217
TGFBR2, 206, 212, 214–215
TGF-β type II receptor (TGF βIIR), 36
T lymphocytes, 85–86
Tol, J., 236
Tomlinson, I., 287–298
TP53, 35, 72, 253
Transforming growth factor beta (TGF-β) signaling pathway
apoptosis, 209–210
cell proliferation, 208–209
family, 202
genome and implications, 256
genomic instability, 210
inhibiting mechanisms
animal model studies, 212, 213

Transforming growth factor beta (TGF-β) signaling pathway (*cont.*)
SMADS, 216–218
TGF-β1, 212, 214
TGFBR1, 215–216
TGFBR2, 214–215
invasive behavior, mechanisms, 211–212
senescence, 210–211
superfamily
activin signaling, 219–220
BMP signaling, 218–219
tumor suppression
deregulation and effects, 204–208
downstream transcriptional targets, 203–204
genes and colorectal cancer risk, 205–206
GWAS, 207
inflammation-associated cancer, 207
intestinal cancer models studies, 205
intestinal epithelial cell behavior, 207–208
mouse models studies, 204
role and elements, 202–203
Trobridge, P., 213
Tumor necrosis factor alpha (TNF-α), 75–76
Tumor suppressor genes
chromosome 17q, 277
focal amplifications, 277
focal deletions, 277, 280
focal genomic amplifications, 276
fragile site deletions, 280
high-density array-methods, 276
SCNA profiling techniques, 276, 280
SMAD3 role, 280
visualized examples, 277–279

V
Van Cutsem, E., 234, 236
Velho, S., 67–92
Virchow, R., 68
Vogelgram, 112
Vogelstein, B., 28, 32, 79, 302, 303
VTI1A-TCF7L2 fusion gene, 280

W
Warburg effect, 167
Warthin, A.S., 3, 7
Wnt signaling and pathway
and APC, 31–33
AXIN2, 33
β-catenin, 33
cancer stem cells, 115–116
familial adenomatous polyposis, 6
genome and implications, 256
intestinal stem cell
β-catenin, 110
gut epithelium, 108
intestinal crypt, 108, 109
villus, 109
therapy and diagnosis, 117
Wood, L.D., 31, 252
WWOX, 280

Y
Yun, J., 157–187

Z
Zhu, Y., 213

MIX
Papier aus verantwortungsvollen Quellen
Paper from responsible sources
FSC® C105338

If you have any concerns about our products, you can contact us on
ProductSafety@springernature.com

In case Publisher is established outside the EU, the EU authorized representative is:
Springer Nature Customer Service Center GmbH
Europaplatz 3, 69115 Heidelberg, Germany

Printed by Libri Plureos GmbH
in Hamburg, Germany